T0252130

DEVELOPMENTS IN LUBRICANT TECHNOLOGY

DEVELOPMENTS IN LUBRICANT TECHNOLOGY

S. P. SRIVASTAVA

Copyright © 2014 by John Wiley & Sons, Inc. All rights reserved

Published by John Wiley & Sons, Inc., Hoboken, New Jersey
Published simultaneously in Canada

No part of this publication may be reproduced, stored in a retrieval system, or transmitted in any form or by any means, electronic, mechanical, photocopying, recording, scanning, or otherwise, except as permitted under Section 107 or 108 of the 1976 United States Copyright Act, without either the prior written permission of the Publisher, or authorization through payment of the appropriate per-copy fee to the Copyright Clearance Center, Inc., 222 Rosewood Drive, Danvers, MA 01923, (978) 750-8400, fax (978) 750-4470, or on the web at www.copyright.com. Requests to the Publisher for permission should be addressed to the Permissions Department, John Wiley & Sons, Inc., 111 River Street, Hoboken, NJ 07030, (201) 748-6011, fax (201) 748-6008, or online at http://www.wiley.com/go/permission.

Limit of Liability/Disclaimer of Warranty: While the publisher and author have used their best efforts in preparing this book, they make no representations or warranties with respect to the accuracy or completeness of the contents of this book and specifically disclaim any implied warranties of merchantability or fitness for a particular purpose. No warranty may be created or extended by sales representatives or written sales materials. The advice and strategies contained herein may not be suitable for your situation. You should consult with a professional where appropriate. Neither the publisher nor author shall be liable for any loss of profit or any other commercial damages, including but not limited to special, incidental, consequential, or other damages.

For general information on our other products and services or for technical support, please contact our Customer Care Department within the United States at (800) 762-2974, outside the United States at (317) 572-3993 or fax (317) 572-4002.

Wiley also publishes its books in a variety of electronic formats. Some content that appears in print may not be available in electronic formats. For more information about Wiley products, visit our web site at www.wiley.com.

Library of Congress Cataloging-in-Publication Data:

Srivastava, S. P. (Som Prakash), 1940–
Developments in lubricant technology / S.P. Srivastava.
     pages   cm
  Includes index.
  ISBN 978-1-118-16816-5 (cloth)
1. Lubrication and lubricants.   I. Title.
  TJ1077.S74 2014
  621.8′9–dc23

                                    2013051266

10 9 8 7 6 5 4 3 2 1

CONTENT

PREFACE

Lubricating oils are extremely important products without which no machinery or engines can run. Modern high-quality industrial products cannot be manufactured without the application of specific lubricants. Each class of equipment needs a distinctive product. Lubricants constitute a group of more than 600 products with different viscosity and quality levels, and hence oil companies manufacturing them continuously strive to develop and upgrade these products through extensive research and development. Lubricant development is a multidisciplinary effort that involves various fields such as chemistry, physics, metallurgy, chemical/mechanical/ automobile engineering, surface science, and polymer science and requires good teamwork for successful production. There are several advanced books that deal with lubricants, lubricant additives, and tribology, but there is a shortage of a simple, concise book that would be useful for scientists and engineers who want to have in-depth knowledge on the subject. Unfortunately, this subject does not form part of a university/college curriculum, mainly because of the fact that this knowledge is regarded as a trade secret, and open literature is not available. During my 40 years of interaction with lubricant users, scientists, engineers, technical service staff, and production and marketing professionals, I have found that there is a considerable gap in knowledge between the users and developers. However, there are some organized industrial sectors, such as the OEMs, where engineers are highly knowledgeable about their equipment and lubricant requirements. If the science of lubrication and its application is understood properly by all users, tremendous benefits can be derived by realizing fuel economy, energy efficiency, reduced wear and tear of equipment, and consequently longer life.

It is with this objective that this concise book has been written, and I am confident that it would be well received by students and all those connected with the development, manufacturing, marketing, and application of lubricating oils. The book covers all the major classes of lubricants such as turbine, hydraulic, compressor, gear, transmission, gasoline engine, diesel engine, two-stroke engine, marine engine, natural gas, and rail road engine oils. However, it has not been possible to cover all the grades of minor lubricants such as specific industry-related products for the textile, cement, paper, sugar mill, and food industry. Nevertheless, it would not be difficult to understand the minor grades of lubricants after going through the major classes covered in this book.

Dr. S. P. Srivastava
Faridabad, India
June 2014

LUBRICANT BASICS

PART 1

LUBRICANT BASICS

CHAPTER 1

INTRODUCTION: LUBRICANT SCENARIO

Lubricants are required in every machinery and engine for reducing friction, wear, and energy consumption. Depending upon the operating and design parameters of the equipment, a properly formulated lubricant can play a major role in extending equipment life and saving energy. For manufacturing modern lubricating oils, lube base oils and chemical additives are required. While base oils are produced in the refineries, chemical additives are manufactured separately in chemical plants, as it involves chemical reactions between several materials and specialized testing facilities. Currently, about 41 MMT (million metric ton) of lubricants are produced globally, and the market is growing slowly at the rate of about 2% per annum. The demand pattern has been described in several publications [1–5]. The growth is mainly in Asia. India and China are the fastest growing countries in this sector (3–5%). Asia Pacific is the largest consumer of lubricants (35%) followed by North America (28%), central and southern America (13%), western Europe (12%), and others (12%). Asian market is dominated by China (4 MMT), Japan (2.8 MMT), India (2.4 MMT), and Korea (1 MMT/year). Asia Pacific countries contribute to about 14 MMT of lubricant business per year.

These 41 MMT of lubricants constitute more than 600 grades of products to meet automotive and industrial requirements. Lubricants for automotive applications constitute the major share of lubricants (55%) followed by industrial oil (30%), process oil (10%), and marine oils (5%). Among industrial oils, turbine, hydraulic, gear, and compressor oils constitute major products (60%). About 15–20% of industrial oils are metal working oils and 5% are greases. The balance constitutes other miscellaneous industrial oils.

There are large numbers of small and major manufacturers of lubricants around the world, but in the last two decades, major consolidation has taken place, and four major global companies—Exxon-Mobil, Chevron-Texaco-Caltex, BP-Amoco-Castrol, and Total-Fina-Elf—operate. Two major regional companies—Chinese Sinopec/CNPC and India's, Indian Oil Corporation—have substantial market share in their respective countries.

Developments in Lubricant Technology, First Edition. S. P. Srivastava.
© 2014 John Wiley & Sons, Inc. Published 2014 by John Wiley & Sons, Inc.

Synthetic lubricants [6] constitute about 3% of the total world lubricants. These are mainly aviation and high-temperature application fluids used in situations where mineral oil-based products cannot provide adequate service. Synthetic lubricants use several additives that are common in mineral-based products but also use special chemicals that provide high-temperature, high-pressure performance. Synthetic lubricants are based on several synthetic materials as base oils such as alkylated aromatics, polyalphaolefins, organic esters, halogenated hydrocarbons, phosphate esters, polyglycols, polyphenyl esters and ethers, silicate esters, and silicones. Synthetic oils are also used when there is a need for longer drain capabilities, lower oil consumption, fuel economy, and environmental issues like biodegradability, emissions, and recyclability. Low-viscosity multigrade engine oils like 0W-30 or 5W-30 also need synthetic base oils to meet the low-temperature viscosity requirements.

Lubricant market is dynamic, and quality levels are continuously changing. Every year, new specifications of automotive lubricants are generated to meet the OEM requirements. The use of multigrade engine oils in both gasoline and diesel engines has given a new dimension to the engine oil formulations. Multifunctional additives like dispersant viscosity modifiers change the ratio of detergent/dispersant. Use of American Petroleum Institute (API) group II, III, and IV base oils also changes the additive requirements. For example, it is possible to formulate multigrade engine oils with polyalphaolefins without or minimal use of viscosity modifiers and pour point depressants. These variations lead to the reformulation of products, and therefore, the additive pattern also changes. The demand pattern provided earlier, therefore, should serve as broad guideline only.

The last two decades have seen a very fast-track upgradation of engines, fuels, and lubricants. U.S., European, and Japanese OEM efforts have resulted in several upgraded engine oil specifications and test procedures. The highest diesel engine oil quality till 1985 was API CD level and gasoline engine oil till 1988 was API SF category. However, after 1988, there has been upgradation every year. Currently, API SN and ILSAC GF-5 for gasoline and API CJ4 for diesel engine are the latest standards for automotive lubricants. There has been a similar trend in the development of automatic transmission fluid specifications. These were, however, heavily driven by two major OEMs, General Motors and Ford, whose Dexron and Mercon fluid specifications are accepted worldwide.

This improvement in oil quality led to higher oil drain intervals, improved fuel economy, and reduced emissions. Simultaneously, the gasoline and diesel fuel quality was also improved to match the emission standards imposed by legislation. From gasoline, lead was phased out, octane number was improved, and benzene content and sulfur content were drastically reduced. Gasoline was reformulated to allow the use of oxygenates and multifunctional additives. Similarly, the diesel fuel quality was improved with respect to improved cetane number, reduced aromatics and olefin content, distillation, and drastic reduction of sulfur content. To formulate improved lubricants, it was also necessary to improve the base oil quality. API responded to this need and came out with its base oil classification, where all base oils were categorized into five groups (groups I–V). In groups II and III, sulfur levels have been reduced to less than 300 ppm and saturate content to minimum 90%. The viscosity index for group III base oil has been specified as 120 minimum. All synthetic polyalphaolefins have been categorized in group IV and remaining synthetic oil of different molecular structure in group V. To match this development in lubricant specifications, fuel quality, and base oil quality, it is obvious that the additive technology has to improve. The new specification of diesel engine oil API CJ-4 imposes restrictions on sulfur, phosphorus, and ash content, which will restrict the use of ZDDP or sulfonate/phenate detergent. Newer additives would therefore be required to formulate these and future lubricants. Biodegradability, environmental friendliness, and toxicity would further impose restrictions on the choice of additives.

Oil is limited and reserves are depleting. The search for alternate fuels is currently at its peak. Following options are currently being considered as an alternative to petroleum fuel:

1. Biofuels such as biodiesel

2. Light gaseous hydrocarbons (CH_4 based) such as CNG, LNG, coal bed methane, gas hydrate, propane, and butane (LPG)

3. Oxygen-containing fuels such as methanol, ethanol, dimethyl ether, and ethers

4. Hydrogen

The increasingly higher cost of crude petroleum and its depleting reserves is driving the development of alternate fuels. In the next few decades, we may witness a shift in the use of alternative fuels depending on the techno-commercial viability of these options. The application of CNG and biodiesel has already taken place in several countries. There is considerable activity in the development and use of biodiesel, which is nontoxic, free from aromatics, low in sulfur, and biodegradable. Biodiesel can be manufactured from renewable sources using varieties of vegetable oils and animal fats through a process of trans-esterification with methyl/ethyl alcohol. These are called fatty acid methyl esters or FAME. It may be necessary to have a separate biodiesel lubricant to take care of specific character of this fuel.

There are several technological issues, like cost of manufacture, energy requirement in the production of hydrogen, and its use in the field that needs to be addressed before hydrogen can be adopted as a transportation fuel. Hydrogen is the most ideal and clean burning fuel. With hydrogen fuel cell-based engines, the crankcase engine oils will not be required. Fuel cell-based engine will have electric motors to drive the wheels, which will require only grease for lubrication. However, hydrogen-fired engine would need special lubricant to meet the changed engine environment.

With these changes, the lubricants and their quality will undergo substantial change, and new innovative technologies need to be developed to meet the challenges lying ahead.

The book discusses various aspects of formulating modern lubricants to meet the modern industrial and automotive vehicle requirements while complying with the environmental regulations. The changes that are taking place in lubricant technology are discussed specifically.

REFERENCES

[1] Srivastava SP. Chapter 1. In: *Modern Lubricant Technology*. Dehradun: Technology Publications; 2007.
[2] Srivastava SP. Chapter 1. In: *Advances in Lubricant Additives and Tribology*. New Delhi: Tech Books International; 2009. p 1–10.
[3] Fuchs M. The world lubricant market, current situation and outlook. 12th International Colloquium Tribology; January 11–13, 2000; Esslingen.
[4] Thomas JA. Change and challenges: the future of Asia-Pacific and Indian lube and additives market. 3rd International Petroleum Conference, Petrotech-99; January 9–12, 1999; p 299.
[5] Srivastava SP. Lubricant and fuel scenario in the new millennium. International Symposium on Fuels and Lubricants, ISFL-2000; March 10–12, 2000; New Delhi. New Delhi: Allied Publisher; 2000. Vol. 1, p 19–27.
[6] Srivastava SP. Synthetic lubricant scenario in 21st century. Proceedings of the 2nd International Conference on Industrial Tribology; December 1–4, 1999; Hyderabad. p 7–17.

CHAPTER 2

CLASSIFICATION OF LUBRICANTS

Lubricants are classified in several ways; these could be liquid, semisolid (greases), and solids such as graphite, molybdenum disulfide, boron nitride, tungsten disulfide, and polytetrafluoroethylene. Majority of lubricants are, however, liquids with different viscosities and other physicochemical characteristics. Semisolid greases are used in several applications where liquid lubricants cannot be used conveniently such as in the antifriction and roller bearings of automotive/rail car wheels and other industrial machinery. Solid lubricants on the other hand are used as coatings in the fine powder form or used as additives in greases and liquid lubricants. Another classification used in the industry is based on the type of base oil utilized for the formulation of products such as the following:

1. Mineral oil-based lubricants
2. Synthetic oils
3. Biodegradable, environmentally friendly oils (based on esters or fatty oils)

Synthetic oils are based on synthetic base oils produced in a chemical or petrochemical plant such as esters, diesters, polyalphaolefins, polyalkeleneglycols, silicones, alkyl benzenes, and polyphenyl ethers. Synthetic oils are used in a wide variety of critical applications such as in engines, turbines, compressors, and hydraulic, gear, aviation, and space equipment. It is possible to formulate biodegradable oils from both selected mineral base oils and synthetic oils. However, vegetable oils and synthetic ester-based products are regarded as highly biodegradable and are preferred in those applications where spillage in soil and water is expected. These applications include products used in agricultural, forestry, outboard motors, snowmobiles, etc.

Lubricants are most conveniently classified according to their applications irrespective of the type of base oil utilized such as the following:

AUTOMOTIVE ENGINE OILS

These are further classified as gasoline engine oils, diesel engine oils, rail road oils, marine oils, two-stroke engine oils, tractor oils, off-highway equipment lubricants, gas engine oils, etc. Rail road and marine oils are also considered as a separate class

Developments in Lubricant Technology, First Edition. S. P. Srivastava.
© 2014 John Wiley & Sons, Inc. Published 2014 by John Wiley & Sons, Inc.

since the chemistry used in formulating these oils is slightly different, but these are basically engine oils. The modern API classification system was established only after 1970. SAE developed viscosity classification of engine oils. Initially, three types of oils were proposed by API: regular, premium, or heavy duty. The regular type was straight mineral oils. The premium type contained antioxidants and were meant for gasoline engines. The heavy-duty oils meant for diesel engines contained both antioxidant and detergent/dispersant. This was a rough classification and did not address issues connected with the fuel differences (such as sulfur content and distillation characteristics) and operating conditions. API later developed a new system including three categories for gasoline engines (ML, MM, and MS) and three for diesel engines (DG, DM, and DS). Finally, the modern classification system was established with sequence testing, standardized testing, and performance requirements agreed by both engine manufacturers and lubricant suppliers. This practice has largely prevented the introduction of multiple OEM specifications. Presently, only Mack Truck Company's EO-X system is prevalent under the modern API licensing system. Gasoline engine oils have been classified as *SX* and diesel engine oils are given classifications in the format of *CX-2/4*. S represents service category for spark ignition engines and C represents commercial for compression ignition engines. The *X* is given in alphabetical order representing the sequence of introduction. For example, *SB* and *CB* come after *SA* and *CA*, respectively. The number 2 or 4 denotes 2T or 4T engine applications. Because of the rapid change in emission regulations and engine technology in the 1990s and onward, a new API classification comes out every 3–4 years. Before this, API-CD remained in place for a long time. The fast change presents a major challenge to the lubricant industry. Sometimes new OEM or industry specifications are reappearing faster than the API standards. For example, Cummins has issued CES 20071 1 year ahead of API-CH-4 and has again issued CES 20076 in 1999 to promote the use of premium oils with better soot-handling capability.

In Europe, engine manufacturers continue to specify their own oil requirements as of today. The specifications issued by Comite des Constructeurs d' Automobiles du Marche Commun (CCMC) have been in place for years but are now obsolete. Mercedes Benz has the most comprehensive testing and approval requirements. Volkswagen, Volvo, MAN, and others all have their own test requirements. Association des Constructeurs Europeens de l' Automobile (ACEA) has replaced CCMC in 1996, and ACEA-comprehensive specifications for gasoline and diesel engine oils applicable from the year 2012 have been issued (A1/B1-12, A3/B3-12, A3/B4-12, A5/B5-12, C1-12 to C4-12, E4-12, E6-12, E7-12, and E9-12).

INDUSTRIAL OILS

Industrial oils constitute large number of products used in a variety of industrial machinery such as in turbine, hydraulic systems, compressor, gear boxes, bearings, refrigeration, machine tools, and other industrial equipment. These are known by their names such as turbine oil, compressor oil, and gear oils.

METAL WORKING FLUIDS

These products are generally referred to as fluids rather than oils, since they are used as emulsions or dispersions in many applications. Metal working fluids constitute cutting, grinding, quenching, honing, broaching, forming, forging, wire drawing, and rolling oils for various ferrous and nonferrous metals.

ISO has evolved a viscosity-grade system of classifying industrial and metal working, and this system is now universally accepted. Earlier, different companies were marketing different viscosity products for individual applications depending on the customer need. These oils were further classified into family L and classes A–Y.

AVIATION OILS

Aviation oils are specifically formulated mineral and synthetic hydraulic and gas turbine oils for aviation industry. These groups of products form a different category and are manufactured by limited companies due to their highly specialized quality control procedures and approval system.

GREASES

Greases are further classified according to the soap used for the manufacture of products such as lithium, lithium complex, calcium, calcium complex, clay, poly urea, sulfonate complex, moly, graphited greases, and mixed soap greases.

In oil industry, lubricants are also sometimes classified based on the additives used in them, such as EP oil (containing extreme pressure additive), detergent oil (containing detergent and dispersant additives), R&O oil (rust and oxidation inhibited), antiwear oil (containing antiwear additive), compounded oil (containing fatty oil to reduce friction), and FM oil (containing friction modifier).

Lubricants are used in equipment under different operating conditions of temperature, speed, and load and require different viscosity grades (VGs) to satisfy performance characteristics. Lubricants are thus classified according to the viscosity and applications by the following organizations:

1. SAE Viscosity Classification of Engine Oils
2. SAE Viscosity Classification of Automotive Gear Oils
3. ISO Viscosity Classification of Industrial Oils
4. National Lubricating Grease Institute (NLGI) Classification of Greases
5. American Gear Manufacturers Association Classification of Industrial Gear Oils
6. U.S. Military Classification of Engine and Gear Oils

Lubricants have been further classified according to their performance characteristics in a series of following documents:

1. API service classification of gasoline engine oils—S (service) category
2. API service classification of diesel engine oils—C (commercial) category
3. API service classification of gear oils—GL category
4. ISO-6743 classification of industrial oils
5. Two-stroke engine oil classification by API, NMMA, JASO, and ISO
6. Automatic transmission oil classification by GM, Ford, Allison, and Caterpillar
7. Rail road oil classification by Locomotive Maintenance Officers Association (LMOA)

Similar to ISO-3448, ASTM D-2422-2007 Standard Classification of Industrial Fluid Lubricants by Viscosity System has also been worked out. There are various other regional or country- or OEM-specific classifications of lubricants, but these generally follow the above described international approach. The classification is then followed by the detailed specifications of the finished products.

ISO 3448 VISCOSITY CLASSIFICATION FOR INDUSTRIAL OILS

The ISO viscosity classification is generally used for industrial lubricants and the VG number from VG 10 grades and above indicates viscosity in centistokes at 40°C. VG 2 to VG 7, however, have different midviscosities (refer to Table 2.1). Each subsequent

TABLE 2.1   ISO viscosity classification of industrial oil

ISO 3448 viscosity grades	Kinematic viscosity at 40°C (mm^2/s = cSt)		
	Midpoint	Minimum	Maximum
ISO VG 2	2.2	1.98	2.42
ISO VG 3	3.2	2.88	3.52
ISO VG 5	4.6	4.14	5.06
ISO VG 7	6.8	6.12	7.48
ISO VG 10	10	9.0	11.0
ISO VG 15	15	13.5	16.5
ISO VG 22	22	19.8	24.2
ISO VG 32	32	28.8	35.2
ISO VG 46	46	41.4	50.6
ISO VG 68	68	61.2	74.8
ISO VG 100	100	90	110
ISO VG 150	150	135	165
ISO VG 220	220	198	242
ISO VG 320	320	288	352
ISO VG 460	460	414	506
ISO VG 680	680	612	748
ISO VG 1000	1000	900	1100
ISO VG 1500	1500	1350	1650

VG has approximately 50% higher viscosity, whereas the minimum and maximum values of each grade range ±10% from the midpoint. For example, ISO VG 100 refers to a VG of $100\,cSt \pm 10\%$ at $40°C$.

ENGINE OIL CLASSIFICATION

The VGs of engine oils are classified by the Society of Automotive Engineers into six winter (W) grades and five mono grades. There can be cross grades as well, for example, SAE-10W-40, which is designated as multigrade oil. Such oils have to satisfy the requirements of both 10W and SAE 40 mono grade. The first number (10W) refers to the VG at low temperatures (W stands for winter), whereas the second number (40) refers to the VG at high temperatures. All automotive oils including rail road, marine, and natural gas engine oils follow SAE viscosity classification (Table 2.2). Performance of individual oil is, however, decided by the actual application and the applicable standard.

TABLE 2.2   Automotive lubricant viscosity grades[a]

Engine oils—SAE J 300, January 2009				
SAE	Low-temperature viscosities		High-temperature viscosities	
Viscosity grade	Cold cranking[b] viscosity (mPa.s) max at temp. (°C)	Pumping[c] viscosity (mPa.s) max at temp. (°C)	Kinematic[d] (mm²/s) at 100°C	High shear[e] rate viscosity (mPa.s) at 150°C
			Minimum Maximum	Minimum
0W	6,200 at −35	60,000 at −40	3.8 —	—
5W	6,600 at −30	60,000 at −35	3.8 —	—
10W	7,000 at −25	60,000 at −30	4.1 —	—
15W	7,000 at −20	60,000 at −25	5.6 —	—
20W	9,500 at −15	60,000 at −20	5.6 —	—
25W	13,000 at −10	60,000 at −15	9.3 —	—
20	—	—	5.6 <9.3	2.6
30	—	—	9.3 <12.5	2.9
40	—	—	12.5 <16.3	3.5[f]
40	—	—	12.5 <16.3	3.7[g]
50	—	—	16.3 <21.9	3.7
60	—	—	21.9 <26.1	3.7

[a]All values are critical specifications as defined by ASTM D-3244.
[b]ASTM D-5293.
[c]ASTM D-4684.
[d]ASTM D-445.
[e]ASTM D-4683, CEC L-36-A-90 (ASTM D-4741), or ASTM D-S481.
[f]Applicable for 0W-40, 5W-40, and 10W-40 grades.
[g]Applicable for 15W-40, 20W-40, 25W-40, and 40 grades.

Selection of proper base oil with satisfactory low pour point and low-temperature rheological properties is necessary to meet the winter grade requirement. The use of low pour synthetic oils such as PAOs and polyol esters is advantageous in formulating lower viscosity multigrade oils.

VISCOSITY

Both industrial and automotive oils use viscosity for classification. It is therefore important to understand various terminologies used in viscosity measurement techniques.

ABSOLUTE AND KINEMATIC VISCOSITIES

Viscosity is a measure of internal friction of the fluid, which is experienced when an external force is applied on the fluid. Kinematic viscosity is a measure of this internal friction under the influence of gravity. In other words, viscosity can be determined by measuring force required to overcome the fluid friction. Absolute viscosity, sometimes called dynamic or simple viscosity, is the product of kinematic viscosity and density of the fluid:

$$\text{Absolute viscosity } (\eta) = \text{Kinematic viscosity} \times \text{density}$$

The dimension of kinematic viscosity is $L2/T$, where L is the length and T is the time. Commonly, the centistoke (cSt) is used. The SI unit of kinematic viscosity is m^2/s, which is 10^6 cSt. The SI unit of absolute viscosity is milli Pascal-second (mPa-s).

NEWTONIAN AND NON-NEWTONIAN FLUIDS

When a fluid follows Newton's viscosity law, that is, its viscosity is independent of shear stress or rate of shear; the fluid is called Newtonian fluid. Most lubricating base oils, solvents, and formulated lubricants without polymers are Newtonian fluids. Thus, Newtonian fluids have constant viscosity under shear stress at a particular temperature. On the other hand, non-Newtonian fluids have different viscosity at different shear stress.

Multigrade engine oils and high VI industrial oils formulated with viscosity modifiers (usually polymers) are non-Newtonian fluids since their viscosity changes with shear rates. The viscosities of such oils decrease with an increase in shear rate.

VISCOSITY MEASUREMENT

Kinematic viscosity of oil is measured by capillary viscometers (ASTM D-445). The time to flow a fixed volume of oil through the capillary orifice at controlled temperature (40 and 100°C) under gravity is measured. Shear rate in this measurement is very low. High-temperature high-shear (HTHS) viscosity is measured by high-pressure capillary viscometer (ASTM D-4624). In this system, a known volume of fluid is forced through a small-diameter capillary by applying gas pressure. The rate of shear can be varied up to $10^6 S^{-1}$. This viscosity is called *high-temperature high-shear* viscosity and is measured at 150°C and $10^6 S^{-1}$.

Rotary viscometers use the torque on a rotating shaft to measure the resistance of the fluid to flow. The cold-cranking simulator (CCS-ASTM D-2602), Mini-Rotary Viscometer (MRV-ASTM D-3829 and D-4684), Brookfield viscometer, and tapered bearing simulator (TBS) are all rotary viscometers. Rate of shear can be changed by changing rotor dimensions, speed of rotation, and the gap between rotor and stator wall. HTHS viscosity is also measured by the TBS by ASTM D-4683 procedure. Very high-shear rates are obtained by using a smaller gap between the rotor and the stator wall in the TBS.

The CCS measures an apparent viscosity in the range 500–200,000 cP in the temperature range of 0 to −40°C. Shear rate ranges between 10^4 and $10^5 S^{-1}$. The CCS data correlate with engine cranking at low temperatures. The SAE J300 viscosity classification specifies the low-temperature performance of engine oils by both CCS and MRV.

The MRV is a low-shear rate measurement and measures apparent yield stress, which is the minimum stress required to initiate the flow of the oil. It also measures an apparent viscosity under shear rates of $1-50 S^{-1}$. MRV measurement correlates with the oil pumpability in the engine. The cooling cycle of ASTM D-3829 is used to measure the borderline pumping temperature.

VISCOSITY INDEX

Viscosity index (VI) is an empirical number indicating the change in viscosity of oil with respect to increased temperature. A high VI signifies a relatively small change in viscosity with increase in temperature, whereas a low VI reflects greater viscosity change with temperature. Most solvent-refined mineral base oils have a VI between 60 and 100. Naphthenic oils may have lower VI between 0 and 50. The API groups II, III, and IV category oils have higher VI. Polymeric viscosity modifier-containing oils have higher VI as compared to their base oil VI. VI is calculated from kinematic viscosities at 40 and 100°C by using tables in ASTM D-2270 or ASTM D-39B. It, however, does not predict low-temperature or HTHS viscosities. These values are derived from CCS, MRV, low-temperature Brookfield, and high shear rate viscometers. The term VI is now obsolete and is not used in characterizing engine oils. SAE classification of engine oils has no requirement of VI. The term is, however, useful in defining viscosity–temperature behavior of oils and is still used to describe industrial oils.

API AND ILSAC CLASSIFICATION OF ENGINE OILS

API and ILSAC have further classified gasoline and diesel engine oils according to their performance to guide users to select proper lubricants. API-S category means service category for gasoline engine. Over a period of time, several categories of API and ILSAC (Table 2.3) have been evolved as follows:

TABLE 2.3   Development of gasoline engine oil specifications

API-SA, SB, SC, SD, SE	1950–1979, all are obsolete
API-SF	1980 obsolete
API-SG	1988 obsolete
API-SH	1993 obsolete
API-SJ	Developed in 1997 and suitable for 2001 and older engines
API-SL	Developed in 2001 and suitable for 2004 and older engines
API-SM	Developed in 2004 and suitable for 2010 and older engines
API-SN	Developed in 2010 and suitable for 2011 and older engines
	Resource conserving grade matches with ILSAC-GF-5
ILSAC-GF1	1990 obsolete
ILSAC-GF2	1996 obsolete
ILSAC-GF3	2000 obsolete
ILSAC-GF4	2004 was valid till September 2011
ILSAC-GF5	2010 current for 2011 and older vehicles

Note: SI category was left out since this denotes spark ignition. SK category was also left out due to its similarity with a country name.

Similarly, diesel engine oils are classified according to API-C category. C denotes commercial. Following categories have been identified in Table 2.4. Combined gasoline–diesel cross lubricants are also permitted and have been marketed widely; for example, SC/CC, SD/CF, or SG/CF. The current high-performance gasoline and diesel engine oils are now highly specialized such as API-SN and API-CJ-4, and the cost of testing is too high to qualify them for both applications. However, technically, it is possible to formulate combined API-SN/CJ-4 oil.

Four-stroke motor cycle oils have been classified by Society of Automotive Engineers of Japan (JASO T 903-2005) into MA, MA1, MA2, and MB categories, specifying dynamic and static coefficient of friction (Table 2.5).

TABLE 2.4   Development of diesel engine oil specifications

API-CA, CB, CC, CD	1950–1985, all are obsolete
API-CE	1985 obsolete
API-CD-II	1988 obsolete
API-CF-4	1991 obsolete
API CF and CF-2	1994 obsolete, CF-2 for two-stroke engines
API-CG-4	1995 obsolete
API-CH-4	1998 current, designed to meet 1998 emission standards
API-CI-4 and CI-4 plus	Introduced in 2002 for 2004 exhaust emission standards
API-CJ-4	For low-emission vehicles of 2010 using 15 ppm sulfur diesel fuel

Note: The numbers 2 and 4 indicate two-stroke and four-stroke engine applications. Each category identifies the oil with several performance characteristics including engine tests.

TABLE 2.5   Four-stroke classification JASO T-903, 2011

JASO T-904	Dynamic friction characteristics index (DFI)	Static friction characteristics index (SFI)	Stop time index (STI)
JASO MA	$1.30 \leq DFI < 2.5$	$1.25 \leq SFI < 2.5$	$1.45 \leq STI < 2.5$
JASO MA1	$1.30 \leq DFI < 1.8$	$1.25 \leq SFI < 1.70$	$1.45 \leq STI < 1.85$
JASO MA2	$1.85 \leq DFI < 2.5$	$1.70 \leq SFI < 2.50$	$1.85 \leq STI < 2.50$
JASO MB	$0.50 \leq DFI < 1.3$	$0.50 \leq SFI < 1.25$	$0.50 \leq STI < 1.45$

SAE CLASSIFICATION OF AUTOMOTIVE GEAR OILS

Eleven grades of auto gear oils have been identified (four winter W grades and seven mono grades) by SAE. For wider temperature applications and energy efficiency, multigrade gear oils are recommended (Table 2.6).

API has further classified automotive gear oils into several categories according to the severity of operation:

API-GL-1: for gears operating under mild conditions, not requiring EP additives.

API-GL-2: for worm gear axles, requiring lubricity or friction modifier additives. GL-1 type oils will not be satisfactory for this application.

API-GL-3: for gears operating under moderately severe conditions.

TABLE 2.6   SAE viscosity of automotive gear oils—SAE J306 June 2005

SAE viscosity grade	Maximum temperature for a viscosity of 150,000 cP (°C)[a] ASTM D-2983	Minimum viscosity[b] at 100°C (cSt) ASTM D-445	Maximum viscosity at 100°C (cSt) ASTM D-445
70W	−55	4.1	—
75W	−40	4.1	—
80W	−26	7.0	—
85W	−12	11.0	—
80	—	7.0	<11.0
85	—	11.0	<13.5
90	—	13.5	<18.5
110	—	18.5	<24.0
140	—	24.0	<32.5
190	—	32.5	<41.0
250	—	41.0	—

[a]Using ASTM D-2983, additional low-temperature viscosity requirements may be appropriate for fluids intended for use in light-duty synchronized manual transmission.

[b]Limit must also be met after testing in CEC l-45-T-93, Method C (20h).

API-GL-4: with antiscore properties. No testing support is now available, but still used commercially. Oils are formulated with half the dosage of GL-5 oil additives.

API-GL-5: for hypoid gears operating under high-speed shock load, high-speed low-torque, and low-speed, high-torque conditions. Performance is described in ASTM STP-512A, requiring CRC L-33, L-60, L-37, and L-42 tests.

API-GL-6: now obsolete.

API-MT-1: issued in 1995 for nonsynchronized transmission of heavy trucks and buses

Proposed API-PG-2: a GL-5 plus oils.

TWO-STROKE ENGINE OILS

API has classified these oils into API-TA and API-TC categories. JASO (Japanese) have categorized into FA, FB, and FC, and ISO have categorized into ISO E-GB, E-GC, and E-GD categories (Table 2.7). For outboard motors, NMMA has categorized oils into TC-W, TC-W-II, TC-W-3, and TC-W-3R.

Test engines:

Honda DIO AF27: lubricity, torque index, detergency, piston skirt varnish.

Suzuki SX800R: exhaust smoke, exhaust blocking.

Piston skirt deposit rating is not required by JASO.

Thailand TISI classification of two-stroke engine oils is provided in Table 2.8. India follows both API and JASO system for 2T and 4T oils for small engines. Most of the motor cycles in India are manufactured with Japanese designs.

RAIL ROAD OILS

These are diesel engine oils recommended for rail road applications. LMOA has classified these oils into five generations (Generations 1–5) based on oil performance.

TABLE 2.7 Two-stroke classification: ISO/JASO

ISO	—	E-GB	E-GC	E-GD
JASO	FA	FB	FC	—
Lubricity	90 min	95 min	95 min	95 min
Torque index	98 min	98 min	98 min	98 min
Detergency	80 min	85 min	95 min	125 min
Piston skirt deposits	—	85 min	90 min	95 min
Exhaust smoke	40 min	45 min	85 min	85 min
Exhaust blocking	30 min	45 min	90 min	90 min

Note: All limits are indices relative to reference oil, JATRE-1.

TABLE 2.8   Two-stroke classification: TISI 1040

Test	Parameter	Limits
Bench tests	Viscosity, 100°C.cSt.	5.6–16.3
	Viscosity index	95 min
	Flash point, °C	70 min
	Pour point, °C	−5 max
	Sulfated ash, % wt	0.5 max
	Metallic element content, % wt	Report
Kawasaki KH 125M	Piston seizure and ring scuffing at fuel oil ratio of 200:1	No seizure
	Detergency (general cleanliness)	
	Ring sticking	8 merit min
	Piston cleanliness	48 merit min
	Exhaust port blocking	None
Suzuki SX800R (JASO M 342-92)	Exhaust smoke	85 min

Note: Since mid-1991, all two-stroke oils used in Thailand are required to meet TISI requirements.

Generation 1

First introduced in 1940, Generation 1 oils were generally straight mineral oils, as well as some containing detergents and antioxidants. Total base number (TBN) of Generation 1 oils was generally below 7.

Generation 2

Introduced in 1964, Generation 2 engine oils contain ashless dispersants and moderate levels of detergency. These oils were developed to reduce engine sludge and extend filter life. Generation 2 engine oils had a TBN of around 7.

Generation 3

Introduced in 1968, Generation 3 engine oils possess improved alkalinity retention, detergency, and dispersancy. With TBN of around 10, Generation 3 oils were introduced to overcome increased piston ring wear.

Generation 4

Generation 4 oils, introduced in 1976, provided added protection under severe operating conditions and were designed to permit 90-day oil change intervals. LMOA set higher base number (13) detergency and dispersancy characteristics for Generation 4 oils. Generation 4 engine oils should also meet API-CD performance level.

Generation 5

Before the introduction of Generation 5 category, oils that eventually were to meet this qualification were described as Generation 4 *Long-Life*.

TABLE 2.9   NLGI classification of greases by cone penetration

NLGI grade	60-stroke worked penetration at 25°C tenth of mm, ASTM D-217
000	445–475
00	400–430
0	355–385
1	310–340
2	265–295
3	220–250
4	175–205
5	130–160
6	85–115

Introduced in 1989, Generation 5 oils are required to meet extended oil drain period of 180-day performance to meet the requirements of new-generation, *fuel-efficient*, and low-oil-consuming diesel locomotive engines. TBN for these oils are not specified, but the products have improved alkalinity reserve, detergency, and antioxidation performance. Oils meeting LMOA Generation 5 must also meet API Service Classification CD and have to be field tested and approved by both GE and EMD. Multigrade versions of Gen IV and Gen V oils (mainly 20W-40) have also been developed, field tested, and approved by OEMs to obtain fuel efficiency and low oil consumption. With the introduction of ULSD having max 15 ppm sulfur, a new category of rail road oil with low SAPS and low TBN of 9 has emerged recently.

NLGI CLASSIFICATION OF GREASES

Greases have been classified according to the worked cone penetration, since these are semisolids and viscosity cannot be determined under normal conditions (Table 2.9). Greases are described by NLGI numbers. NLGI 000 grade is the thinnest grease and No 6 grade is the thickest.

METAL WORKING OIL CLASSIFICATION

ISO and DIN have classified metal working oil into several groups according to the application. Table 2.10 provides ISO system, and Table 2.11 shows DIN classification.

DIN 51385 has identified metal working fluids into seven categories.Industrial oils have been systematically classified by ISO under class L and family A–Y. The following ISO documents have been issued. Readers are advised to refer to the original documents for details. This classification is then followed by the ISO specifications of individual classes.

TABLE 2.10   ISO 6743/7 metal working lubricant classification L—Family M and applications

Neat oils	Aqueous fluids
MHA—refined mineral oil or synthetic fluid	MAA—milky emulsions with anticorrosion properties
MHB—with friction reducing properties in MHA	MAB—emulsions with friction reducing properties
MHC—with EP properties in MHA (noncorrosive)	MAC—emulsions with EP properties
MHD—with EP properties in MHA (corrosive)	MAD—emulsions with friction reducing and EP properties
MHE—with friction reducing properties in MHC	MAE—microemulsions with anticorrosion properties
MHF—with friction reducing properties in MHD	MAF—microemulsions with friction reducing and/or EP properties
MHG—grease type	MAG—solutions with anticorrosion properties
MHH—soap type	MAH—solutions with friction-reducing and/or EP properties
	MAI—greases and pastes blended in water

Applications

Cutting: MHA to MHF, MAA to MAF, and MAH.
Abrasion: MHC, MHE, MHF, MAG, and MAH.
Rolling: MHA, MHB, and MAG.
Sheet metal forming: MHB to MHG and MAA, MAB, MAD, MAI.
Forming and stamping: MHB, MAG, and MAH.
Wire drawing: MHB, MHG, MHH, MAB, and MAI.
Power spinning: MHB, MHE, and MAD.

TABLE 2.11   DIN 51385 classification of metal working fluids

Term	Fluid type	Code letter
0	Metal working fluids	S
1	Nonwater-miscible MWF	SN
2	Water-miscible metal working concentrate	SE
2.1	Emulsifiable metal working concentrate	SEM
2.2	Water-soluble metal working concentrate	SES
3	Diluted metal working fluids	SEW
3.1	Metal working emulsion—oil-in-water	SEMW
3.2	Metal working solution	SESW

1. ISO 6743-1:2002: Lubricants, industrial oils and related products (class L)—Classification—Part 1: Family A (Total loss systems).

2. ISO 6743-2:1981: Lubricants, industrial oils and related products (class L)—Classification—Part 2: Family F (Spindle bearings, bearings and associated clutches).

3. ISO 6743-3:2003: Lubricants, industrial oils and related products (class L)—Classification—Part 3: Family D (Compressors).

4. ISO 6743-4:1999: Lubricants, industrial oils and related products (class L)—Classification—Part 4: Family H (Hydraulic systems).

5. ISO 6743-5:2006: Lubricants, industrial oils and related products (class L)—Classification—Part 5: Family T (Turbines).

6. ISO 6743-6:1990: Lubricants, industrial oils and related products (class L)—Classification—Part 6: Family C (Gears).

7. ISO 6743-7:1986: Lubricants, industrial oils and related products (class L)—Classification—Part 7: Family M (Metalworking).

8. ISO 6743-8:1987: Lubricants, industrial oils and related products (class L)—Classification—Part 8: Family R (Temporary protection against corrosion).

9. ISO 6743-9:2003: Lubricants, industrial oils and related products (class L)—Classification—Part 9: Family X (Greases).

10. ISO 6743-10:1989: Lubricants, industrial oils and related products (class L)—Classification—Part 10: Family Y (Miscellaneous).

11. ISO 6743-11:1990: Lubricants, industrial oils and related products (class L)—Classification—Part 11: Family P (Pneumatic tools).

12. ISO 6743-12:1989: Lubricants, industrial oils and related products (class L)—Classification—Part 12: Family Q (Heat transfer fluids).

13. ISO 6743-13:2002: Lubricants, industrial oils and related products (class L)—Classification—Part 13: Family G (Slide ways).

14. ISO 6743-14:1994: Lubricants, industrial oils and related products (class L)—Classification—Part 14: Family U (Heat treatment).

15. ISO 6743-15:2007: Lubricants, industrial oils and related products (class L)—Classification—Part 15: Family E (Internal combustion engine oils).

16. ISO 6743-99:2002: Lubricants, industrial oils and related products (class L)—Classification—Part 99: General.

17. ISO 7745:2010: Hydraulic fluid power—Fire-resistant (FR) fluids—Requirements and guidelines for use.

Following ISO specifications have been worked out, and others are in the process of being finalized. Details of some of these specifications have been discussed in individual chapters dealing with the subject. However, there are several OEMs and national and international standards of lubricants. ISO efforts are to consolidate these different standards into unified internationally accepted ISO standards.

1. ISO 8068:2006: Lubricants, industrial oils and related products (class L)—Family T (Turbines)—Specification for lubricating oils for turbines.

2. ISO 10050:2005: Lubricants, industrial oils and related products (class L)—Family T (Turbines)—Specifications of triaryl phosphate ester turbine control fluids (category ISO-L-TCD).

3. ISO 11158:2009: Lubricants, industrial oils and related products (class L)—Family H (hydraulic systems)—Specifications for categories HH, HL, HM, HV and HG.

4. ISO/FDIS 12922: Lubricants, industrial oils and related products (class L)—Family H (Hydraulic systems)—Specifications for hydraulic fluids in categories HFAE, HFAS, HFB, HFC, HFDR and HFDU.

5. ISO 12922:1999: Lubricants, industrial oils and related products (class L)—Family H (Hydraulic systems)—Specifications for categories HFAE, HFAS, HFB, HFC, HFDR, and HFDU.

6. ISO 12924:2010: Lubricants, industrial oils and related products (Class L)—Family X (Greases)—Specification.

7. ISO 12925-1:1996: Lubricants, industrial oils and related products (class L)—Family C (Gears)—Part 1: Specifications for lubricants for enclosed gear systems.

8. ISO 13738:2011: Lubricants, industrial oils and related products (class L)—Family E (Internal combustion engine oils)—Specifications for two-stroke-cycle gasoline engine oils (categories E-GB, E-GC and E-GD).

9. ISO 15380:2011: Lubricants, industrial oils and related products (class L)—Family H (Hydraulic systems)—Specifications for categories HETG, HEPG, HEES and HEPR.

10. ISO 19378:2003: Lubricants, industrial oils and related products (class L)—Machine-tool lubricants—Categories and specifications.

11. ISO 24254:2007: Lubricants, industrial oils and related products (class L)—Family E (internal combustion engine oils)—Specifications for oils for use in four-stroke cycle motorcycle gasoline engines and associated drive trains (categories EMA and EMB).

CHAPTER 3

MINERAL AND CHEMICALLY MODIFIED LUBRICATING BASE OILS

Lubricating base oils are higher-molecular-weight, high-boiling, high-viscosity, and refined crude oil products from refinery, which form the basis of finished lubricating oil. Lubricants are manufactured from these base oils by blending two or more base oils to obtain the desired viscosity. Most lubricants contain some amount of chemical additives to improve certain performance characteristics. There are, however, many products that are straight base oil blends of specific viscosity. Base oils thus constitute the backbone of any lubricant. It is therefore necessary to understand base oil properties and the need to modify certain characteristics of base oil through additives. Base oils are both petroleum derived and synthetic. Only about 3% of total world lubricants are based on synthetic base oils and are used in specialized applications such as in aviation or high-temperature industrial applications. The earliest methods of producing lubricating base oils were based on acid and clay treatment and SO_2 treatments to remove aromatic and undesirable products from vacuum distillates. In the second phase, solvent extraction and solvent dewaxing followed by hydrofinishing were used for several decades. Mineral base oils are produced in the refinery by processing several fractions of high boiling vacuum distillates and residue (for bright stock (BS)). A typical composition [1] and properties of the raw distillates are shown in Table 3.1.

The high pour point (above 40°C), low viscosity index (VI), and high aromatic content of these distillates make these streams unsuitable for lubricant application. The aromatics make them unstable and are also responsible for poor viscosity–temperature characteristics. Waxes create fluidity problems and the oils have high pour points. These raw vacuum lube distillates are, therefore, further processed to reduce aromatics and waxes to make them suitable as lubricant base oils. Various types of compounds present in vacuum distillate have different characteristics:

1. Linear paraffins have high VI, high pour point, and high oxidation stability.

2. Isoparaffins have high VI, low pour point, and high oxidation stability.

3. Naphthenes have low VI, low pour point, and moderate oxidation stability.

Developments in Lubricant Technology, First Edition. S. P. Srivastava.
© 2014 John Wiley & Sons, Inc. Published 2014 by John Wiley & Sons, Inc.

TABLE 3.1   Typical properties of raw vacuum distillates

Properties	Inter neutral	Heavy neutral
Density	0.9215	0.9467
Pour point (°C)	42	41
Kinematic viscosity 100°C	9.04	22.17
VI	60	56
Wax content (%)	8.1	4.7
Saturates (paraffins+naphthenes) (%)	43.5	35.9
Aromatics (%)	56.5	64.1

4. Aromatics have low VI, low pour point, and low oxidation stability.

5. Long-chain alkyl aromatics have low VI, low pour point, and medium oxidation stability.

6. Polycyclic aromatics have low VI, low pour point, and high oxidation stability.

Therefore, in a refinery, various processing steps are selected to remove undesirable molecules and enrich the vacuum distillate with the desirable compounds. In a conventional refinery, the following steps are followed [2]:

1. The vacuum distillates are subjected to solvent extraction by sulfur dioxide or phenol or furfural or n-methyl pyrrolidone (NMP) to remove mixture of aromatics (mainly bi- and polycyclic aromatics). Aromatics have higher affinity with these solvents. NMP is used up to 400% of the feed and is the preferred solvent for modern refineries. This process upgrades the VI of the feed by removing low VI aromatics.

2. This extract is called aromatic extract and is used as rubber processing oil. The dearomatized oil may now have VI of 95–115, but the pour point is generally not influenced by this process. The product still contains high amount of wax.

3. The removal of wax is carried out by solvent dewaxing process (by cooling with a mixture of toluene and methyl ethyl ketone) and removing the precipitated wax. The produced wax is called slack wax and is further processed to yield wax products. Depending on the severity of dewaxing process, base oils of pour points lower than −6°C can be produced. There is, however, some loss of VI due to the removal of short-chain paraffins. The desired VI is therefore controlled by the VI of solvent-extracted dearomatized oil. Dewaxing process thus controls the pour point of the feed by removing high-pour waxes.

4. Heaviest or high-viscosity base oils called BS (Bright stocks) are produced from the vacuum residue. The residue contains substantial amount of asphaltenes and resins. These must be removed before subjecting the deasphalted oil to the usual solvent extraction and solvent dewaxing process. The asphaltenes and resins are generally removed by propane deasphalting process, where propane to feed ratio is maintained at 4–10 vol/vol and at a pressure of 37–44 kg/cm². Propane dissolves asphaltenes and resins, and the deasphalted oil can now be subjected to usual solvent extraction and dewaxing processes to produce BS.

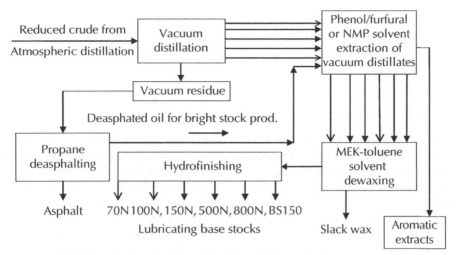

FIGURE 3.1   Conventional refinery production of lubricating base oil.

5. Finally, the solvent-extracted and solvent-dewaxed products are subjected to the hydrofinishing treatment to stabilize the product with respect to color, color stability, and thermal and oxidative stability. In the earlier processes, this was achieved by clay treatment.

6. Hydrotreating process for base oil was developed in the 1950s by Amoco and others as an additional finishing step at the end of a conventional solvent refining process. It is a process for adding hydrogen to the base oil at elevated temperatures in the presence of catalyst to stabilize the most reactive components in the base oil, improve color, and increase the useful life of the base oil. This process removed some of the nitrogen- and sulfur-containing molecules. This process is, however, not severe enough to remove a significant amount of aromatics. Hydrotreating was only a small improvement in base oil technology.

These processes are based on separation, and therefore, the crude oil processed must have the potential to produce lubricating base oils. The production of lube base oils depends on the specific crudes. All crudes cannot produce lubricating base oils economically. The conventional lube refining process flow diagram is provided in Figure 3.1. The lubricating base oils produced in such a configuration are called neutral oils, and different viscosity grades can be produced depending on the market need (from 70 neutral to BS). These are also sometimes referred to as solvent-refined, dewaxed oils. Different refineries produce different types of neutral base oil (referred to as solvent neutral (SN)), depending on the crude oil quality and market need. Some of the typical oils produced are as follows:

SN-70

SN-100

SN-150

SN-350

SN-500

SN-650

SN-800

SN-1300

BS 150 and 200

Just few decades back, most of the lubricants were manufactured [3] by using solvent-extracted and dewaxed oils produced in a refinery using the processes described earlier. Any need to improve base oil quality was addressed by the incorporation of suitable additives such as antioxidants to improve life and pour point depressant and VI improvers to improve rheological properties. Refiners have been resisting lube quality upgradation by conventional processes due to yield losses, and additive treatment was the only choice to improve finished lubricant quality. However, the finished lubricant quality was continuously improved to meet the new equipment requirements and environmental regulations, and thus, a shift toward the use of synthetic lubricating base oil in critical applications was witnessed. The emergence of all-hydroprocessing routes [4] and gas-to-liquid (GTL) technology to produce base oils of higher quality levels has permitted oil producers to meet the modern lubricant requirements at a lower cost.

Currently, several choices are available to formulate different quality-level products using a variety of base oils consisting of synthetic and mineral oils. These are as follows:

1. API (American Petroleum Institute) group I oils, solvent extracted, dewaxed, and hydrofinished
2. API group II oils, using hydroprocesses with some conventional process
3. API group III oils, through hydrocracking processes or chemically modified oils
4. GTL base oils of API group III, produced by chemical synthesis from syngas
5. Synthetic polyalphaolefins (PAOs), API group IV oils
6. Synthetic polyinternalolefins (PIOs)
7. Synthetic esters, alkyl aromatics, PAG, phosphate esters, and all other synthetics, API group V oils

HYDROPROCESSES FOR LUBE PRODUCTION

New refinery processes have now been developed to produce improved quality of base oils in higher yield by using hydroprocessing [5–7]. Both dearomatization and dewaxing can be carried out by hydroprocesses, without actually removing molecules. Aromatics and olefins can be saturated by hydrogen and wax can be isomerized to isoparaffins or *n*-paraffins can be selectively hydrocracked without much loss of yield. Sulfur, nitrogen, and oxygen can be removed from the molecules by hydrogen treatment. Newer hydroprocessing technologies also allow low-viscosity base oil production from paraffin wax cracking and also from GTL technology. Hydrocracking is a severe form of hydroprocessing at higher temperatures and pressure. In this process, molecules are restructured and cracked into smaller molecules by breaking of C–C bond, scission of condensed ring structure, and restructuring of molecule by isomerization. This also removes most of the sulfur and nitrogen compounds. Aromatics are saturated and naphthenic rings opened up and paraffin isomers are redistributed. Hydrocracking technology continued to improve [8, 9]. In 1969, the first hydrocracker for base oil manufacturing was commercialized by Idemitsu Kosan in Chiba Refinery, Japan, using technology licensed by Gulf [10].

These improved hydroprocessed base oils containing lower amount of aromatics, sulfur, nitrogen, and waxes and having high VI are required for the formulation of premium low-viscosity and energy-efficient multigrade oils like 10W-30, 5W-30, or 0W-30 oils. Modern engine oils meeting API SM/SN and ILSAC GF-4/GF-5 specifications require base oils of low volatility and higher VI to achieve higher fuel economy and lower oil consumption. For 5W-XX engine oil, typical Noack volatility of 15% is required, while API group I base oils have volatility of the order of about 25%. If base oils of low volatility are produced in a conventional refinery, the yield will go down considerably. The new hydroprocessing routes were, therefore, developed [11–16] to meet these requirements. The first catalytic dewaxing and wax isomerization technologies were introduced in 1970. Shell utilized hydroisomerization with solvent dewaxing to produce high-viscosity-index base oils in Europe. Mobil developed catalytic dewaxing process and combined it with solvent dewaxing unit to produce improved neutral oils. Chevron combined catalytic dewaxing with hydrocracking and hydrofinishing to develop an all-hydroprocessing route for the manufacture of lubricating base oils. The modern hydroprocesses convert polyaromatics into naphthenes by saturating the aromatic ring. Isomerization converts high-pour linear paraffins into branched molecules that have lower pour points. Some of the saturated compounds can also undergo ring opening and get converted to low pour point alkanes. Sulfur and nitrogen are also removed during the processes in the form of hydrogen sulfide and ammonia. These changes can be depicted by the following reactions:

1. Sulfur removal reactions—hydrodesulfurization

$$R - SH + H_2 \rightarrow RH + 2H_2S$$
Thiol

$$R - S - S - R + 3H_2 \rightarrow 2RH + 2H_2S$$
Disulfide

$$R - S - R + 2H_2 \rightarrow 2RH + H_2S$$
Sulfide

$+ 3H_2 \rightarrow C_4H_8 + H_2S$

Thiophene

R can be an aromatic, alkyl, or heterocyclic group.

2. Nitrogen removal reactions—hydrodenitrogenation

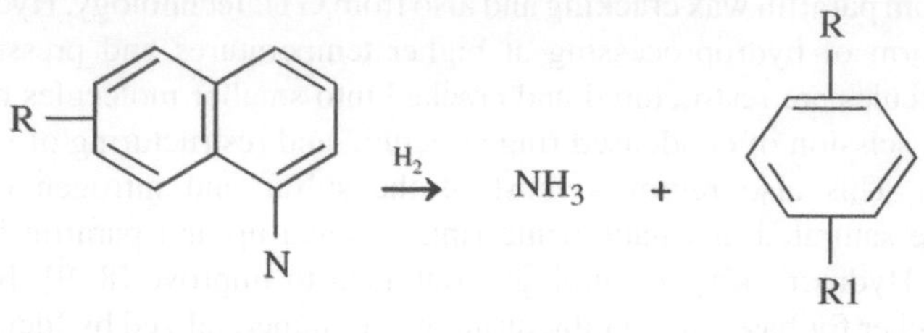

Alkyl quinoline   Ammonia   Alkyl benzene

3. Polyaromatic saturation

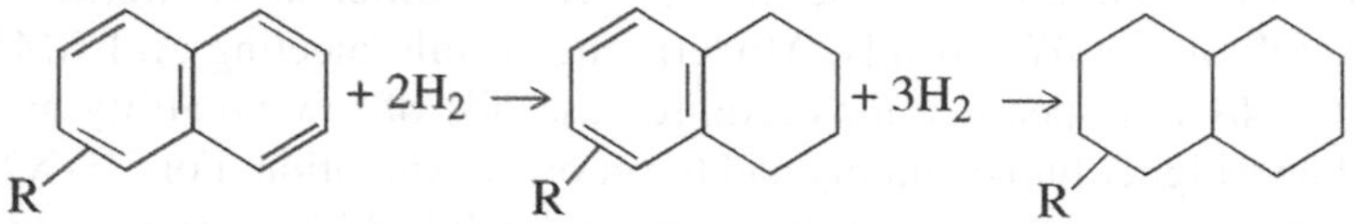

4. Isomerization reaction

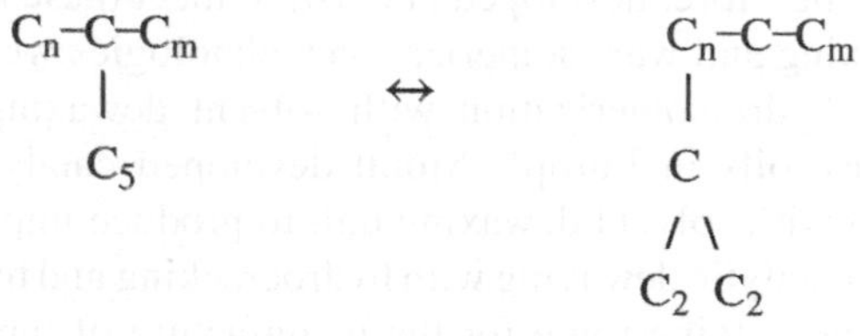

Paraffin   Iso-paraffin (lower pour point and viscosity)

5. Hydrocracking

$$C_nH_{2n+2} + H_2 \rightarrow C_aH_{2a+2} + C_bH_{2b+2} \quad \text{where } a+b=n$$

6. Olefin hydrogenation

Olefins are unstable molecules and can be completely hydrogenated to yield stable molecules. The reaction is highly exothermic. The addition of H_2 to alkenes yields an alkane:

$$RCH = CH_2 + H_2 \rightarrow RCH_2CH_3 \left(R = \text{alkyl, aryl}\right)$$

These reactions can be conveniently utilized to get rid of undesirable molecules and generate most desirable molecules in the lubricating base oils. Modern refineries utilize hydroprocesses to produce API group II and III oils. Typical operating conditions of hydroprocesses are provided in Table 3.2.

TABLE 3.2   Typical operating conditions of hydroprocesses for lube base oils

Parameters	Hydrocracking	Hydrorefining	Catalytic dewaxing	Hydroisomerization or isodewaxing	Hydrofinishing
Pressure (psig)	2500–3000	1500–3000	250–3000	500–2500	500–1000
Temperature (°C)	385–440	260–315	275–370	315–370	260–315
Space velocity (h^{-1})	0.5–1.0	0.5–1.0	0.5–5.0	0.3–1.5	1.0–1.5
Hydrogen recycle, SCFB	3500–5000	1000–3000	500–5000	—	300–1000

In a refinery, these hydroprocesses can be incorporated in different combinations with conventional processes to obtain operational and cost benefits. In one of the combination, the hydroprocessing is incorporated in the existing lube refinery along with the solvent extraction and solvent dewaxing processes. API group II base oils can be produced by subjecting the raffinate from solvent extraction unit to hydroconversion followed by solvent dewaxing (Fig. 3.2, Scheme 3). Solvent dewaxing route is preferred, when wax production is desired. However, if wax is not required, the streams can be hydrodewaxed to increase the yield and VI. This approach is quite cost-effective since the preextraction allows the hydroprocessing at a relatively mild temperature and pressure with lower hydrogen consumption. Through this process, improved API group I base oils can also be produced by resorting to milder extraction, which offset the yield loss due to hydroprocessing. The lube yield can be further improved by introducing hydroisomerization process, which converts waxes into isoparaffins. This combination provides both API group II and group III base oils. Hydroisomerization also permits the raffinate hydrotreater to operate at lower severity, which further increases the yield of base oils.

CATALYTIC HYDROPROCESSING/DEWAXING PROCESS

Catalytic hydrodewaxing is an alternate process to the conventional solvent dewaxing, which selectively cracks long-chain paraffin wax molecules into light petroleum gases and naphtha. This process is quite cost-effective and yields low-pour-point products. Modern catalytic dewaxing process combines both catalytic cracking and wax isomerization.

Exxon Mobil catalytic isomerization and Chevron–Texaco isodewaxing processes are the two main commercial processes available, and several plants are in operation based on these technologies. This process in combination with conventional solvent extraction unit produces good-quality API group I base oils (Scheme 1 of Fig. 3.2).

Further developments in catalyst technology led to the hydroisomerization process, which selectively dewaxes the streams from lube hydrocracker to yield

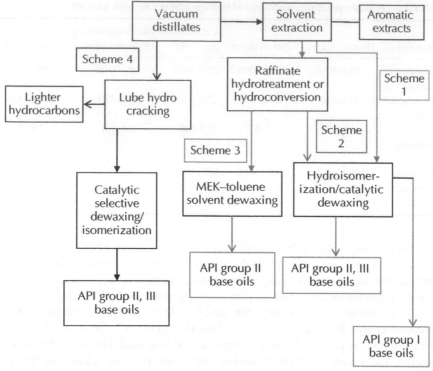

FIGURE 3.2   Integration of hydroprocessing with conventional extraction and dewaxing processes to produce API group I, II, and III base oils.

good-quality API group II and III base oils (Scheme 2 of Fig. 3.2). This configuration, however, consumes more hydrogen but is more flexible in the selection of crudes, and even nonlube crudes can be processed. Selective dewaxing process also yields higher-viscosity-index oils.

The process flow diagram of lube base oils produced through hydrocracker and selective dewaxing processes is provided in Figure 3.2 Scheme 4. This process is also being used commercially to produce good-quality base oils [17]. The lube hydro-cracking steps convert aromatic molecules into saturated compounds having higher VI and also simultaneously remove undesirable elements such as sulfur and nitrogen compounds. The loss of yield in the hydrocracker unit due to higher conversion into lighter compounds is made up by the selective dewaxing process. This process does not remove wax physically but converts them into isoparaffins having low pour points.

WAX HYDROISOMERIZATION PROCESS

Slack wax containing about 70% wax can be hydroisomerized to produce API group III base oils. This process can produce product of low pour point up to −50°C. Jet fuel and diesel fuel are obtained as by-products. This approach has been utilized in several refineries in Europe. However, only lower-viscosity base oils can be produced through this route.

Fuel hydrocracker bottoms also contain high amount of waxes that can be conveniently converted to high-quality lubricating base oils by catalytic hydrodewaxing followed by hydrofinishing/hydrotreatment. This process also yields low-viscosity base oils of API group II and III quality.

Modern refineries thus have options to combine conventional solvent extraction or solvent dewaxing processes with different hydroprocesses to obtain a cost-effective solution to produce desired quality of lubricating base oils. All-hydroprocessing route can also be utilized to produce higher-quality base oils while processing different crude oils. Various options are depicted in Figure 3.2.

High-viscosity high-quality group II lube base oils of 4–6 cSt at 100°C have been produced [18] by an integrated hydrocracking and dewaxing process with improved properties.

UNCONVENTIONAL OR CHEMICALLY MODIFIED LUBRICANT BASE OILS

According to API-1509 (Table 3.3), the difference between group II and group III base oils is only in terms of the VI. Base oils with VI 80–120 are group II and base oils with VI above 120 are group III. Group III oils are also referred to as unconventional base oils (UCBOs) or very high VI (VHVI) base oils. Solvent-dewaxed group III base oils have been produced in Europe for more than 10 years, primarily by Shell and BP. Group III base oils are now produced in North America through all-hydroprocessing routes. These modern group III oils have improved oxidation stability and low-temperature performance as compared to the solvent-refined group III oils. This has resulted in the wider application of this technology to produce group III base oils. Modern group III base oils have properties superior to the conventional base oils and approach close to the properties of PAOs.

The base oils produced through lube hydrocracker or fuel hydrocracker bottoms are now called as chemically modified base oils, since most of the chemical compounds present in the original vacuum distillates have been modified. Nitrogen, sulfur, and oxygen from the molecules are removed, aromatics have been saturated into naphthenes, long-chain paraffins have been isomerized, and rings have been opened up. With such drastic changes, the molecules have been virtually reformu-

TABLE 3.3   API-1509: Base oil categories

Base oil category	Sulfur (%)	Saturates (%)	VI
Group I	>0.03	≤90	80–120
Group II	≤0.03	≥90	80–120
Group III	≤0.03	≥90	≥120
Group IV	All PAOs		
Group V	All others not included in groups I, II , III, and IV		

lated, and it no longer qualifies to be called just mineral oil. These oils have actually become synthetic and are also very close to PAOs in properties at a very low cost. There is a convention to call such oils as API group II+ or API group III+ oils depending on the properties. Base oils produced through GTL technology are actually synthetic, since these are synthesized from CO and hydrogen only.

GTL BASE OILS

GTL base oils are produced from carbon and hydrogen containing gaseous products such as hydrogen, carbon monoxide, carbon dioxide and gaseous hydrocarbons by various synthetic routes. The process basically involves production of syngas from coal/natural gas/ heavy residue/petroleum coke or biomass and then synthesizing hydrocarbon molecules through Fischer–Tropsch (FT) process. In FT process, synthesis gas, a mixture of H_2 and CO, is catalytically converted into liquid hydrocarbons. The end products are mixtures of hydrocarbons, which on distillation yield high-boiling waxy products. These can be further subjected to catalytic hydrodewaxing or isomerization to yield good-quality lubricating base stocks having above 99% saturates. Such base oils are generally of API group III quality level and are virtually free from sulfur (<5 ppm) and nitrogen (<25 ppm) and have very low aromatic content. VI of such oils is of the order of 130. Syngas is a mixture of hydrogen and carbon monoxide that is used to reconstruct organic molecules through FT reactions. German researchers Franz Fischer and Hans Tropsch, working at the Kaiser Wilhelm Institute in the 1920s, discovered this process, and since then, many modifications have been made, and the term "Fischer–Tropsch" now applies to a wide variety of similar processes. The process was discovered in petroleum-deficient but coal-rich Germany to produce liquid fuels. It was used by Germany and Japan during World War II to produce fuels.

Fisher and Tropsch reacted carbon monoxide with hydrogen (syngas) in the presence of a catalyst to obtain products having more than one carbon atom. The syngas can be produced from coal or natural gas. Modern GTL process typically produces a waxy syncrude consisting of straight-chain hydrocarbons in the range of C5–C20. The FT reaction is carried out at a temperature of 250–350°C and a pressure of about 100–3500 psi in the presence of an iron or cobalt catalyst. FT synthesis requires hydrogen/carbon monoxide in the ratio of about 3:1, and the syngas is, therefore, subjected to water–gas shift reaction to enrich it with the desired level of hydrogen.

Water–Gas Shift Reaction

$$CO + H_2O \rightarrow CO_2 + H_2$$

Overall FT reactions can be represented by the following simplified equation. This reaction, however, takes place through a series of discrete endothermic reactions. The molecules are formed by the reaction of CO and carbon dioxide with hydrogen in several steps:

$$nCO + (2n+1)H_2 \rightarrow C_nH_{2n+2} + nH_2O$$

FT reactions are carbon chain building reactions where methylene groups are attached to the carbon chain. The actual mechanism is not fully understood. The following endothermic reactions are believed to take place in the presence of iron or cobalt catalyst:

$$CO + 2H_2 \rightarrow -CH_2 - + H_2O$$

$$2CO + H_2 \rightarrow -CH_2 - + CO_2$$

$$CO + H_2O \rightarrow H_2 + CO_2$$

$$CO_2 + 3H_2 \rightarrow -CH_2 - + 2H_2O$$

$$3CO_2 + H_2 \rightarrow -CH_2 - + 2CO_2$$

The hydrocarbon mixture thus obtained can be fractionated in the usual petroleum refinery to get fuels and other higher products like lubricating base oils after further processing through usual refinery hydroprocesses. These products are free from sulfur and nitrogen and would meet the most recent specifications. Lubricating base oils produced by FT synthesis are generally of API group III quality.

Petroleum coke, tar sands, and heavy residues can also be subjected to gasification in a similar manner for the production of value-added fuels and other products. As the gasification technology has matured, a low-emission integrated gasification combined cycle (IGCC) technology has been evolved, which generates electricity at a higher efficiency as compared to the conventional thermal power plants. This process also produces hydrogen that can be used in direct fired engines or in fuel cell-powered vehicles.

The IGCC plant has a gasifier, a gas purifier, a gas turbine, and a heat recovery system from the gasifier and gas turbine. The heat thus recovered produces steam, which drives a steam turbine to produce electricity. The combined efficiency of IGCC plant is over 40% as compared to about 37% for the thermal power plant. A large size of IGCC plant can produce fuels and lubricating base oils through FT synthesis. A typical complex with IGCC and FT synthesis is depicted in Figure 3.3.

NATURAL GAS AND GTL

Natural gas can also be converted to liquids directly or by following FT synthesis after gasification. This process called GTL technology is useful where large reserves of gas are available. Gas can be conveniently converted to liquids and then stored and transported to the marketplace. All GTL fuels and lubricants are free from sulfur and nitrogen and meet the ultralow sulfur specifications. Methane or other hydrocarbons

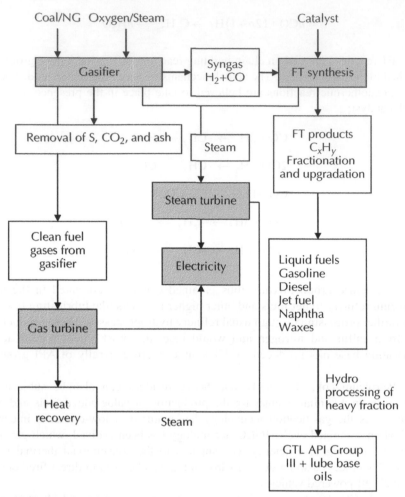

FIGURE 3.3   IGCC and FT synthesis to produce power, fuels, and base oils.

are easily converted to syngas by steam reforming. The reaction is exothermic and can be depicted by the following equation:

$$CH_4 + H_2O \rightarrow CO + 3H_2 \quad \text{or} \quad C_nH_m + nH_2O \rightarrow nCO + \left(\frac{n+m}{2}\right)H_2$$

The thermal cracking reaction may also take place, producing carbon and hydrogen. Carbon may again react with steam to produce CO and hydrogen:

$$CH_4 \rightarrow C + 2H_2$$

$$C + H_2O \rightarrow CO + H_2$$

Produced gases are deficient in hydrogen. Water–gas shift reaction is used to produce more hydrogen. This reaction is endothermic:

$$CO + H_2O \rightarrow CO_2 + H_2$$

Syngas thus produced from hydrocarbon can be used in FT process to get liquid products.

Sasol in South Africa uses coal and natural gas as a feedstock and produces a variety of synthetic petroleum products including country's most of the diesel fuel requirement. Sasol plant has more than 50 years of experience converting syngas from coal to liquid fuels and claims that syngas from any source can be converted to clean fuels. These processes are advantageous when the gas reserves are at a far-off location and its transportation to the users poses difficulties. GTL base oils are of API group III quality.

CHARACTERIZATION OF LUBRICATING BASE OILS

Lubricating base oils are mixtures of a large number of chemical compounds and are therefore characterized by the following properties related to their end performance:

Viscosity

VI (indicates viscosity–temperature relationship)

Specific gravity

Pour point

Acidity

Aniline point (indicates aromaticity)

Saturate content

Carbon residue

Flash point

Sulfur content

Volatility

Air release value

API (API-1509) has categorized all the base oils into five groups. These are called as groups I–V. All base oils are defined by their sulfur content, saturate content, and VI. Table 3.3 provides the API-1509 classification of base oils. This is quite useful for the selection of base oils for the formulation of engine oils and industrial oils.

Among the synthetic oils, API group IV base oils, POAs, are the largest selling products [19]. These are available in various viscosities of 2, 4, 6, 8, 10, 40, and 100 cSt at 100°C and are hydrogenated oligomers of an olefin, generally decene. Decene can be obtained by controlled polymerization of ethylene. PAOs are essential in formulating modern high-performance low-viscosity multigrade engine oils, transmission oils, compressor oils, and aviation oils.

TABLE 3.4   Some typical characteristics of API Group I, II, III, PAOs and GTL base oils

	Specific Gravity at 60 °F	Sulfur (%wt.)	Viscosity Index	Kinematic Viscosity 40°C	Kinematic Viscosity 100°C	Pour Point (°C)
	API Group 1					
150 neutral	0.861	0.120	98	24.38	4.55	−15
500 neutral	0.878	0.150	96	107.0	11.0	−12
150 BS	0.895	0.260	95	438.0	29.46	−6
	Hydro treated API Group II					
150 neutral	0.862 Saturates 99%	2ppm	105	31.3	5.38	−21
500 neutral	0.867 Saturates 99%	2ppm	106	89.59	10.85	−15
	Hydro treated API Group III					
150 neutral	0.861 Saturates 99.6%	2ppm	121	29.70	5.41	−15
	GTL base oils API Group III					
GTL base oils	Saturates >99%	S,< 5 ppm N,< 20 ppm	130	—	2,4,6 Cst	−10 to −30
	API group IV base oils—PAOs					
PAOs,2,4,6,8,10,40 and 100 cSt	>99%	S and N ND	120–170	5.5–1400	2–100	−27 to −72

PIOs (polyinternalolefins) have been produced in Europe by polymerizing C15–C16 internal olefins with BF3 and a proton source as catalyst. The products thus obtained are similar to regular PAOs in most properties. These have been extensively evaluated in gasoline and diesel engine tests to replace PAO, and European ATIEL EELQMS has recognized PIOs as a new category group VI oils allowing full interchangeability [20] with group IV PAOs, without the need to carry out additional engine tests. API has, however, not included this category in their base oil interchangeability guidelines.

Typical properties of some neutral oils (produced by conventional method), GTL base oils and hydroprocessing route are provided in Table 3.4. For comparison, PAO properties are also given in the end of the table. Noticeable differences are in VI, pour point, saturate content, and sulfur content. As the degree of refining is increased, other properties like emulsion characteristics, air release value, and carbon residue are also improved. The following inherent properties of base oils are related with the molecular structure and cannot be modified by the use of additives:

Volatility

Flash point

Specific gravity

Air release value

Cloud point

Thermal stability

Carbon residue

Aniline point

Biodegradability

These properties depend on the molecular structure of large number of compounds present in the base oil [21, 22]. However, several important properties can be conveniently modified by the use of chemical additives. These properties are also important for the lubricant performance in the equipment. API group IV oils are synthetic PAO and have much superior properties as compared to mineral oils. However, the API group III and GTL base oils have come quite near to the PAO properties and can be used in many applications where earlier PAOs were only required.

Base oil quality has considerable impact on the oxidation stability in turbine and hydraulic oils. The Turbine Oil Stability Test (TOST), or ASTM D-943, measures the time required for a turbine oil to oxidize to the point where the total acid number reaches 2.0 mg KOH/g. API group I base oil without additives have only about 200 h of TOST life. A modern high-quality VG 32 turbine oil formulated with group I base oil typically will have a TOST life of maximum 5,000 h, while a high-quality formulated turbine oils based on group II oils can run for 8,000–10,000 h in this test. Oils formulated with group III base oils will have TOST life comparable with the oils formulated with PAO of similar viscosity. With greater supply of higher-quality oils (groups II and III), long-life energy-efficient products now utilize such oils for formulating these oils [23, 24].

Chemical additives can modify or improve several existing properties of base oils. Some new properties such as antiwear and extreme pressure (EP) can also be imparted by additives. The following properties can be influenced by chemical additives:

Viscosity

VI

Pour point

Rheological properties at low and high temperatures

Friction properties

Detergency

Dispersancy

Oxidation stability

Antiwear

EP

Load-carrying capacity

Foaming

Water separation characteristics

Rusting

Corrosion

Some properties like viscosity, VI, pour point, and water-separating characteristics are indicative of degree of refining in the refinery through processes like dewaxing, extraction, and hydrofinishing. It is, therefore, desirable to obtain these properties to the desired level in the refinery itself so that the oils can be used for the major applications in automotive and industrial oils. Usually, a pour point of −6 to −12°C and VI of 95–98 are adequate for API group I oils. Base oil properties influence several lubricant performance characteristics. For example, viscosity and VI influence oil film thickness, wear, pumpability, fluidity, and energy consumption. Oil consumption, deposit formation, and oil thickening are influenced by volatility and flash point. Presence of surface compounds in base oils influence foaming, air release, emulsification, and demulsification. Aniline point/solvency/presence of aromatics effects seal compatibility, engine cleanliness, and additive solubility. Oxidation stability is one of the most important properties influencing the life of lubricants, acidity buildup, oil thickening, deposit formation, and metal corrosion. A similar correlation has been provided by Webster model with the base oil properties and performance characteristics.

Lubricating base oils produced by hydrodewaxing, hydroisomerization, or base oils obtained through the GTL technology are highly paraffinic in nature and have low solvency properties. This property can be improved by the incorporation of a synthetic ester in limited amount. This approach also improves the seal swell characteristics of GTL base stocks or PAO/PIOs, or group III base oils.

Lubricants are used in a variety of equipment and engines operating in different conditions of temperatures and pressures and need a set of properties so that the equipment provides trouble-free service for a longer time. Each lubricant specification has been carefully defined to meet these requirements. This is only possible through the use of suitable chemical additives. A careful selection of the additive combination is therefore necessary in a finished lubricant. Additives exhibit both synergistic and antagonistic effects when used in combination, and this requires careful selection and evaluation. For example, the antioxidant property of an antioxidant is enhanced when used in combination with a metal deactivator. However, the antioxidant properties are reduced when active EP and antiwear compounds are also present. The dosage of each additive in a particular product is of great importance, since this decides the cost and performance of the product.

Lubricant technology is thus the science of combining appropriate base oils with additives in a most cost-effective manner to obtain desired viscosity and satisfactory performance in the intended equipment.

Gasoline engine oils will typically contain the following groups of additives:

Detergent, dispersant, antioxidant, antiwear, corrosion inhibitor, pour point depressant, antifoam, viscosity modifiers (for multigrade oils), and friction modifiers (for energy efficiency).

Diesel engine oils will also contain these additives, but their ratio and nature of additives will vary; for example, diesel engine oils will require more of detergent/dispersant to take care of deposits and soot dispersion. Diesel engine oils would also require a different type of friction modifier to withstand higher temperatures as compared to gasoline engine oil.

Marine oils, on the other hand, will need high TBN detergent and dispersant of specific quality to combat higher sulfur and high-boiling fractions in the residual fuels.

Industrial oils generally do not need detergents and dispersants, since these do not come in contact with the fuel combustion products, but need antirust, antifoam, and oxidation inhibitor along with antiwear and EP additives (wherever these properties are required). The nature of these additives and dosage vary according to the equipment requirement and need to be optimized in every grade. Emulsifying metalworking fluids will require an appropriate surface active compound.

Thus, the lubricant formulation methodology involves a very careful selection of base oils and chemical additives along with elaborate testing facilities in the laboratory followed by field testing in the actual equipment for the desired performance.

BASE OIL TOXICITY

Long-term carcinogenicity studies have been reported for different base oils. The toxicity testing has consistently shown that lubricating base oils have low acute toxicities. Several tests have shown that lubricating base oil's mutagenic and carcinogenic potential correlates with its 3–7-ring PAC content and the level of DMSO extractable (e.g., IP346 assay). Base oil toxicity is also related to the degree of refining (World Health Organization (WHO) [25]; International Agency for Research on Cancer (IARC) [26], API [27]; CONCAWE [28]).

Unrefined and mildly refined base oils have LD_{50}s of greater than 5000 mg/kg (bw) and greater than 2 g/kg (bw) for the oral and dermal routes of exposure, respectively, which have been observed in rats dosed with an unrefined light paraffinic distillate (API, 1986d). The same material was also reported to be "moderately irritating" to the skin of rabbits (API, 1986d). When tested for eye irritation in rabbits, the material produced Draize scores of 3.0 and 4.0 (unwashed/washed eyes) at 24h, with the scores returning to zero by 48h (API, 1986d). The material was reported to be "not sensitizing" when tested in guinea pigs (API, 1986d).

Highly and Severely Refined Base Oils

Multiple studies of the acute toxicity of highly and severely refined base oils have been reported. Irrespective of the crude source or the method of refining, the oral LD_{50}s have been observed to be greater than 5 g/kg (bw), and the dermal LD_{50}s have ranged from greater than 2 to greater than 5 g/kg (bw) (API, 1986c; CONCAWE [28]). The LC_{50} for inhalation toxicity ranged from 2.18 to greater than 4 mg/l (API,

1987a; CONCAWE [28]). When tested for skin and eye irritation, the materials have been reported as "nonirritating" to "moderately irritating" (API, 1986c; CONCAWE [28]). Testing in guinea pigs for sensitization has been negative (API, 1986c; CONCAWE [28]).

The base oil quality and their processing techniques have undergone substantial changes during the last two decades. All hydroprocessed base oils are approaching the quality levels of PAOs and other synthetic oils and are virtually free from aromatics, sulfur, oxygen, and nitrogen compounds. Thus, the modern highly refined base oils such as hydroprocessed oils do not pose any toxicity problems. However, some of the additives used in formulating finished products have higher toxicity, and due care is warranted in using fully formulated oils.

REFERENCES

[1] Singh H, Anwar M, Chaudhary GS, Kaushik RS. Hybrid technology—an innovative approach to make HVI base oils. Proceedings of the International Symposium on Production and Application of Lube Base Stocks; November 23–25; Indian Institute of Petroleum, Dehradun, India. New Delhi: Tata McGraw Hill; 1994. p 149–155.

[2] Gupta SC, Handa S. Lube base oil production technologies, present status and future trend. Proceedings of the International Symposium on Advances in Production and Applications of Lube Base Stocks; November 23–25; New Delhi. New Delhi: Tata McGraw Hill; 1994. p 138–148.

[3] National Petroleum Refiners Association. *1999 Lubricating Oil and Wax Capacities of Refiners and Re-Refiners in the Western Hemisphere*. Washington, DC: NPRA; January 1999.

[4] Sequeira A Jr. *Lubricant Base Oil and Wax Processing*. New York: Marcel Dekker; August 1994. Chemical Industries Series.

[5] Sequeira A. *Lubricating Base Oil and Wax Processing*. New York: Marcel Dekker; 1994.

[6] Qureshi W. Recent advances in lube hydroprocessing technology. 2nd Annual Fuels and Lube Asia Conference; January 29–31, 1996; Singapore.

[7] Howell RL, Srinivas B, Mayer JF. Manufacturing base oil for the 21st century. International Conference, Petrotech; January 9–12, 1999; New Delhi.

[8] Stormont DH. New process has big possibilities. Oil Gas J 1959;57 (44):48–49.

[9] Zakarian JA, Robson RJ, Farrell TR. All-hydroprocessing route for high-viscosity index lubes. Energy Prog 1987;7 (1):59–64.

[10] Wilson MW, Mueller TA, Kraft GW. Commercialization of isodewaxing—a new technology for dewaxing to manufacture high quality lube base stocks. Presented at 1994 NPRA, National Fuels and Lubricants Meeting; November 3–4, 1994; Houston, FL-94-112; 1994.

[11] Miller SJ, Shippey MA, Masada GM. Advances in lube base oil manufacture by catalytic hydro processing. NPRA National Fuels and Lubricating Meeting; November 5–6, 1992; Houston.

[12] Jacob SM. Lube base oil processing for the 21st century. 4th Annual Fuels and Lubes Asia Conference; January 14–16, 1998; Singapore.

[13] Cohen SC, Mack PD. HVI and VHVI base stocks. The World Base Oil Conference; December 2–3, 1996; London.

[14] Moon WS, Cho YR, Yoon CB, Park YM. VHVI base oils from fuel hydro cracker bottoms. The World Base Oil Conference; December 2–3, 1996; London.

[15] Mukhopadhyay PK, Singh H. Recent trend in production of lube base oils. Proceedings of the International Symposium on Production and Application of Lube Base Atocks; November 23–25, 1994. p 7–23; New Delhi.

[16] Kramer DC, Lok BK, Krug RR. The evolution of base oil technology. In: Herguth WR, Warne TM, editors. *Turbine Lubrication in the 21st Century*. West Conshocken: ASTM; January 1, 2001. ASTM Special Technical Publication 1407.

[17] Howell RL. Hydroprocessing routes to improved base oil quality and refining economics. 6th Annual Fuels and Lubes Conference; January 25–28, 2000; Singapore.

[18] Holtzer GL, Dandekar AB, Baker CL, Fingland BR, Hagee BE, Yung C, Wang FC, Sanchez E, Pathare RP. High viscosity high quality group II lube base stocks. US patent 2013,0264,246. October 10, 2013.

[19] Bui K. Synthetics II. *Lubricants World*, November 1999.

[20] Navarrini F, Ciali M, Cooly R. Polyinternalolefins. In: Rudnick LR, editor. *Synthetics, Mineral Oils, and Bio-Based Lubricants—Chemistry and Technology*. Boca Raton: CRC Press; 2006. p 41–43.

[21] Sarpal AS, Sastry MIS, Bansal V, Inder Singh, Mazumdar SK, Basu B. Correlation of structure and properties of groups I to III base oils. Lubr Sci August 2012;24 (5):199–215.

[22] Kapur GS, Sarpal AS, Bhatnagar AK. Temperature-dependent effects in base oils: Carbon-13 NMR spin-lattice relaxation time and viscome try studies. Lubr Sci May 2002;14 (3):287–302.

[23] Min PY. VHVI base oils: supply and demand. 4th Annual Fuels and Lubes Conference; January 14–16, 1998; Singapore.

[24] Lehmus M, Cabanel C. Top tier base oils for high performance lubricants with reduced environmental impact. In: Srivastava SP, editor. Proceedings of the International Symposium on Fuels and Lubricants (ISFL-2000); March 10–12, 2000; New Delhi. New Delhi: Allied Publisher; p 45–52.

[25] World Health Organization (WHO). Selected petroleum products, Environmental Health Criteria 20. Geneva: WHO; 1982.

[26] International Agency for Research on Cancer (IARC). *IARC Monographs on the Evaluation of Carcinogenic Risks of Chemicals to Humans*. Volume 33, Polynuclear aromatic hydrocarbons. Part 2. Mineral oil. Lyon: IARC; 1984.

[27] American Petroleum Institute (API). Mineral oil review, health and environmental science, API Department Report No. DR 21. Washington, DC: API; January 1992.

[28] CONCAWE's Petroleum Products and Health Management Groups. Lubricating oil and base stocks. Product Dossier No. 97/108. Brussels: CONCAWE's Petroleum Products and Health Management Groups; 1997.

CHAPTER 4

SYNTHESIZED BASE OILS

Conventional crude oil-derived lubricating base oils always had the problem of consistent quality, since each crude oil is unique and the groups of chemical compounds present in them differ considerably. Consequently, each lubricant formulation in different lube base oils had to be reformulated or rebalanced to take care of the molecular composition variations. Certain lubricants such as turbine oils and compressor oils had to be formulated with specific base oil from a particular crude source. Additive response in these different base oils was different, especially with respect to oxidation stability [1–4]. This problem has been greatly resolved with the advent of hydrotreated or hydrocracked base oils of API group II and III quality levels [5–7], since during hydroprocessing several structural changes take place such as aromatic and unsaturated compounds are saturated and sulfur and nitrogen are removed. In API group III oils, saturated compounds become more than 99% irrespective of the crude source. Base oils produced by hydrocracking process are referred to as chemically modified mineral oils (CMMO).

There are still several applications of lubricants where even API group III base oils are not adequate to operate the equipment to its optimal level such as in aviation, where very-low- and high-temperature conditions are encountered and lubricant must perform with a high safety factor. Also, several industrial applications such as high-pressure compressors and sealed-for-life equipment need synthetic products. Synthetic lubricants are formulated with synthetic base oils, which are produced in a chemical plant and have well-defined molecular structures and properties. This allows the selection of particular group of synthetic base oils for an individual application to provide superior performance as compared with the mineral oil-based products. Technically, synthetic base oils can be used to formulate all of the existing mineral base oil-based products, but the only limitation is the high cost of synthetic oils (generally two to four times and higher) as compared to crude oil-based products. Therefore, synthetic oils are presently used in those situations where performance is the main concern and cost is of secondary importance. Such applications include aviation, space, military, sealed-for-life, and high-performance industrial equipment. During the last 70 years, several synthetic products have been developed. The most useful class of compounds falls into following categories:

Synthesized hydrocarbons

1. Polyalphaolefins (PAOs)
2. Polyinternalolefins (PIOs)

Developments in Lubricant Technology, First Edition. S. P. Srivastava.
© 2014 John Wiley & Sons, Inc. Published 2014 by John Wiley & Sons, Inc.

3. Alkylated aromatics/benzenes

4. Polyisobutylenes (PIBs)

Esters

1. Organic diesters

2. Organic polyolesters

3. Phosphate esters

4. Silicate esters

Others

1. Perfluoropolyalkyl/aryl ethers

2. Polyphenyl ethers

3. Chlorofluorocarbons

4. Polyalkylene glycols (PAGs)

5. Silicones

Currently, about 80% of synthetic lubricants are based on synthetic hydrocarbons (55%) and organic esters (25%). PAOs are the most popular synthetic hydrocarbons that have found extensive applications in all fields of lubrication. High-performance energy-efficient low-viscosity engine oils, transmission oils, compressor oils, hydraulic oils, gear oils, gas turbine oils, and aviation hydraulic oils are based on PAOs. The use of PAO in wind power plant as gear oil is increasing due to higher power generation through wind energy. The only shortcoming of PAOs is their inferior seal swell characteristic, which is modified by the incorporation of certain amount of organic esters. The other major group is PAGs (about 13%) for coolant, brake fluids, metalworking, gas compressors, gear oils, hydraulic fluids, and phosphate esters for fire-resistant hydraulic and compressor oils. Currently, the total demand for synthetic lubricants is only about 3% of the total lubricant demand of about 41 MMT/year. But based on the value, it would be about 10% of the lubricant value, due to the higher cost of synthetics.

The superior properties of synthetic lubricants over mineral oils arise out of the following characteristics:

1. Improved oxidation and thermal stability leading to longer oil life

2. Lower pour points that provide easy engine startability under colder climates

3. High viscosity index (VI) and good viscosity–temperature relationship providing wide temperature operability

4. Good lubricity properties leading to longer engine life due to less wear

5. Wide operating temperature range (low to high)

6. Lower volatility providing reduced oil consumption

It must, however, be remembered that synthetic oils are of several different molecular structures and not all synthetic oils have superior properties as compared to mineral oils. For example, phosphate esters have inferior hydrolytic stability, corrosion protection, and toxicological properties but have far superior fire-resistant properties as compared to mineral oils. Similarly, other synthetics will also have certain deficiency. In general, all synthetic oils are costlier than mineral oils. Therefore, the use of synthetic oils is restricted to the applications where mineral oils cannot perform due to their limitations. This is the reason why synthetic oils constitute only 3% of the total world lubricants. The following classes of synthetic oils are used in different applications.

POLYALPHAOLEFINS

PAOs are produced by first controlled polymerization of linear alpha olefins derived from ethylene to get decene. Polymerization is carried out in the presence of BF_3 and protic cocatalyst ROH or water and carboxylic acid [8]. Earlier attempts to obtain consistent quality of decene by using aluminum chloride as catalyst have been unsuccessful [9]. Decene is then oligomerized to obtain PAOs. PAO thus obtained have some unsaturation which is saturated with hydrogen to get stabilized PAOs. At this stage, these are mixtures of several grades of PAOs that on fractionation yield different grades of PAO.

$$R - CH = CH_2 \xrightarrow[\text{BF}_3 + \text{ROH}]{\text{Polymerization}} \text{Decene–I} \xrightarrow[\text{Catalytic hydrogenation}]{\text{oligomerization}}$$
$$\alpha - \text{Olefin}$$

$$\downarrow$$

$$CH_3 - CH - CH_2 - CH - CH_2 - CH_2$$
$$\quad\quad | \quad\quad\quad\quad | \quad\quad\quad\quad |$$
$$\quad C_8H_{17} \quad\quad C_8H_{17} \quad\quad C_8H_{17}$$

Polyalphaolefin

Commercially, several grades of PAO are produced, the most common being PAO having K, viscosity of 2, 4, 6, 8, and 10 cSt at 100°C. Higher grades of PAO such as 40 and 100 grades are produced selectively by using different catalyst systems [10, 11] other than BF_3 + ROH system. PAOs and their branch structures have been characterized by NMR spectroscopy [12]. The even-numbered PAOs are produced from the C10 alpha olefins. However, the odd-numbered PAOs such as PAO 2.5, PAO 5, PAO 7, and PAO 9 are produced from C12 alpha olefins. Low-viscosity PAO 2 cSt is used in industrial and automotive applications, such as hydraulic, transmission, and heat transfer fluids. The PAO 2.5, PAO 4, PAO 5, PAO 6, PAO 7, PAO 8, PAO 9, PAO 40, and PAO 100 are also used in various industrial, automotive, and aviation lubricants such as gear oils, compressor oils, hydraulic fluids, greases, engine oils, aviation lubricants, and other fluids. These lubricants based on PAOs provide specific characteristics such as high-pressure stability, high VI, low toxicity, low volatility, high oxidative stability, low-temperature fluidity, low flammability, high hydrolytic stability, and high-temperature stability.

Physicochemical characteristics of some of the commercially available PAOs are provided in Table 4.1. Actual values may differ slightly depending on the production processes and from company to company. Commercial companies manufacturing PAOs provide details of their product specifications and guidelines for using such products (e.g., Chevron Phillips PAO 2008 Rev. O.doc).

TABLE 4.1 Typical properties of some commercial PAOs

Properties	PAO 2	PAO 4	PAO 6	PAO 8	PAO 10	PAO 40	PAO 100
Kinematic viscosity at 100°C, cSt	1.7–1.8	3.8–3.9	5.8–5.98	7.74–7.8	9.87	40–42	103–110
Kinematic viscosity at 40°C, cSt	5.1–5.54	16.68–16.8	30.5–30.89	46.3–46.7	64.50	400–423	1,260–1,400
Kinematic viscosity at –40°C, cSt	255–306	2390–2432	7749–7830	18,200–19,877	34,600	39,000–41,000 at –18°C	175,000–203,000 at –18°C
VI	—	124	138–143	136–137	137	145–147	160–170
Pour point (°C)	–3 to –73	–69 to –72	–61 to –64	–55 to –57	–53	–36 to –45	–21 to –27
Flash point (°C)	161–165	213–226	235	258	270	275–280	280–290
Noack, % loss at 250°C	99.5	11.8–13	6.1	3.1	1.8	0.8–1.4	0.6–1.1

Very low pour point, high VI, good low-temperature rheological properties, high thermal stability, and low evaporation loss make PAOs the best choice for formulating a variety of high-performance industrial, automotive, transmission, and aviation lubricants. In most applications, these are often combined with about 10% organic ester so that the elastomer compatibility or seal swell properties of the formulated product are obtained. The percentage of ester could, however, vary depending upon the seal swell requirements. Low-viscosity energy-efficient engine oils such as 0W-30 or 5W-30 grades are best blended with PAO due to their superior low-temperature and low-evaporation-loss properties requiring no pour point depressant additive and viscosity modifiers. Only API group III base oils can offer some competition to PAO in these applications. PAOs' cost has always been higher than the mineral oils, and efforts have been made to reduce the cost by oligomerizing 1-dodecene [13] in place of decene. Such PAOs have 10–15°C higher pour point but have higher flash point and lower evaporation losses as compared to the decene-based PAO. 1-Tetradodecene may also be oligomerized to obtain PAOs of different characteristics. 1-Dodecene-based PAOs are distilled to odd-numbered viscosities as compared to even-numbered PAOs from decene. These odd-numbered PAOs (3, 5, 7, and 9 cSt) can be utilized in a number of industrial and automotive applications where very low pour point is not the requirement, but evaporation loss and/or cost is more important. A high-viscosity 25-cSt PAO has also been developed for industrial oils and grease application with −45°C pour point and 155 VI [14]. A highly shear-stable high-viscosity PAO of 135 cSt or greater at 100°C has been reported to be produced by mechanical breakdown of a high-viscosity PAO or by a selective catalyst system used in oligomerization or polymerization of the feedstock [15].

POLYINTERNALOLEFINS

To further optimize cost and performance over PAOs and hydrocracked mineral base oils, PIOs have been developed from internal n-olefins (C15–C16) that are obtained by the cracking of paraffins. These n-olefins have been oligomerized in the presence of BF_3 complex with a proton source [16, 17]. The products are then neutralized, hydrogenated, distilled, and fractionated to obtain stable products of 4-, 6-, and 8-cSt viscosity at 100°C. PIOs have VI of the order of 122–127 and pour point in the range of −45 to −51°C. These products are presently manufactured and marketed by Sasol Italy S.p.A. In Europe, ATIEL has approved PIO interchangeability with PAOs (API group IV) without further testing. API in North America has, however, not yet granted such an interchangeability approval.

ALKYLATED AROMATICS

Alkylated aromatic compounds are present in mineral base oils. Some of these aromatics are removed during the solvent extraction process. However, these are complex compounds consisting of mono-, di-, and polynuclear aromatic rings, some with heteroatoms. Such compounds are not considered advantageous from lubrication point of view. In the modern hydroprocesses, it is aimed to convert these aromatics into saturated compounds or paraffins. Thus, the naturally occurring aromatic compounds are not very useful, but the synthetic alkyl aromatics with well-defined structure have been found to be useful in several applications. The long-chain alkylated aromatics have very low pour point (−27 to −50°C) and low VI (0–10). Their oxidation stability in the presence of antioxidant is quite good and comparable to PAOs [18]. Products in the lubricating oil viscosity range are therefore suitable for low-temperature applications such as in refrigeration oil. During World War II, Germany used alkyl aromatics as base oils due to the shortage of other mineral oil-based products. Mixtures of heavier alkyl aromatics are available from linear alkyl benzene (LAB) production plant as by-product. In LAB plants, benzene is alkylated with normal C10–C14 paraffins using HF or BF_3 as catalyst. The product contains LAB and some amount of dialkyl benzene mixture. After separating out LAB, the residue is generally known as heavy alkylate. The properties of the final product can be altered by changing the structure and position of the alkyl groups [19]. Dialkylated aromatics are in the lubricating oil viscosity range. In place of benzene, naphthalene can also be alkylated to produce base oil quality fluids. Strong acidic catalysts like BF_3 or HF or $AlCl_3$ have been used but are known to produce a complex mixture of isomers.

Monoalkylated   Dialkylated benzene

Recently, solid catalysts based on zeolites [20] have been used, which can control isomer distribution. Commercially, several alkylated aromatic-based base oils are available in different viscosity grades (VG) to formulate a variety of lubricants:

1. Synesstic 5 and 10 from Exxon Mobil based on alkylated naphthalene
2. Dialkyl benzenes of 4 and 5 cSt from Sasol and Conoco
3. Zerol fluids of 4, 5, 6, and 8 cSt from Chevron–Texaco based on branched alkyl benzene

Alkylated aromatics can also be used as the base fluid in combination with mineral oil for formulating engine oils, gear oils, hydraulic fluids, and grease for subzero applications. They are also used as the base fluid in power transmission fluids and gas turbine, air compressor, and refrigeration compressor lubricants.

POLYISOBUTYLENES

These are produced by the polymerization of C4 olefins obtained either from refinery catalytic cracker or steam naphtha crackers producing ethylene. Dehydration of tertiary butyl alcohol also yields isobutylene. The C4 olefin cut obtained from naphtha cracker usually contains a mixture of isobutylene, *n*-butene, butane, and butadiene. Butadiene is extracted and is utilized in the production of synthetic rubber. The isobutylene-rich fraction is selectively polymerized in the presence of Lewis acid (anhydrous $AlCl_3$) catalyst to yield PIB of desired molecular weight and structure. Other catalysts and cationic polymerization have also been used [21]. The unreacted butane and butenes are recovered and processed further. PIB polymers of 10,000 molecular weight and above are rubbers. Similarly, polybutene-1 polymers are thermoplastic in nature. The lower-molecular-weight materials produced by this process have lubricating properties, while medium-molecular-weight materials, usually referred to as PIBs, are used as VI improvers and thickeners. PIB of 1000 molecular weight are also used for the manufacture of engine oil dispersant. High-molecular-weight PIB is used as tackiness additive. Usually, 1000–1500-molecular-weight PIB is useful in 2T oils since this replaces bright stock in this formulation and therefore provides smokeless product. Now, since the population of two-stroke engines is declining, PIB use in this application has also come down. A typical simplified structure of PIB is provided below. There may be other vinylidene and butane structures associated with it as well. The carbon–carbon double bond may be predominantly present at the end group, but it is possible to have the double bond at other positions also. *Cis*- and *trans*-structures are determined by this position of C=C.

Typical polyisobutylene structure

PIBs are hydrocarbons and are fully compatible and miscible with mineral oils, PAOs, hydrocracked base oils, and alkyl aromatics. In polar solvents and glycols, their miscibility is poor. PIBs are available in low viscosities (1 cSt at 100°C) to very high viscosities of 45,000 cSt. One of the most striking properties of PIBs is that they have a very low carbon residue (CCR) value, less than 0.01%. It means that PIBs are clean-burning compounds and do not leave CCR. This property makes them most suitable for applications where low CCR values are required such as in high-temperature compressor oils, where deposit formation is a fire and safety hazard. In two-stroke engine oils, also PIBs have completely replaced bright stocks to formulate low deposit forming smokeless 2T oils. PIBs can also be used as blending stocks with mineral and synthetic oils to adjust viscosity and VI of the finished product. British Petroleum has been marketing several grades of PIBs under the brand names of Hyvis, Nepvis, and Indopol. From 2001 onward, Indopol is their international brand name for polybutenes. These are available in different viscosities from 2 cSt (molecular weight of 260) at 100°C to 40,000 cSt (molecular weight of 5900) at 100°C. Several additive companies like Lubrizol, Infineum, Chevron, and others have been producing PIBs for their internal consumption for the manufacture of lubricant dispersant and fuel detergent additives.

Organic Esters

The earliest lubricants known to humans were natural triglyceride fatty acid esters of vegetable and animal origin such as castor, rapeseed, sperm whale, and lard oil. Organic esters have been an important class of synthetic base fluids from the time of World War II. In Germany, these were used in mineral oil blends to improve low-temperature properties of formulated products.

Simple organic ester compounds are formed by the reaction of an acid and an alcohol.

$$\underset{\displaystyle R-\overset{\textstyle O}{\overset{\textstyle \|}{C}}-OH}{} + HO-R_1 \;\rightleftharpoons\; R-\overset{\textstyle O}{\overset{\textstyle \|}{C}}-O-R_1 + H_2O$$

Synthetic ester lubricant base stocks can be synthesized from a variety of alcohols and acids to obtain suitable products of desired composition and properties. The acids and alcohols are reacted in the presence of an acidic catalyst such as sulfuric acid or p-toluenesulfonic acid or phosphorus oxides. The mixture is refluxed and water of reaction is continuously removed. The acidity of the reaction mixture is measured to monitor the progress of the reaction. For diesters, 5–10% excess alcohol is taken, while for polyolesters, 5–10% excess acid is used. At the end of the reaction, the products are neutralized with alkali, filtered and distilled if necessary. Product color can be improved with activated charcoal. Polyolesters can also be prepared by transesterification of monomethyl esters with neopentyl alcohol in the presence of anhydrous sodium sulfate.

Commonly used esters are:

1. Monoesters
2. Dibasic esters
3. Polyolesters

Monoesters are generally not used as lubricant base stocks, but products like glycerol monooleate and sorbitan monooleates have been used as friction modifiers and emulsifiers in engine oils and metalworking fluids as additives. Monoesters are produced by the reaction of a monofunctional acid such as oleic or isostearic acids with a monofunctional alcohol. These fluids are characterized by low and high VI.

Dibasic Acid Esters

These esters are prepared by the reaction of a dibasic acid with a monohydric alcohol containing one reactive hydroxyl group. The backbone of the structure is formed by the acid, with the alcohol radicals joining at its ends. The physical properties of the final product can be varied by using different alcohols or acids. Usually, adipic, azelaic, and sebacic acids are used for diester synthesis. Among the commonly used alcohols, 2-ethylhexanol, trimethylhexyl alcohol, isodecyl alcohol, and isotridecyl alcohol are utilized.

The use of diesters is now mainly restricted to older military jet engines and some jet engines in industrial service. They are also used as the base fluid in low-temperature aircraft greases and some industrial oils.

$$
R-OH + (H_2C)_n
\begin{matrix}
\overset{O}{\underset{\parallel}{C}}-OH \\
\\
\underset{\parallel}{C}-OH \\
O
\end{matrix}
\longrightarrow
(H_2C)_n
\begin{matrix}
\overset{O}{\underset{\parallel}{C}}-OR \\
\\
\underset{\parallel}{C}-OH \\
O
\end{matrix}
+ 2H_2O
$$

Alcohol   Acid   Dibasic Ester

When $n = 4$, adipates; $n = 7$, azealates; $n = 8$, sebacates; and $n = 10$, dodecanedioates.

Dibasic esters are characterized by following typical properties due to their "dumbbell"-type configuration:

1. Good oxidation/thermal stability
2. High VI
3. Excellent low-temperature properties
4. Low volatility
5. Excellent solvency due to their polar nature
6. Good shear stability
7. High biodegradability

Phthalates

Phthalates are also diesters prepared from phthalic acid/anhydride and a monohydric alcohol of C7–C18 linear or branched alkyl group. These are commercially available from VG 2 to VG 220 ISO viscosity grades and are used in selected industrial oils and greases. Trimellitate and pyromellitates have also been synthesized with good thermal stability, but their higher costs limit their applications. Structure of phthalate can be represented by the following diagram, where R could be C7–C18 linear or branched alkyl group.

Polyolesters

These esters are formed by the reaction an alcohol having two or more hydroxyl groups (polyhydric alcohol), with a monobasic acid. In the polyolesters, the polyol forms the backbone of the structure with the acid radicals attached to it unlike the dibasic ester where the acid is the backbone of the structure. These are very stable esters and used in aviation oils. Neopentyl glycol, trimethylolpropane, and pentaerythritol or dipentaerythritol are the common polyhydric alcohols to produce polyolesters.

Polyolester's high thermal stability is due to the absence of beta hydrogen and is also called hindered ester.

$$
\underset{\text{Polyhydric Alcohol}}{R-\overset{\overset{\displaystyle CH_2OH}{|}}{\underset{\underset{\displaystyle CH_2OH}{|}}{C}}-CH_2OH} + \underset{\text{Acid}}{3R_1-\overset{}{\underset{\underset{\displaystyle O}{\|}}{C}}-OH} \longrightarrow \underset{\text{Polyol Ester}}{R-\overset{\overset{\displaystyle CH_2-O-\overset{\overset{\displaystyle O}{\|}}{C}-R_1}{|}}{\underset{\underset{\displaystyle \underset{O}{\overset{\|}{CH_2-O-C-R_1}}}{|}}{C}}-CH_2-O-\underset{\underset{\displaystyle O}{\|}}{C}-R_1} + 3H_2O
$$

There are other complex polyolesters, which are prepared by the reaction of polyhydric alcohols and a dibasic acid and end capped with an alcohol. PAG esters are also prepared by reacting PAG hydroxyl group with an acid.

The structure of polyolesters provides the following exceptional properties to the fluid:

1. Very high thermal/oxidation stability
2. High VI
3. Good low-temperature properties
4. High flash/fire points
5. Very low volatility
6. Moderate to high biodegradability

The new-generation jet engine oils used in commercial and military jet aircrafts are formulated with polyolester base oils. These lubricants are also used in many industrial and marine gas turbines that use aircraft jet engines as power generators.

Diesters, polyolesters, and complex esters are also highly biodegradable and are utilized for formulating biodegradable oils for forestry and agricultural applications. Biodegradability test data as per CEC L-33-T-82 test method for 21 days show that most diesters, polyolesters, and complex esters are biodegradable in the range of 70–100%. Phthalates are biodegradable in the range of 40–90%. Trimelliates are biodegradable to a lesser extent in the range of 40–70% only. Mineral oils are not readily biodegradable, but higher-performance base oils such as VG 32 grade turbine oil is biodegradable to the extent of 59% [22].

Polyglycols

PAGs are produced by reacting compounds containing active hydrogen atoms (e.g., alcohols) with alkylene oxides, usually in the presence of a basic catalyst (e.g., sodium or potassium hydroxide, tertiary amines, or Lewis acids and double metal cyanide complexes). However, alkali metal hydroxides are the most preferred catalyst system. The commonly used alkylene oxides are ethylene oxide (EO), propylene oxide (PO), and butylene oxide, which are produced by the oxidation of an olefin. The basic reaction is the addition of alkylene oxide to a hydroxyl group.

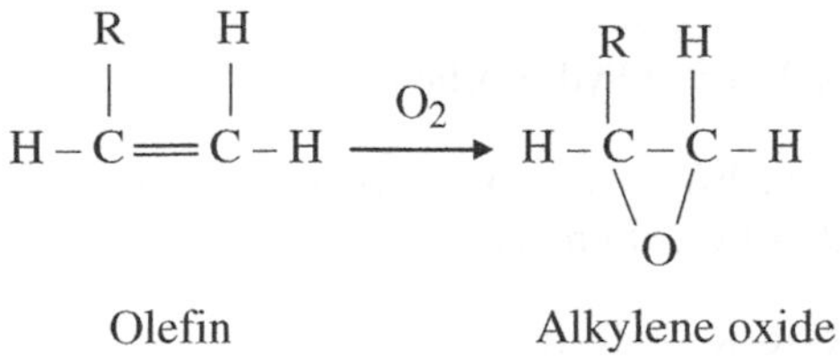

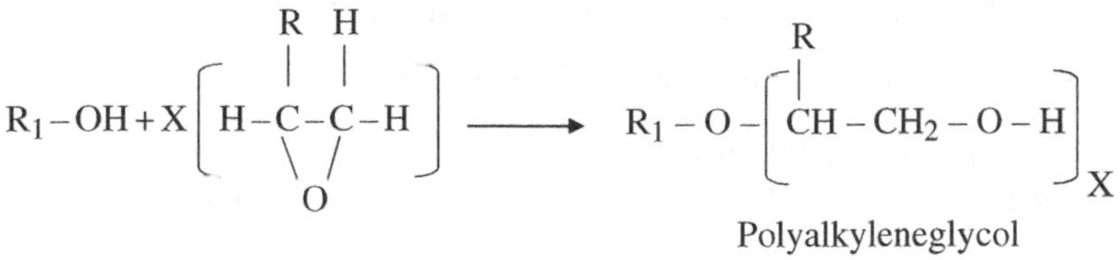

$$
\underset{\text{Olefin}}{H-\underset{|}{\overset{\overset{\displaystyle R}{|}}{C}}=\underset{|}{\overset{\overset{\displaystyle H}{|}}{C}}-H} \xrightarrow{\ O_2\ } \underset{\text{Alkylene oxide}}{H-\underset{\diagdown\ \ \diagup}{\overset{\overset{\displaystyle R}{|}}{C}}\underset{O}{-}\underset{}{\overset{\overset{\displaystyle H}{|}}{C}}-H}
$$

$$
R_1-OH + X\left[H-\overset{\overset{\displaystyle R}{|}}{C}-\overset{\overset{\displaystyle H}{|}}{\underset{\diagdown O \diagup}{C}}-H \right] \longrightarrow R_1-O-\left[\overset{\overset{\displaystyle R}{|}}{CH}-CH_2-O-H \right]_X
$$

Polyalkyleneglycol

R and R1 could be hydrogen, or an alkyl group, or an aryl group.
When R is H, EO.
When R is CH_2, PO.
When R is CH_3CH_2, butylene oxide.
These reactions take place in following three steps:

Step 1   ROH $\xrightarrow{\text{KOH}}$ RO$^-$M$^+$

Step 2   Alkoxylation

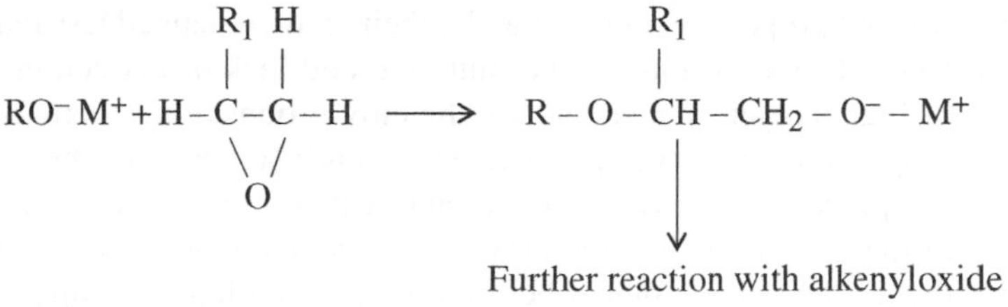

$$
RO^-M^+ + H-\overset{\overset{\displaystyle R_1}{|}}{C}-\underset{\diagdown O \diagup}{\overset{\overset{\displaystyle H}{|}}{C}}-H \longrightarrow R-O-\overset{\overset{\displaystyle R_1}{|}}{CH}-CH_2-O^--M^+
$$

Further reaction with alkenyloxide

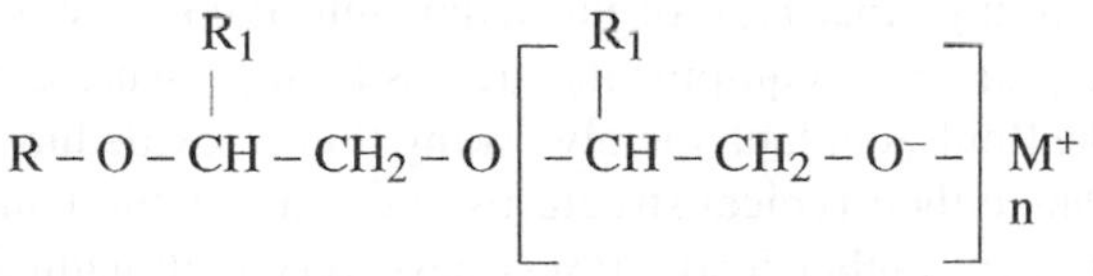

$$
R-O-\overset{\overset{\displaystyle R_1}{|}}{CH}-CH_2-O\left[-\overset{\overset{\displaystyle R_1}{|}}{CH}-CH_2-O^- - \right]_n M^+
$$

Step 3   Catalyst removal

In step 3, the catalyst is removed, replacing M^+ with H in the aforementioned molecule. Further treatment may be necessary to neutralize alkali and remove metal content in the products.

The type of alkylene oxide used also influences the properties of final product. For example, PO homopolymers are water insoluble. Introducing some quantity of EO into the PO polymer will make it water soluble to the desired level:

EO:PO = 0:1 insoluble in water and hydrocarbons

EO:PO = 1:1 soluble in cold water, soluble in alcohol, but not soluble in hydrocarbons

EO:PO = 4:1 water soluble, not soluble in hydrocarbons

EO and PO polymerized together in different ratios can produce homopolymer, block copolymer, and copolymers with different properties:

Homopolymer: R–R–R–R–R–R–R–R–R–R–R

Copolymer or random polymer: R–R$_1$–R–R$_1$–R–R$_1$–R–R$_1$–R

Block copolymer: R–R–R–R–R–R$_1$–R$_1$–R$_1$–R$_1$–R$_1$

where R is EO and R1 is PO.

Linear polymers of EO and PO can be represented by the following general structure:

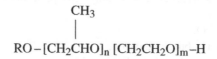

$$RO-[CH_2CHO]_n \ [CH_2CH_2O]_m-H$$
with CH_3 on the CHO carbon

PAGs are an interesting class of compounds, where the polar and nonpolar groups could be balanced in such a manner that the product properties are controlled according to the requirements. Such a choice is not available with the hydrocarbon-based synthetic oils and mineral oils. It is possible to formulate PAG-based lubricants for majority of applications with proper understanding of PAG structure and properties. Unfortunately, their share remained less than 10% of total industrial oils due to conventional thinking and lack of marketing support. PAGs possess high oxygen content and are the only groups of synthetic compounds that offer opportunity to develop water-based multifunctional lubricants with good antiwear properties for wide range of applications in metalworking and industrial applications. The only limitation PAGs have is that mineral oils are not compatible with them. Thus, the system where these are used has to ensure that there is no mixing up or leakage of any other type of lubricants into PAGs. PAGs, thus, find applications in a variety of select equipments such as high-pressure compressors, gears, hydraulics, brake fluids, coolant, metalworking fluids, textile lubricants, and heat transfer fluids. Due to their typical structures, they have natural antiwear and EP properties and can carry higher loads. PAGs have very high natural VI in the range of 200 and have good lubricity properties. Most PAGs are also nontoxic and can be used in food grade lubricants.

The applications for the polyglycols are divided into those for the water-soluble types and those for water-insoluble types. The largest volume application of the water-soluble polyglycols is in hydraulic brake fluids. Water-insoluble polyglycols are used as heat transfer fluids due to their high thermal stability and as the base fluid in certain types of industrial hydraulic fluids and high-temperature bearing oils. They also possess low pour point depending on the structure. Water-soluble polyglycols are used in the preparation of water–glycol fire-resistant hydraulic fluids. Product details for different applications can be found in the technical bulletin of Cognis Co. [23] and Dow Chemical Co. [24].

Phosphate Esters

There are large numbers of organophosphorous compounds used in a variety of industries. Neutral, metal-free, and fully substituted esters of orthophosphoric acids (H_3PO_4) and alkyl alcohols or phenols have been used in lubricants. Some are used as antiwear/EP; anticorrosion additives and tertiary neutral esters have been used as fire-resistant hydraulic fluids and compressor oils. Tricresyl phosphate (TCP) has been a well-known plasticizer. Further development of phosphate esters took place after World War II, when military and commercial aircraft control system required a safe and fire-resistant hydraulic fluid. Extensive work was carried out on both trialkyl phosphate and triaryl phosphates [25, 26]. Soon, TCP, tributyl phosphate (TBP), and trixylenyl phosphate (TXP) were marketed for hydraulic and compressor lubrication. U.S. Navy specification MIL-H-19457 is based on trixylenyl phosphate. The phosphate esters of phenols and alcohols have the general formula $O{=}P{-}(O{-}R)_3$, where R represents aryl, alkyl, or a mixture of alkyl and aryl components. The physical and chemical properties of phosphate esters can be varied considerably depending on the choice of substituents, and these are selected to provide optimum performance for a particular application. Phosphate esters are produced by reaction of phosphoryl chloride with phenols or alcohols:

$$3ROH + POCl_3 \longrightarrow O = P(RO)_3 + 3HCl$$

Several categories of fluids can be produced through this reaction:

1. Trialkyl phosphates
2. Triaryl phosphates
3. Mixed alkyl/aryl phosphates

Triaryl phosphates are most important from the point of lubrication. TCP and TXP have been produced by using cresols and xylenols from coal tar distillation. These contain several isomers including *o*-cresol, which is neurotoxic. Thus, *ortho*-isomer has to be reduced to a minimum level from cresol feed. Xylenol also contains several isomers such as 2,4-, 3,4-, and 3,5-xylenols. 3,5-Xylenol is most important to obtain hydrolytically stable product [27]. Triaryl phosphates have the following structure:

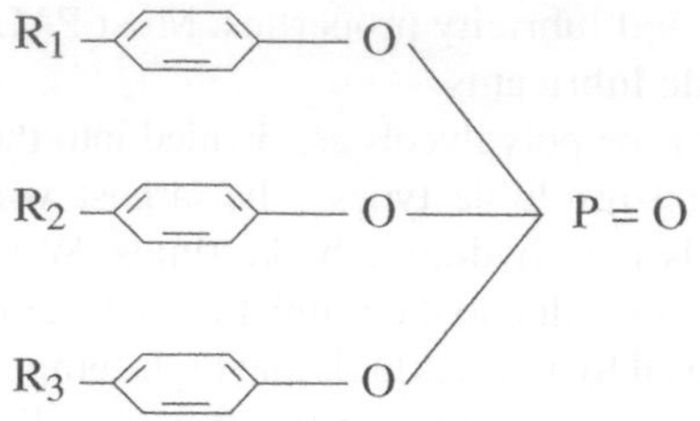

where R_1, R_2, and R_3 may be methyl or higher alkyl group.

By varying R_1, R_2, and R_3, fluids of different viscosity and stability can be prepared. In 1960, synthetic alkyl phenols were produced by the alkylation of phenol, which solved the toxicological problems. Isopropylphenyl and tertiary butyl phenyl phosphates are now commercially available in different viscosity range. Some of the important commercially available phosphate esters are described along with the common abbreviations used:

Trialkyl phosphate esters

Tributyl phosphate (TBP)

Tri-isobutyl phosphate (TIBP)

Trioctyl phosphate (TOP)

Triaryl phosphates

Triphenyl phosphate (TPP)

Tricresyl phosphate (TCP)

Trixylenyl phosphate (TXP

Tripropyl phosphate (TPP, propylated)

Trisisopropylphenyl phosphate (TIPPP)

Cresyl diphenyl phosphate (CDP)

Alkyl/aryl phosphates

Dibutyl phenyl phosphate (DBPP)

2-Ethylhexyl diphenyl phosphate (EHDPP)

Silicones

Silicones are wide-temperature functional fluids with a very high VI and highly resistant to thermal and oxidative degradation. But they are poor lubricants for steel-on-steel application, and this has restricted their use. Silicones are used in specialty hydraulic fluids for applications such as liquid springs and torsion dampers where their high compressibility and minimal change in viscosity with temperature are beneficial. Silicone fluids (polydimethylsiloxane) are used in formulating DOT 5 and SAE 1705 brake fluids. U.S. military uses DOT 5 fluids due to the superior properties exhibited by these fluids with respect to the wet equilibrium reflux boiling point and low viscosities at lower temperatures.

Silicate Esters

Silicate esters have excellent thermal stability and with proper inhibitors show good oxidation stability. They have excellent viscosity–temperature characteristics. Silicate esters are used as heat transfer fluids and dielectric coolants.

Silahydrocarbon

Close competitors to the PAOs as hydraulic fluids in aerospace applications are silahydrocarbons, a new class of thermally stable wide-liquid-range functional fluids. Silahydrocarbon or tetra-alkyl silane is used as wide-temperature-range, high-temperature, and fire-resistant hydraulic fluid. Their excellent stability at temperatures up to 370°C permits their extended use at elevated temperatures. Another very important application is as liquid space lubricants. Their excellent viscosity–temperature characteristics permit the selection of extremely high-molecular-weight (1000–1500) trisilahydrocarbon fluids to be used. These fluids have extremely low volatility, which makes them excellent for long-life, noncontaminating liquid lubricants for space.

Perfluropolyalkylethers

The perfluropolyalkylethers are nonflammable liquids. Because of their exceptional stability over a wide-temperature corrosive environment, radiation resistance, and ability to function at very high vacuum, they are suitable for use in rockets, missiles, and manned spacecraft.

Unconventional Synthetic Oil

A natural algae-based synthetic lubricant derived from harvesting kelp from the ocean has a composition of polyols (mannitol about 70% and mannose about 0.4%) and about 28% by weight of a high-molecular-weight polymer. The synthetic lubricant may be used as a drag reducing agent and additive for existing lubricants and also further reacted with fatty esters to form a hybrid lubricant that may serve as a total replacement for existing lubricants [28].

Aviation Oils

The aviation environment offers a very tough condition for the lubricants such as high temperatures, low pressure, very low temperatures, high load, etc. Therefore, special products are required for these applications. Synthetic oils are usually preferred for severe application, but some mineral oil-based products with very strict controls are also used. Aircraft generally uses two types of engines:

1. Piston engines
2. Turbojets

For piston engines, mineral or semisynthetic oils with additives are sufficient. However, turbojet or turboprop engines require synthetic ester-type lubricants.

TABLE 4.2   Synthetic jet engine oils

Specifications	JSD	NATO code	Viscosity at 100°C
MIL-PRF-7808 K grade 3	OX-9	0–148	3 cSt
MIL-PRF-7808 K grade 4	—	0–163	4 cSt
MIL-PRF-23699 F	—	0–156	5 cSt
DEF STAN 91-101/2	OX-27	0–156	5 cSt
DEF STAN 91-98/2	OX-38	0–149	7.5 cSt
DEF STAN 91-100/2	OX-26	0–160	5 cSt
AIR 3514/A (French)	—	0–150	3 cSt

Aircrafts also have sophisticated hydraulic systems, which require clean or superclean mineral or synthetic oil.

There are also several varieties of greases and specialty products that are required.

Specifications

The United States, the United Kingdom, France, and Russia have developed their own specifications and designations for aviation oils. Most of these have been developed by military authorities and are followed by civil aviation authorities. However, in certain cases, engine builders like Rolls-Royce, General Electric, and Pratt and Whitney have their own additional requirements.

The UK Ministry of Defense has developed series of DEF STAN standards and also evolved Joint Service Designations covering products required by Navy, Air Force, and Army. The United States has covered these oils under MIL designations. NATO has also developed a code system where oils developed from different countries can be considered as equivalent and interchangeable. Table 4.2 provides the details of UK, U.S., and French specifications of jet engine oils.

Formulating Jet Engine Oils

Two widely used synthetic oils are covered under DEF STAN 91-98/2 (DERD 2487) JSD (OX-38), and MIL-PRF-23699 F specification (JSD-OX-27). These oils are formulated with polyolesters and other additives to meet various characteristics including seal compatibility. The approval procedures are lengthy with various bench tests and flight tests and require several years before an oil company can hope to get an approval.

REFERENCES

[1] Murray DW, MacDonald JM, White AM. The effect of base oil composition on lubricant oxidation performance. Pet Rev February 1982:36–40.
[2] Saxena D, Mooken RT, Pandey LM, Gupta A, Mishra AK, Srivastava SP, Bhatnagar AK. Long life anti-wear turbine oils. Petrotech 1999; January 9–12, 1999; New Delhi. p 203–208.

[3] Gatto VJ, Grina MA. Effect of base oil type, oxidation and test conditions and phenolic antioxidant structure. Lubr Eng January 1999;50 (1):11–20.

[4] Mooken RT, Saxena D, Basu B, Satpathy S, Srivastava SP, Bhatnagar AK. Dependence of oxidation stability of steam turbine oils on base stock composition. 52nd Society of Tribologist and Lubrication Engineers (STLE), USA Annual Meeting; May 18–22, 1997; Kansas City. p 19–24.

[5] Okazaki ME, Militante SE. Performance advantages of turbine oils formulated with Group II base oils. In: Herguth WR, Warne TM, editors. *Turbine Lubrication in the 21st Century*. West Conshocken: American Society for Testing and Materials; 2001. ASTM Special Technical Publication 1407.

[6] Schwager BP, Hardy BJ, Aguiilar GA. Improved response of turbine oils based on Group II hydro-cracked base oils compared with those based on solvent refined oils. In: Herguth WR, Warne TM, editors. *Turbine Lubrication in the 21st Century*. West Conshocken: American Society for Testing and Materials; 2001. ASTM Special Technical Publication 1407.

[7] Irvine DJ. Performance advantages of turbine oils formulated with group II and group III base stocks. In: Herguth WR, Warne TM, editors. *Turbine Lubrication in the 21st Century*. West Conshocken: American Society for Testing and Materials; 2001. ASTM Special Technical Publication 1407.

[8] Shubkin RL. Synthetic lubricants by oligomerization and hydrogenation. US patent 3,780,128. 1973.

[9] Sullivan FW Jr, Voorhees V, Neeley AW, Shankland RV. Synthetic lubricating oils relation between chemical constitution and physical properties. Ind Eng Chem 1931;23 (6):604–611.

[10] Brennan JA. Polymerization of olefins with BF_3. US patent 3,382,291. May 1968.

[11] Loveless FC. Method for the oligomerization of olefins. US patent 4,041,098. 1977.

[12] Kapur GS, Sarpal AS, Sarin R, Jain SK, Srivastava SP, Bhatnagar AK. Detailed characterisation of polyalphaolefins and their branched structures using multi-pulse NMR techniques. J Syn Lubr October 1998;15 (3):177–191.

[13] Kumar G, Shubkin RL. New polyalphaolefins fluids for specialty application. Lubr Eng 1993; 49:723–725.

[14] Hope KD, Driver MS, Harris TV. High viscosity polyalphaolefins prepared with ionic liquid catalyst. US patent 6,395,948. 2000.

[15] Wu MMS, Stavens WWH, Abhimanyu OP. Shear-stable high viscosity polyalphaolefins. US patent 2013,0210,996. August 15, 2013.

[16] Priola A, Farrore V, Arrighetti S, Mancini G. Heterogeneous cationic catalytic systems suitable for the olegomerization of linear non-terminal olefins and a process for preparing said oligomers using said system. European patent 0068554 A1. January 1983.

[17] Priola A, Farrore V, Arrighetti S, Mancini G. Heterogeneous cationic catalytic systems suitable for the olegomerization of linear non-terminal olefins and a process for preparing said oligomers using said system. European patent 0068554 B1. November 1985.

[18] Gschwender LJ, Snyder CE Jr, Driscol CL. Candidate high temperature hydraulic fluids. J Soc Trib Lubr Eng 1990;46:377–381.

[19] Kapur GS, Sarpal AS, Jain SK, Srivastava SP, Bhatnagar AK. Detailed characterization of heavy alkylated benzene fluids by multipulse one- and two-dimensional NMR spectroscopy. J Syn Lubr April 2000;17 (1):41–54.

[20] Le QN, Marler DO, MacWilliams JP, Ruben MK, Shim J, Wong SS. Process for preparing long chain alkyl aromatic compounds. US patent 4,962,256. 1990.

[21] Kennedy JP. *Cationic Polymerization of Olefins*. New York: Wiley Interscience; 1975.

[22] Singh MP, Ali N, Tyagi BR, Srivastava SP, Bhatnagar AK. Biodegradable lubricants. In: Srivastava SP, editor. Proceedings of the 1st International Symposium on Fuels and Lubricants (ISFL-1997); December 8–10, 1997; New Delhi. New Delhi: Tata McGraw Hill; p 369–374.

[23] www.ilco.chemie.de/downloads/os/_Breox.pdf. Accessed March 26, 2014.

[24] Technical Bulletin, *UCON Fluids and Lubricants*. Dow Chemical Company. Available at www.dow. com/ucon. Accessed March 26, 2014.

[25] Morton FI. Development and testing of fire resistant hydraulic fluids. SAE paper 490229. Society of Automobile Engineers; 1949.

[26] Gamrath HR, Hatton RE, Wessner WR. Chemistry and physical properties of alkyl and aryl phosphates. Ind Eng Chem 1954;46:208–212.

[27] Vilyanskaya GD, Vainshtein AG, Razarenova MM, Salnikova GK, Lysko VV, Bakai VS. Procédé pour la préparation d'un liquide incombustible. European patent 0382178. 1990.

[28] Copp EA, Glantz D. Environmentally-friendly kelp-based energy saving lubricants, biofuels and other industrial products. US patent 2012,0190,600 A1. July 26, 2012.

CHAPTER 5

LUBRICANT ADDITIVES AND THEIR EVALUATION

The basic function of a lubricant is to reduce friction between the two moving parts under all operating conditions of load, speed, torque, pressure, and temperatures encountered in the equipment. There are several applications where a simple mineral oil of certain viscosity will be adequate to lubricate under the hydrodynamic conditions. However, as the severity of operation in the equipment increases, mineral base oils alone are no longer able to provide protection, and hence chemical additives are required. The appearance of petroleum lubricant dates back to the late 1880s when early distillation and fractionation techniques were used to produce a range of petroleum products, some of which were suitable as lubricants. By the 1920s, good-quality lubricants could be produced by vacuum distillation. Between the 1930s and the 1950s, the use of additives has become common. The use of zinc dialkyl dithiophosphates (ZDDPs) and pour point depressants (PPDs) started the modern additive technology in the 1930s. Antioxidants (to increase oil life), detergents (to keep engine clean), and viscosity index improvers (VIIs) found their uses in the 1930s and 1940s. Over-based detergents appeared in the 1950s. By the 1960s, most of the basic additive compounds used presently in lubricants have been developed and introduced. These additives are oil-soluble chemicals that improve or modify or impart desired properties in the base oils. There are a large number of such compounds, but these generally fall under the following categories:

1. Antioxidants
2. Metal deactivator
3. Corrosion inhibitor
4. Rust inhibitor
5. Friction modifiers/lubricity/film strength/oiliness additives
6. Antiwear compounds
7. Extreme pressure (EP)/load-carrying compounds
8. VIIs or viscosity modifier
9. Tackiness compounds
10. PPD
11. Antifoam compounds

Developments in Lubricant Technology, First Edition. S. P. Srivastava.
© 2014 John Wiley & Sons, Inc. Published 2014 by John Wiley & Sons, Inc.

12. Detergents
13. Dispersants
14. Emulsifiers
15. Demulsifier
16. Biocides
17. Dyes
18. Seal swell compounds
19. Solid additives [1] such as MoS_2, graphite, boron nitride, PTFE, and their nano-forms

Industrial lubricants would typically use two or more additives. Simple turbine, compressor, or hydraulic oils would contain rust inhibitor, antifoam, and antioxidants and are thus known as R&O oils. In most applications, lubricants come in contact with moisture or water and thus need rust protection. Antioxidants are required to increase the life of lubricants. Antiwear version of these oils will additionally contain an antiwear additive. Commonly employed additives are triaryl phosphate and ZDDP. Certain hydraulic oils that have to operate under extreme temperature conditions may additionally contain a polymeric viscosity modifier and a PPD to modify the viscosity–temperature relationship of the lubricant so that it can function over a wide temperature range.

Lubricants operating under severe load and high torque conditions such as gear oils shall require EP additives. These EP additives function through chemical reaction with metal surfaces and contain active sulfur and phosphorous compounds.

In engine oils, an additional parameter influences the lubricant composition. Lubricants come in contact with the fuel combustion products such as carbon deposits, soot, lacquer, and oxides of sulfur and nitrogen. This severity also depends on the nature of fuel and its quality. Engine oil compositions, thus, become more complex, containing detergent, dispersant, antioxidant, anticorrosion, antifoam, antiwear, viscosity modifier, PPD, and friction modifier.

These compositions become even more complex when transmission oils and universal tractor oils are formulated, which require a very balanced additive treatment having optimum friction, seal swell, and viscometrics in addition to the other described properties. The environmental regulations also put strict control on the choice of additives in lubricants. The selection of right base oil combination and chemical additives becomes extremely important in lubricant formulation. The dosage of additives is the next important parameter to obtain a cost-effective product having desired performance characteristics. Lubricant technology is all about combining right type of additives in right quantities in right base oils. Since the additives are chemical molecules, care is required to select compatible molecules so that undesirable effect of a possible chemical and physical interaction between the additives is avoided.

BASE OIL QUALITY AND ADDITIVE TREATMENT

There are several properties of base oils such as flash point, volatility, vapor pressure, thermal conductivity, and air release value that cannot be modified by any known chemical additives. Such properties need to be controlled in the refinery. During the

TABLE 5.1   Typical properties of different category of base oils

	API Gr I, solvent refined and dewaxed	API Gr II, hydrotreated	API Gr III, two-stage hydrotreated	GTL base oils API Gr III level	Polyalphaolefin 4 to 6 cSt
K. Viscosity (40°C, cSt)	32–500	32–500	32–100	3.5–30	16.7–30.89
Viscosity index	95	105	121	130	124–143
Pour point (°C)	−6 to −12	−18	−21	−10 to −30	−64 to −70
Saturates (%)	80	99	99.6	>99	—
Sulfur (ppm)	0.05–1.0	5	2	<5 ppm	ND
Aromatics (%)	12–15	0.5	0.2	0.1	ND
Nitrogen (ppm)	50	5	2	<20	ND

last 20 years, newer refining technologies based on hydroprocesses have been developed, which provide improved base oils with respect to lower sulfur, nitrogen, aromatics, volatility, and higher viscosity index (VI). Conventional base oils produced by solvent extraction, solvent dewaxing, and hydrofinishing contain several sulfur and nitrogen compounds that are natural antioxidants and lubricity compounds. Aromatics present in these base oils also provided good solvency power and some amount of surface activity. A typical comparison of these base oils along with PAO 4 and 6 are shown in Table 5.1. These data are based on several commercial base oils available from different sources and are compiled to demonstrate the difference in properties. Other synthetic oils such as ester and polyol esters are also available in different viscosities.

These different base oils may need different levels and types of additives to obtain desired performance levels in lubricants. The higher VI and low pour oil will need less of viscosity modifiers and pour depressant additives. The lower aromatics in higher-quality base oils reduce their solvency power and would need additional seal swell compound to avoid leakages in equipment through seals. Lower sulfur and nitrogen compounds reduce the natural antioxidants, and such oils need to be stabilized with antioxidants. However, the higher saturate content improves their additive response. Thus, in lubricant formulation, base oil properties need to be considered carefully while selecting additives and their dosages. Similarly, due consideration has to be applied for synthetic base oil properties. Polyalphaolefins (PAOs), alkyl aromatics, esters, diesters, polyol esters, phosphate esters, and glycols have all different properties and require specifically tailored additive system. For example, PAOs and polyol esters may not require any PPD and viscosity modifiers in view of their low pour points and high VI.

Common additives used in lubricants are described in the following sections.

Antioxidants

Industrial equipments, including diesel and gasoline engines, offer ideal environment for the lubricant hydrocarbons to undergo oxidation during their service. Higher temperatures, presence of air (oxygen), and metals accelerate the oxidation process through free radical mechanism. The free radical formation takes place by hydrogen extraction at higher temperature. These are also formed by the reaction of hydrocarbon molecule RH with oxygen at higher temperature. The formation of free radical is followed by a series of chain reactions, which converts hydrocarbon molecules into several new compounds leading to oil degradation.

Initiation

$$RH \rightarrow R*+H*$$

$$R-H+O_2 \rightarrow R*+HOO*$$

Chain Propagation

This chain reaction continues to propagate to form peroxide radical ROO* by the reaction of alkyl radical with oxygen, which further generates free radicals and hydroperoxides ROOH.

$$R*+O_2 \rightarrow ROO*(peroxide)$$

$$ROO*+RH \rightarrow ROOH+R*$$

$$HOO*+RH \rightarrow H_2O_2+R*$$

The initial free radical R* can react with large number of hydrocarbon molecules RH to produce equal number of hydroperoxide molecules ROOH. These are unstable molecules, where O–O bond splits to produce two new free radicals, and the chain reaction branches.

$$ROOH \rightarrow RO*+HO*$$

$$RO*+RH \rightarrow ROH+R*$$

These radicals further propagate the chain reactions to produce a variety of oxidation products like aldehydes, ketones, and carboxylic acids resulting in the development of acidity and sludge in oil [2, 3].

Chain Termination

These chains can be terminated by the combination of free radical itself.

$$2R* \rightarrow R-R$$

$$2ROO* \rightarrow ROOR+O_2$$

$$ROO*+R* \rightarrow ROOR$$

During the combustion of fuels in the engine in the presence of air (containing nitrogen and oxygen), especially in diesel engines, oxides of nitrogen (NOx) are also produced, which can react with hydrocarbons and free radicals to form corrosive products. NOx can also initiate the formation of free radicals like R* and HO*, which can further propagate the chain reactions.

Oxidation of lubricants can be prevented under these conditions by the use of antioxidant that can deactivate free radicals and decompose hydroperoxides at both moderate and high temperatures. It would also be desirable to deactivate catalytic metal surface by metal passivators/deactivators to retard the formation of free radicals. Chemical compounds that deactivate or scavenge free radicals are known as primary antioxidants.

Hindered phenols and alkyl-aromatic amines that react with free radicals by donating hydrogen atom are primary antioxidants. Sulfur- and phosphorous-containing compounds like organic phosphites, disulfides, polysulfides, xanthates, dialkyl dithiocarbamates, and thioethers destroy peroxides and are called secondary antioxidants.

Similarly, there are two types of metal deactivators: one which forms chelate with metals and the other which forms a protective film on the metal surface. Disalicylidene propylene diamine acts as a chelating agent, while benzotriazole and alkyl (butyl) benzotriazole passivate metal surface by forming a protective film on it.

Some of the important classes of compounds used in lubricants for inhibiting oxidation are described as follows.

HINDERED PHENOLS

Hindered phenols are the largest group of compounds used as antioxidants in a variety of applications. 2,6-Di-*tert*-butyl *p*-cresol (DBPC or BHT) is the most preferred compound. Other variations are 2,6-di-*tert*-butyl phenol (DBP), mixture of DBP, 4,4-methylene bis-2,6-di-*tert*-butyl phenol, and 2,6-di-*tert*-butyl-alpha-diamino *p*-cresol. DBPC is manufactured by the reaction of isobutylene with *p*-cresol in the presence of an acidic catalyst. Most of the hindered phenols except the 4,4-methylene-bis-DBP are effective at temperatures below 110°C since these tend to volatilize above this temperature. Hindered phenols are universally used in industrial lubricants operating at moderate temperatures.

ALKYLATED AROMATIC AMINES

Alkylated aromatic amines are most effective at moderate to high temperatures encountered in several industrial equipment. Commercially available compounds are as follows:

Octylated phenyl naphthylamine

Di-octyl/dinonyl-diphenylamine

N,N'-di-*sec*-butyl-*p*-phenylene diamine

Phenothiazines [4, 5]

2,2,4-trimethyl dihydroxy quiniline

Diphenyl amines or substituted diphenyl amines are prepared [6] by reacting a phenol with an aniline in the presence of a hydrogen transfer catalyst. Phenol is converted to cyclohexanone by accepting protons. Aniline is then added in the presence of alkali metals and a noble metal of group VIII as hydrogen transfer catalyst. Diphenylamine can be alkylated with di-isobutylene or nonane or with other alkyl groups using a Friedel–Crafts reaction and a Lewis acid catalyst [7, 8].

PHENOTHIAZINES

Phenothiazines are tertiary aromatic amines and are good high-temperature antioxidants for synthetic lubricants used in aircraft engines. The oil-soluble phenothiazine can be synthesized from aromatic secondary amine, which can be conveniently prepared from the condensation of aryl amine with aryl hydroxy compound.

METAL DIALKYLDITHIOCARBAMATE

These compounds are highly effective as high-temperature antioxidants. Metal dithiocarbamates react with peroxide radicals without the formation of hydroperoxides. Zinc dialkyl dithiocarbamates [9] have very high thermal stability up to 300°C. Following is the general structure of metal dithiocarbamates:

$$
\begin{array}{c}
R_1 \qquad\quad S \\
\diagdown \qquad \|\; \\
N \!-\! C \!-\! S \!-\! M_1 \\
\diagup \\
R_2
\end{array}
$$

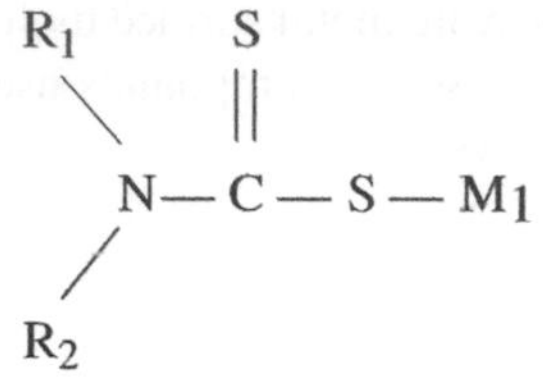

where R_1 and R_2 are alkyl group or hydrogen, and M_1 is a monovalent metal. Metals could also be multivalent (e.g., Cu, Zn, Sb, Cd, and Mo) also, and in that case, several structures are possible. Molybdenum dialkyldithiocarbamate complexes possess antioxidant properties in addition to antiwear and friction-reducing properties. These products are extensively used in energy-efficient industrial and automotive oils.

BENZOTRIAZOLE AS METAL DEACTIVATOR

Benzotriazole and alkyl [butyl] benzotriazole form a layer over the copper surface and thus do not allow it to catalyze the oxidation reaction. Benzotriazole can be easily prepared by the reaction of ortho phenylenediamine [10], sodium nitrite, and acetic acid in aqueous medium. Benzotriazole solubility in mineral oil is low, and therefore, alkylated (butylated) benzotriazole having better solubility is generally used.

EVALUATION OF ANTIOXIDANTS

Following parameters influence the oxidation of lubricants and the formation of free radicals:

Time duration

Temperature

Supply of air/oxygen and its pressure

Presence of metals

All laboratory oxidation tests have been designed by varying these parameters to simulate equipment operating conditions. For example, for evaluating turbine and hydraulic oils, moderate temperatures and long test duration are the main factors. For compressor oils, high test temperature and moderate test duration are desirable. Both metal-catalyzed and noncatalyzed tests have been developed. Higher pressures of oxygen have also been used to accelerate the test. The common laboratory test involves bubbling of air or oxygen through the lubricant in a tube at a specified temperature and time duration. After the test, viscosity, acid number, and carbon residue are measured in the oil. Some of the important oxidation tests widely used in the petroleum industry are provided in the following text. Many of these tests are required by the lubricant specification and have been well documented in the ASTM, IP, DIN, ISO, and other national standards.

INDUSTRIAL TURBINE, HYDRAULIC, AND CIRCULATING OILS

ASTM D-943, ASTM D-4310, ASTM D-4871, ASTM D-5846, ASTM D-2272 IP-280, ISO-4263, IP-306, IP-48

COMPRESSOR OILS

DIN-51352 Parts 1&2 (Pneurop oxidation test)

GEAR OILS

ASTM D-2893, U S Steel S-200

Tests like ASTM D-943, ISO-4263, ASTM D-4310, ASTM D-4871, and ASTM D-5846 are based on this scheme. In the most popular ASTM D-943 test, oxygen is bubbled at 3 l/h for 1000 h through 300 of ml oil sample containing 60 ml of water and iron and copper coiled wires as catalyst. Temperature is maintained at 95°C. After the test duration, oil acidity in mg/KOH/g is measured and reported. Good oils will have low acidity. ASTM D-4310 is slightly modified in that the oil is filtered through a 5 μm filter, and insolubles are also reported. ASTM D-4871 and ASTM D-5846 are universal oxidation tests, where 100 ml of oil is oxidized at 135°C. The test is terminated when the oil acidity is increased by 0.5 mg KOH/g.

IP-48, IP-280, IP-306, and Pneurop oxidation tests are carried out in a similar apparatus. In IP-48, air is bubbled through 40 ml of oil at the rate of 15 l/h for 2 × 6 h of duration, at 200°C. Viscosity increase and carbon residue are determined after the test. In IP-306 test, oxygen is bubbled through 25 ml of sample at 135°C for 48 h in the presence or without the presence of copper wire catalyst. At the end of the test, insolubles and oil acidity are measured. All these tests are designed for individual application of lubricants depending upon the temperature, pressure, air/oxygen flow rate, and duration. Attempts to find a correlation between different oxidation tests have not been very successful [11].

For engine oils, the oxidation properties are evaluated by engine tests. Usually, Sequence III G/F tests are used to evaluate engine oil stability, and viscosity increase at 80–100 h is measured.

CORROSION INHIBITORS

In lubrication systems, there are many situations where corrosion of metals can take place due to the presence of acidic components. In gasoline and diesel engines, sulfuric acid is formed during the combustion of fuels containing sulfur. Corrosion by such acid in engines is prevented by using an over-based detergent containing dispersed calcium carbonate, which continuously neutralizes the formed acid. In most systems, the presence of trace amount of water cannot be avoided that can lead to electrochemical processes of corrosion. In several applications, such as in metal working fluids and water/glycol systems, large amount of water is present in the lubricants, and metals need to be protected against corrosion. Water-based systems are protected from corrosion by using water-soluble corrosion inhibitors such as sodium salts of benzoic acid, sebacic acid, sodium nitrite, and triethanol amines.

RUST INHIBITOR

Rusting of ferrous metal components can take place in the presence of trace amount of water in lubricants. This can be prevented by forming a protected layer on the metal surface. Following polar compounds are used as rust inhibitors in lubricant formulations:

1. Calcium/magnesium/barium petroleum sulfonates or synthetic sulfonates
2. Ethoxylated alkyl phenols
3. Dodecyl succinic ester (widely used in industrial oils)
4. Oleyl sarcosine/imidazolines in greases and rust preventive oils

Rust inhibitors are evaluated by ASTM D-665 A and B procedures for 24 h or more. Procedure A utilizes distilled water, while procedure B uses salt water and is more severe. The steel pin immersed in oil/water is evaluated for any rust formation. Rust preventive oils are, however, evaluated by tests like salt spray corrosion or humidity cabinet corrosion tests.

FRICTION MODIFIERS/LUBRICITY/FILM STRENGTH/ OILINESS ADDITIVES

The basic function of a lubricant is to reduce friction between the moving metal surfaces, however, as the load increases; friction goes up due to metal to metal contact. At moderate loads, friction can be reduced by some polar compounds that get adsorbed at the metal surfaces, and the oil film is capable of carrying higher loads. These chemicals at further higher loads are, however, squeezed out of the metal surface and do not provide protection. Such compounds are vegetable/animal fats, fatty acids, and fatty acid esters. To reduce friction at higher loads and temperature, EP additives are required.

A blend of carboxylic esters derived from the reaction of mono-, di-, and/or poly alcohols with mono- and/or dicarboxylic acids with alkylated phosphorothionates provides energy efficiency to the engines [12]. Alkenyl succinimide is reacted with an acidic molybdenum compound to provide the liquid molybdated succinimide complex suitable as dispersant and friction modifier [13]. Branched-chain alkenyl succinimide is reacted with acidic molybdenum compound to obtain a liquid molybdated succinimide complex, which is useful as a dispersant and friction modifier [14]. A multifunctional additive has been prepared based on boron salt of an oil-soluble hydroxylated amine and a hindered phenolic acid, which imparts improved oxidation resistance and frictional properties to a lubricant formulation [15].

Friction characteristics of lubricants can be determined by several tribological test machines such as four-ball, Amsler, Falex, SRV, SAE, LFW-1, and low-velocity friction apparatus.

Lubricating oil of reduced friction has been prepared by dissolving carbon nanomaterials functionalized with ester or amide [16]. These have been prepared by dropwise additions of reactants to the carbon nanomaterials. Self-dispersing cerium oxide nanoparticles [17] are prepared by reacting a mixture of organo-cerium salt, fatty acid, and amine (in 1:1:1 to 1:2:2 ratio) in the organic solvent at a temperature between 150°C and 250°C. These nanoparticles have friction-reducing properties in lubricants. Recently, there has been a tremendous interest in the application of nanotechnology in lubricant additives, and large numbers of publications have appeared in the literature [18–28]. It is expected that this technology has the potential to bring in some of the new-generation lubricant additives with greatly improved performance.

Recently, ionic lubricants and additives have been described in the literature, which are claimed to provide fuel efficiency. These additives are based [29] on imidazoline, pyridinium, ammonium, BF4-, and PF6- compounds.

ANTIWEAR AND EP ADDITIVES

When load between the two moving surfaces rises, metal to metal contact takes place, and friction increases leading to higher temperatures and loss of metal from surfaces. This is a condition when the lubricant film is not able to carry the load. To prevent wear, specific chemicals with active sulfur, phosphorous, or chlorine are required. These antiwear materials are temperature sensitive and release active S, P, or Cl at the metal surfaces and generate a completely new surface that is easily sheared at the applied load. Chlorine-containing additives are now phased out due to toxicity problems. The higher load is thus carried by the regenerated surface, reducing the friction and temperature at the surfaces. The difference between antiwear additives and EP additives is very little; both additives contain a reactive atom like S or P. Antiwear additives are effective at milder loads, while EP additives are effective at higher loads and temperatures. Both, however, function through a similar mechanism of forming a reactive film at the surface. Phosphate esters and ZDDP are examples of antiwear additives, while sulfurized hydrocarbons or esters are EP additives.

Some of the following sulfur- and phosphorous-containing organic compounds exhibit antiwear and EP characteristics.

SULFUR COMPOUNDS

Sulfurized isobutylene, sulfurized esters, sulfurized fats
Dibenzyl disulfides
Antimony/molybdenum dialkyldithiocarbamate

SULFURIZED HYDROCARBONS

Unsaturated compounds like fats, some esters, terpenes, and isobutylene can be reacted with elemental sulfur (up to 40% sulfur) to produce sulfurized compounds. Sulfurized isobutylene is a well-known gear oil EP additive and is used in combination with phosphorous compounds.

Sulfur is a good EP additive but is corrosive and cannot be used in elemental form. However, in many metal working fluids, sulfurized base oils are used as EP additive. Sulfur is easily incorporated in base oil (about 1–1.5%) by slowly adding sulfur powder into heated base oil. Dibenzyl disulfide releases sulfur at a relatively moderate temperature and is used in many applications requiring EP protection.

SULFUR–PHOSPHOROUS ORGANO-COMPOUNDS

The most effective multifunctional antioxidant, antiwear, EP, and corrosion inhibitor compound discovered in 1944 is zinc dialkyl/aryl dithiophosphates. Using different primary/secondary alcohols and phenols, large number of products can be derived to meet the antiwear performance requirements. These compounds are widely used in engine oils, industrial oils, and greases. Molybdenum dithiophosphates also exhibit antioxidant properties in addition to antiwear and antifriction properties. Combination of sulfur and phosphorous compounds in gear oils is used to provide EP and antiwear properties.

ZINC DIALKYL DITHIOPHOSPHATES

ZDDP is produced by a reaction of an alkyl/aryl alcohol with phosphorous pentasulfide to form dithiophosphoric acids. This is then neutralized with zinc oxide to produce ZDDP. Performance characteristics of ZDDP can be varied by using primary or secondary alkyl alcohol or the phenol. ZDDPs, like hindered phenols and amines, are mainly free radical scavengers.

$$4\, ROH + P_2S_5 \longrightarrow 2\,(RO)_2\, PSSH + H_2S$$

$$2(RO)_2\, PSSH + ZnO \longrightarrow (RO)_2 - \overset{\overset{\displaystyle S}{\|}}{P} - S - Zn - S - \overset{\overset{\displaystyle S}{\|}}{P} - (RO)_2 + H_2O$$

Dithiophosphoric acid   Zinc dialkyl dithiophosphate

Similarly, zinc diaryl dithiophosphates can be prepared by taking alkyl phenol in place of alkyl alcohol. Aryl ZDDPs are more thermally stable than the alkyl ZDDP and are preferable in high-temperature applications. ZDDP reacts with alkyl free radicals as well as destroys alkyl hydroperoxides. The antioxidant behavior of the secondary alkyl ZDDP is better than that of the primary alkyl ZDDP. ZDDP in combination with hindered phenols and amines provides synergistic effect.

PHOSPHOROUS COMPOUNDS

Alkyl aryl phosphates such as tricresyl and tri-isopropyl phenol phosphates have been extensively used as antiwear additives in hydraulic, gear, and compressor oils. Some mono- or dialkyl phosphate esters can be neutralized by alkylamines to yield a multifunctional antiwear and anticorrosion additive. Triaryl phosphate esters have the following structure:

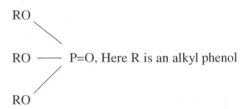

ALKYL PHOSPHITES

These compounds are known for their antiwear and EP properties, but they can also destroy hydroperoxides and peroxy and alkoxy free radicals. Alkyl phosphate thus behaves like secondary antioxidants and can be conveniently used as synergistic additives in combination with hindered phenols and amines. Hydrolytic stability of phosphates/phosphites is however poor, and the acidic compounds formed may lead to the corrosion of metals.

PASSIVE EP ADDITIVES

Highly over-based calcium petroleum sulfonate (400 and above TBN) have been found to possess good EP and antiwear properties and have been successfully utilized in metal working fluids in combination with another sulfur-containing compound such as sulfurized fat. Their action appears to be due to the nanosize calcium carbonate particles used for over-basing.

Wear and EP properties determinations
Wear and EP properties of lubricants are determined by four-ball wear, four-ball EP, Timken, and FZG machines. These are the general laboratory tests for evaluating antiwear and EP additives. There are individual test procedures for hydraulic, gear, engine, and transmission fluids and will be discussed in respective chapters discussing these oils. These tests are also called rig tests and are discussed in the following section.

RIG TESTS FOR THE EVALUATION OF ANTIWEAR AND EP ADDITIVE-CONTAINING OILS

In lubricant terminology, rigs are small machines on which the performance evaluation of lubricant additive and finished products is carried out. Details of some of these rigs are provided in the following text. It is useful to understand these tests since they are referred in many lubricant specifications.

FOUR-BALL WEAR TEST (ASTM D-4172 FOR OILS AND ASTM D-2266 FOR GREASES)

Three 12.7 mm diameter steel balls are clamped together and covered with the lubricant under test. A fourth half-inch steel ball is pressed with a force of 147 or 392 N into the cavity formed by the three balls for the *three-point contact*. The temperature of the test lubricant is regulated at 75°C, and the top ball is rotated at 1200 rpm for 60 min. Lubricants are compared by using the average size of the scar diameters worn on the lower three clamped balls.

FOUR-BALL EP TEST (ASTM D-2783)

The four-ball EP tester is operated with one steel ball rotating at 1780 rpm against three steel balls dipped in test oil and held stationary in the form of a cradle. A series of tests of 10 seconds duration are made at increasing load until welding occurs. The lowest load in kilograms at which the rotating ball welds to the three stationary balls is called weld point. Load wear index is the ability of the lubricant to prevent wear at applied loads. The equation for load wear index reflects the ability of a lubricant to carry a high load without welding and to allow only relatively small wear scars at loads below the weld point.

TIMKEN EP TEST (ASTM D-2782)

Timken OK load test is widely used for specification purpose only and is used to differentiate between lubricants having low, medium, or high EP characteristics. The results may not correlate with the actual performance of the oil in service.

In the Timken Wear machine, a test block and a rotating cylinder (cup) at 800 rpm are used with a loading system. The minimum load that will rupture the lubricant film and cause scoring or seizure and the maximum load (OK load) that will not rupture the lubricant film are determined. Test is initiated at an applied load of 30 lbs and increased in increments of 10 lbs until scoring occurs. The load is then reduced by 5 lbs to determine the final score load and OK load values. Each load stage is run for 10-min duration with oil temperature at 38°C.

FZG LOAD-CARRYING CAPACITY TEST (IP334/DIN 51354)

The FZG spur gear test rig consists of a closed power circuit with drive and test gears connected by two torsion shafts. One of the shafts has a positive clutch for the application of load. Special gear wheels are run in the lubricant under test at a constant speed for a fixed time. The initial oil temperature is controlled but allowed to rise freely during each stage of the test. Loading is raised in stages. The test is continued until the damage load stage is reached, but if no damage occurs at load stage 12, the test is terminated. The gears are inspected visually, without removal, at the end of each load stage. The failure load stage is determined by the summation of deep scoring, seizure lines, or seizure areas on any of the gear teeth. The load stage in which failure occurs is reported together with the test conditions, for example, A/8.3/90, where A is gear type, 8.3 is pinion speed at pitch circle in m/s, and 90 is the initial temperature in oil sump in degree Celsius. These are the usual standardized conditions, but can be changed as required such as A/16.6/90 or A/16.6/140. Results are reported in terms of the highest pass stage for the IP method or the first fail load stage.

FZG MICRO PITTING TEST

This test evaluates the ability of gear lubricants to resist micro pitting and is carried out in two stages. First the load stage test is carried out, and then this is followed by an endurance test. During the load stage test, the ability of the gear lubricant to resist micro pitting is determined. The endurance test provides information on the progress of the damage after a higher number of load cycles. The gears are examined for weight loss and area of micro pitting.

DENISON T6C HYDRAULIC VANE PUMP TEST (DENISON SPECIFICATION TP-30283)

The pump test is carried out to evaluate the wear and filter blocking performance of hydraulic fluids in Denison T6C Vane Pump Rig with and without water contamination. The same fluid is used for two 305 h test phases, first with less than 0.05% water and then with 1% water. A new pump cartridge is used for each phase. Pump rating is carried out by Denison at their factory in Verizon, France.

Vickers's vane pump tests are also used in the industry to evaluate wear in hydraulic fluids.

POUR POINT DEPRESSANTS

The presence of long-chain paraffin in lube base oils tends to limit low-temperature fluidity of these oils. In refinery, the dewaxing process can lower the pour point to a limited extent usually in the range of −6 to −12°C. For lower-temperature applications, polymeric additive called PPD is used to bring down oil fluidity to the desired temperature. The pour point of the oil is defined as the lowest temperature at which the oil can be poured. The pour point of base oil is controlled by the nature and amount of wax present, which are crystallized as the temperature is lowered. The polymeric pour depressant additives interact with wax crystals and modify their structure. There are several other mechanisms through which these additives function such as by preventing nucleation and retarding crystallization. Copolymers of alkyl methacrylate, poly alkyl acrylamides, alkylated naphthalene, and styrene–maleic ester copolymers are some of the examples of PPDs. Large numbers of commercial products are available for specific base oils.

VISCOSITY MODIFIERS (VISCOSITY INDEX IMPROVERS)

These are polymeric organic compounds and were earlier called as VIIs since they were used to modify viscosity–temperature properties of lubricants. VI is an arbitrary number, indicating the tendency of the lubricant to change viscosity with temperature. High VI lubricants have better resistance to thicken at lower temperatures and thin out at higher temperatures. Ideal lubricant should undergo least change in viscosity with respect to temperatures. Different API group base oils have different VI. VI can be calculated from viscosity measurement at 40 and 100°C. This property is important for industrial equipment operating under hydrodynamic regime, where the lubricant must have viscosity within a limit at all operating temperatures. Many industrial lubricant specifications have VI stipulations, but engine oil specifications do not have this requirement; instead, they specify viscosities at different temperatures and shear. Due to this reason, the polymeric VI improvers are now called viscosity modifiers. Following groups of polymeric compounds of appropriate molecular weight are used as viscosity modifiers:

Polymethacrylates

Polyisobutylene

Hydrogenated styrene–isoprene/butadiene copolymer

Ethylene–propylene copolymer (better known as OCP)

The molecules of polymers are so designed that these are oil soluble and remain in coiled form at lower temperatures (say at 40°C) and do not increase oil viscosity to a great extent. As the temperature rises, the coils open up and take up large amount of oil thereby increasing the oil viscosity. This phenomenon increases the VI of oil. Shear stability of polymer molecules is another important property that is considered while selecting a polymer for an application. The polymer chains break down under shear, and this property is determined by measuring shear stability index of polymers. Thus, the selection of a viscosity modifier for industrial or automotive oil requires a very careful analysis and measurements of oil-thickening property, VI increase, and shear stability (under different shear rates). Often, suppliers of viscosity modifiers provide the polymer as oil solution with their recommendation for formulating various grades of oils. Shear stability is determined by a diesel injector test called Kurt Orban test or shear stability test. The molecular structure and molecular weight of the polymer play important role in providing VI improvement, low-temperature flow properties, and shear stability [30]. Amine-functionalized polymeric additive having both dispersant and viscosity modification properties suitable for internal combustion engine has been prepared [31].

SHEAR STABILITY TEST—KURT ORBAN (CEC-L-14-A-48)

Shear stability of polymers used in multigrade oils or in hydraulic oils is determined by this diesel injector test. A sample of oil is subjected to 30 or 250 cycles in a two-cylinder diesel injection pump and injector nozzle set to a pressure of 175 bar. Viscosity of the oil is determined after the test, and viscosity loss is reported. Shear stability index is calculated from the formula.

TACKINESS ADDITIVES

Higher-molecular-weight polyisobutylene is used as tackiness additive and increases oil viscosity substantially. Other oil-soluble high-molecular-weight polymers such as OCP may also be used to a limited extent. There is no standard test available for evaluating this property, but it is often tested by quickly drawing a glass rod from the oil and observing the length of the unbroken oil thread.

ANTIFOAM COMPOUNDS

Most oils on vigorous shaking will produce foam. Oils that have higher viscosity and polar compounds will tend to form higher amount of foam due to surface activity. Lubricants during the use are subjected to shaking, and it is therefore necessary to use antifoam additives to control foam. Foam, if not controlled, can damage equipment due to false lubricant level and lead to high wear due to compressible air bubbles. Foaming characteristics of oil are measured by ASTM D-892 test method under three sequences. Silicon polymer is the most common antifoam compound. These are generally dispersed in oil in low concentrations. In several applications where air release value is important, such as turbine and hydraulic oils, acrylate-based antifoam compounds are used, since these do not affect the air release characteristics of the oil.

DETERGENTS AND DISPERSANTS

In both gasoline and diesel engines, unburnt fuel combustion products and thermo-oxidative products of oil generate soot, sludge, and deposits. For optimum performance, the engine must be kept clean. This cleaning is carried out by incorporating detergent and dispersants into the crankcase oil. Both detergents and dispersants are oil-soluble surface-active compounds having a polar head and a nonpolar hydrocarbon chain. These additives form major portion of engine oil additives and are required to be carefully selected. In an engine, sludge, deposit formation, ring sticking, oil thickening, wear control, bore polish, and acid neutralization (due to SOx and NOx) play major role. Both detergents and dispersants in combination with antioxidant/antiwear (through ZDDP) can control these parameters. Detergents keep the parts clean, which are subjected to high temperatures, such as piston rings, skirts, and undercrown. In these parts, oil decomposes to form deposits, due to localized high temperatures. If these deposits are allowed to build up, heat transfer from piston to oil will be affected, leading to still higher temperatures. Dispersants on the other hand keep the sludge, soot, and deposits dispersed in oil phase. Sludge is a complex material of fuel combustion products, water, carbon, and oxidized oil, which on agglomeration becomes insoluble in oil. Thus, both detergents and dispersants play complimentary role. In diesel engines, soot formation is higher, and therefore, dispersant level also has to be higher for these oils as compared to gasoline engine oils. Detergents are neutral or over-based calcium or magnesium sulfonates, phenates, phosphonates, carboxylates, or salicylates. Over-basing is generally carried out with carbonate or hydroxide, which can neutralize acidic compound formed during oxidative combustion. Phenates can also be sulfurized to impart additional antiwear properties. Dispersants on the other hand are long-chain alkyl amine, amide, and imides. Polyisobutylene succinimide (PIBS) has been one of the most popular dispersants for engine oils. Both detergents and dispersants have a polar group and a nonpolar long hydrocarbon chain, but in dispersants, the two are linked through another polar molecule, that is, the imide group is linked to PIB group through succinic anhydride group.

SYNTHETIC OR PETROLEUM SULFONATES

These could be Ca/Mg/Ba salts of alkyl aryl sulfonic acids. Ba salt is, however, not used now due to toxicity considerations. Sulfonic acids derived from white oil plants have undefined structures and can be represented by RSO_3H. These on neutralization with CaO/MgO or hydroxides yield neutral sulfonates, $(RSO_3)_2Ca/Mg$. These neutral sulfonates can be over-based by reaction with metal hydroxide in the presence of a promoter alcohol and carbon dioxide. Calcium/magnesium carbonate formed is attached to the molecule, providing basicity to sulfonate. Basicity could range from 30 to 400 mg KOH/g or more depending upon the reaction conditions. Calcium carbonate is held within the invert micelle structure of sulfonate molecule [32] $(RSO_3)_2$ Ca—$xCaCO_3$.

PHENATES

When alkyl phenols react with calcium hydroxide, metal phenates are produced. These phenates on sulfurization and over-basing become multifunctional additives, possessing detergent, antioxidant, anticorrosion, and antiwear properties. These are important additives for diesel engine oils due to their versatile properties and high-temperature stability. Like sulfonates, calcium carbonate is also held within the invert micelle structure of phenate molecule. Over-based sulfurized phenates can be represented by the following structure:

$$nMCo_3$$
$$OMO$$

$$R_1 —\bigcirc— Sx —\bigcirc— R_1$$

Over-based calcium sulfurized phenate

M may be calcium or magnesium.

Alkyl salicylates are prepared from alkyl phenol by first converting phenols into salicylic acid and then neutralizing salicylic acid with metal hydroxides of Ca or Mg to obtain neutral metal alkyl salicylates. These can be converted to over-based products by passing carbon dioxide in the presence of excess metal hydroxides. Over-based calcium/magnesium alkyl salicylate has the following structure:

$$OH \quad (MCO_3)_m \quad OH$$

$$R_1 —\bigcirc— C\text{-}O\text{-}M\text{-}O\text{-}C —\bigcirc— R_1$$
$$\quad\quad\quad \| \quad\quad \|$$
$$\quad\quad\quad O \quad\quad O$$

Over-based metal (M is calcium/magnesium) alkyl salicylate detergent

R1 is an alkyl group, and in place of calcium metal, magnesium can be taken to obtain magnesium alkyl salicylate. Salicylates are claimed to be very effective detergents in several specialized diesel engine oil applications. Their good antioxidant properties are due to the presence of phenolic groups.

ALKENYL SUCCINIMIDE DISPERSANTS

These compounds do not contain metal and are ashless. Succinimides are characterized by a linkage group between the polar amine group and a nonpolar hydrocarbon group. The nonpolar hydrocarbon group is usually provided by a 1000-molecular-weight polyisobutylene, and the polar group is a polyamine. These two are connected through a succinate group. Thus PIB succinimides are produced by first reacting PIB with maleic anhydride to obtain PIB succinic anhydride. Both mono- and bis-succinic anhydrides can be produced by this reaction by changing mole ratio of reactants. The succinic anhydride is now reacted with polyamine to get PIB-succinimide. PIB-succinate esters can be produced by using polyols in place of polyamines. If amino alcohols are used in place of polyamine, then hydroxyl alkyl succinimide structures are obtained.

EVALUATION OF DETERGENTS AND DISPERSANTS

Laboratory tests to evaluate detergents and dispersants have limitations since these have to actually work in the engines, and engine test conditions cannot be simulated in these tests. However, the following two tests can be carried out to find out the relative performance of various molecules:

1. Panel Coker test for detergency characteristics according to FTMS 791 B-3462 procedure.

2. Lamp black dispersancy test, where activated lamp black is dispersed in oil and viscosity increase is measured. Less increase means better dispersancy.

EMULSIFIERS AND DEMULSIFIERS

Emulsification and demulsification are two opposite processes and are both utilized in lubricants. Several lubricating oils such as cutting fluids, fire-resistant hydraulic fluids, and steel-mill rolling oils are used as water emulsions or dispersions, and these products need emulsifiers or surface-active compounds. Emulsion formation is undesirable for other industrial and automotive oils and could be a cause of equipment failure. In a centralized lubrication system, the oil is continuously subjected to centrifugation and filtration to remove water, loose emulsion and other impurities. In engine oils, water produced by the fuel combustion can get into the crankcase and form emulsion since engine oil contains polar detergents and dispersants. To control emulsion formation tendencies, high-quality base oils and a demulsifier additive combination is most useful. However, both emulsifiers and demulsifiers are surface-active compounds having a polar group and capable of concentrating at the surface or interface. Water separation characteristics of lubricating oils are determined by ASTM D-1401/ISO-6614 or ASTM D-2711 methods. Common emulsifiers used in lubricants are as follows:

Sorbitan mono-oleate

Triethanolamine oleate

Calcium petroleum sulfonate

Ethylene oxide/propylene oxide condensates with alcohols and phenols

Soaps and carboxylates

Demulsifiers are also surfactants with slightly different structures. Some of the following groups of compounds have been used as demulsifiers in lubricants:

Aliphatic amines such as octadecyl amine

Long-chain imidazolines

Barium dinonyl sulfonate

Polyalkylene glycols and EO/PO condensates

Detailed discussion on lubricant additives can be found in books exclusively dealing with additives [2, 3, 32].

COMPATIBILITY OF ADDITIVES AND LUBRICANT BLEND STABILITY

Different lubricants like engine oils, transmission oils, and industrial oils use a combination of several chemical additives in different ratio to obtain desired performance. Additives such as over-based detergents are basic in nature; some rust inhibitors, friction modifiers, and antiwear additives are acidic and can thus react slowly to form another compound that may not have the required properties. Such reaction products may sometimes form an insoluble compound, forming sediments in the lubricants. EP additives are chemically and thermally unstable. During the application of lubricants, some of these additives undergo decomposition or transformation into another compound, which in turn can interact and produce entirely different performance characteristics. These changes could be synergistic (where the performance of either additive is enhanced due to the presence of the other additive) or antagonistic (where the performance of one or both additives is reduced due to the presence of the two additives together). These interactions can also sometimes alter the solubility of a particular additive. The interactions could at times be quite slow, so that the effect is manifested after sometime. These interaction are, therefore, quite important from the point of view of formulating lubricating oil, and the additive combinations must be examined for these aspects. In lubricant industry, this aspect is important while selecting the additive combination. Due to this reason, the mixing up of different brands of lubricating oil is not suggested in the field.

The simplest method to test the compatibility of additives or finished lubricating oil is to subject the blended product (in different ratio) to a hot and cold cycle (100 to −9°C for 3–7 days of each cycle) in a glass container and then examine the formation of sediment in the bottom or layer separation in the blend. Any separation or layer formation indicates additive incompatibility and needs further investigations. Interaction is still possible, even if there is no sedimentation, since the reacted products could also be oil soluble. This can be monitored by taking an FTIR or FTNMR spectrum of individual additives and also the blended lubricant. Presence of any additional peak or disappearance of an existing peak of absorbance in the spectrum of the blend could be the sign of interaction. Positive interaction, that is, synergism is desirable. Negative interaction, that is, antagonism is undesirable. In synergism, a combination of two or more similar additives together provides improved properties, for example, phenolic and amine antioxidant combination gives better oxidation stability. More than one antiwear and EP additives are often used in lubricants, but these have to be carefully selected to avoid any adverse interaction. Several interactions within the additives have been reported. These are as follows:

1. ZDDP interacts with large number of additives such as EP additives, detergents, dispersants, rust inhibitors, VI improvers, and friction modifiers based on molybdenum.
2. Over-based detergent interaction with acidic components like oleic acid/stearic acid.
3. Antioxidant and EP additive interaction. This interaction could be antagonistic.

4. In metal working fluids, ethanolamine interacts with sodium nitrite and various fatty acids and should be studied carefully.

Even base oils interact differently with antiwear and EP additives. The effect of different types of base oil of API groups I–V, such as mineral based (solvent refined, hydrofinished, hydrocracked/wax isomerized) and synthetic based (PAO and ester), on the interaction of ZDDP with PIBS has been studied [33] by variable-temperature IR and ^{31}P NMR spectroscopic techniques. ZDDPs are known to exist in monomeric and polymeric forms and as neutral and basic salts, and their functioning is temperature dependent. The structural changes observed in the NMR and IR spectra of the ZDDP–PIBS system in different base oils have been explained in terms of the degree of solvation provided by the base oil in which they are present. The results indicate that the complexes are stronger and stable in polar base oils, such as from groups I and V. The equilibrium shifts toward the neutral form of ZDDP in polar base oils. Polar solvents, such as methanol and tetra hydrofuran, also favor the formation of the neutral form of ZDDP and stronger complexes between PIBS and ZDDP, in a manner similar to that observed for polar base oils. ZDDP is present in combination with several other additives in engine oils, transmission fluids, and hydraulic oils, and its interaction with other molecules has been investigated by several authors [34–39]. ZDDP structure is such that it can form complexes with many compounds, especially with amines and PIBSs, and may affect the antiwear performance. ZDDP's main function is to provide antiwear protection. For this function, it has to reach the metal surface asperities and provide sulfur and phosphorous at the contact point to generate new surfaces of sulfides or phosphides under the frictional heat. If the other polar additives or EP/antiwear additives are present in the lubricating oil, these will compete with ZDDP to reach the surface, and its antiwear properties would be affected. This can explain the antagonistic effect of two or more similar EP/antiwear additives. However, in majority of the cases, the combination can also create synergy if proper combinations are selected. In gear oils, sulfur and phosphorous compounds are combined in specific ratio to obtain higher EP and antiwear properties.

Compatibility of primary C4–C5–ZDDP with neutral, over-based sulfonate (25 and 300 TBN), over-based phenate of 200 TBN, PIB succinimides, and antioxidants in a diesel engine oil formulation has been investigated [40, 41]. The interaction has been monitored by conducting panel coker test for deposit-forming tendency, lamp black dispersancy test, and four-ball wear test. From these studies, it has been concluded that the best optimized combination of ZDDP, over-based sulfonate, neutral sulfonate, over-based phenate, PIBS in a lubricant formulation is in the ratio of 1:5:7:10:6. The probable mechanism of interaction in ZDDP–PIBS is through asymmetrical complex formations. Similarly, with bis-succinimides, strong bidentate complexes are formed. With antioxidants, the interaction is through hydrogen bonding. It has been reported that surface-active rust inhibitors [42] also interact with secondary ZDDP and promote severe metal attack in the hydrolytic stability test.

Interaction between ZDDP, molybdenum dithiocarbamate (MoDTC), ashless alkyl phosphorodithioate, PIBS, and metal sulfonates/phenates has been investigated using ^{31}P NMR, electron spectroscopy for chemical analysis (ESCA), and thermogravimetric analysis (TGA) techniques [43–45]. ESCA, which is a surface analytical technique, has been used to provide basic evidence for the formation of various complexes through interactions occurring in the electronic binding energies of orbitals of various atoms of the additives. The ESCA studies have also revealed the actual atomic sites of interaction between the additives responsible for the formation of adduct or complexes. The shifts in the ^{31}P NMR signals; the changes in the binding energies of the s, p, and d orbitals of additive elements; and the multistage decomposition profiles in the TGA thermograms of interacting systems due to complexion and adduct formation have enabled to propose the mechanism of interaction.

An interesting study shows that Mo-compounds alone do not show good antiwear properties. However, when MoDTC is used in combination with ZDDP, a synergistic behavior is observed, and the antiwear and friction properties are improved considerably. The synergistic activity of MoDTC with ZDDP has been studied by several authors [46, 47], and it is assigned due to the functional group exchange between ZDDP and MoDTC. Due to this exchange, the mixture of ZDDP and molybdenum dithiophosphate (MoDTP becomes a mixture of four species in (ZDDC + MoDDP + ZDDP + MoDTC) place of the original two species, and thus, better antiwear and friction properties are exhibited. This interaction has been confirmed by recent studies [45] conducted on the blends of ZDDP and MoDTC by using ^{31}P NMR, ESCA, and TGA.

ZDDPs, although, are good multifunctional additives, but their interactions with other additives in lubricants are quite extensive, and these need to be investigated properly in each formulation.

SYNERGISM OF METAL DEACTIVATORS WITH ANTIOXIDANTS

Metal deactivators do not react with antioxidants directly, but these indirectly supplement the role of antioxidants by passivating catalytic action of metals like copper or brass. Metal deactivators and antioxidants thus act synergistically and would be useful in increasing the life of lubricant in the system. Thus, metal deactivator and antioxidant synergism is not regarded as an interaction, but the combination provides improved antioxidant response.

ANTIOXIDANT–METAL DEACTIVATOR–EP AGENTS

It has been reported [48, 49] that when a commercial sulfur-/phosphorous-based gear oil additive is used in combination with an amine antioxidant and benzotriazole-based metal deactivator, both antioxidant properties and EP properties are greatly enhanced. Oxidation properties have been evaluated by ASTM D-943 or TOST test, and EP properties by four-ball and Timken tests. Table 5.2 provides data obtained from these two additive systems. This combination is quite useful in formulating high-performance products with minimum additive dosage.

TABLE 5.2 Effect of amine antioxidant and an MDA in a gear oil

Properties	Commercial S/P industrial gear oil A	A plus amine antioxidant and metal deactivator
Oxidation ASTM D-943, TAN after 1000 h	2.4 mg KOH/g	0.9 mg KOH/g
Timken Ok load (lbs)	55	80
Four-ball weld load (kg)	310	440
Load wear index	48	60

Amine phosphates are extensively used in gear oils and other lubricants to impart antiwear and low-friction properties. These, however, adversely interact with phenolic and amine antioxidants and reduce the effect of antioxidants [50].

DETERGENTS, DISPERSANTS, AND ANTIWEAR ADDITIVES

The interaction of succinimides and calcium salicylate/sulphonate has been investigated by Vuk and others [50, 51]. They found that the viscosity of blends containing succinimides and calcium salicylates increases to as high as 45%, due to some kind of interaction. The interaction was more intense when the TBN of detergent and nitrogen content in the succinimides was higher. In these mixtures, formations of amine salt/adduct or complex have been postulated. These authors also showed that the incorporation of an antiwear agent into detergent–dispersant mixture weakened the antiwear action of the additive due to stronger surface interactions of detergents/dispersants. The interaction between ZDDP and succinimides could be seen in the IR spectrum as a shift in the P=S doublet, suggesting N–P–S linkage.

ACIDIC AND BASIC ADDITIVES

Acidic additives like oleic acid, stearic acid, half ester of succinic acid (rust inhibitor), and also ZDDP can react with basic materials such as over-based sulfonates, phenates, salicylates, triethanolamine, and diethanolamine and can form insoluble salts. Mixtures of these chemicals, if used in lubricants, could be the source of incompatibility, and additive separation can take place.

Most lubricating oils contain multiple additives to improve the performance of base oils. It is therefore important to understand the mechanism of additive interaction properly to formulate a balanced product. Lubricants containing large number of additives such as engine oils, transmission oils, and universal tractor oils need specific attention with respect to additive compatibility and require a balanced formulation for good performance. The use of API group III, GTL base oils, and synthetic oils such as PAOs brings in additional solubility issues of additives in them and needs to be handled carefully. Alternately, the base oil blend has to be such that the additives are fully miscible in them.

SOLID LUBRICANT ADDITIVES

Solid compounds such as MoS_2, graphite, PTFE, and others provide [1] low-friction characteristics due to their structures, and these have been extensively used in greases and high-viscosity gear oils where their separation does not pose problems. However, their use in other industrial and engine oils is restricted due to the separation or incompatibility problem. Instead, soluble molybdenum compounds have been developed, which provide energy efficiency in both industrial and automotive engine oils. The application of nanotechnology offers advantage, and in future, several nano solid materials might be used in lubricating oils.

REFERENCES

[1] Clauss FJ. *Solid Lubricants and Self-Lubricating Solids*. New York: Academic Press; 1972.
[2] Rasberger M. Oxidative degradation and stabilization of mineral oil based lubricants. In: Motier RM, Orszulik ST, editors. *Chemistry and Technology of Lubricants*. London: Blackie Academics & Professional; 1997. p 98–143.
[3] Migdal CA. Antioxidants. In: Rudnick LR, editor. *Lubricant Additive Chemistry & Applications*. Boca Raton: CRC Press; 2003. p 1–27.
[4] Salomon MF. N-substituted thio alkyl phenothiazine. US patent 5,034,019. 1991.
[5] Germanaud L, Arozin P, Turelo P. Nitrogen containing antioxidants for lubricating oils. French patent 2639956. 1990.
[6] Nagata T, Kusuda C, Wada M. Process for the preparation of diphenylamine or nucleus substituted derivatives thereof. US patent 5,545,752. 1996.
[7] Zhu PY. Synthesis of alkylated aromatic amines. US patent 5,734,084. March 1998.
[8] Gatto VJ, Elnagar HY, Moehle WE, Schneller ER. Redesigning alkylated diphenylamine antioxidant for modern lubricant. Lubr Sci 2007;19 (1):25–40.
[9] Parenago OP, Bakumin VN. Problem of inhibiting the high temperature oxidation of hydrocarbon, additive 97. In: Kovacs A, editor. Proceedings of Hungarian Chemical Society; May 21–23, 1997; Sopron. p 81–88.
[10] Chan MS, Hunter WE. Preparation of benzotriazole. US patent 4,299,965. 1981.
[11] Jay Prakash KC, Srivastava SP, Anand KS, Goel PK. Oxidation stability of steam turbine oils and laboratory method of evaluation. J ASLE February 1984;40 (2):89–95.
[12] Rinklieb R, Rettemeyer D, Scherer M. Lubricant compositions. US patent 2012,0220,508 A1. August 30, 2012.
[13] Nelson KD, Harrison JJ, Rogers P, Hosseini M. Process for preparation of low molecular weight molybdenum succinimide complexes. US patent 8,476,460. July 2, 2013.
[14] Nelson KD, Harrison JJ, Rogers P, Hosseini M. Process for preparation of high molecular weight molybdenum succinimide complexes. US patent 8,426,608. April 23, 2013.
[15] Suen YF, Ward J, Miller T. Lubricating composition containing multifunctional borated hydroxylated amine salt of a hindered phenolic acid. US patent 8,334,242. December 18, 2012.
[16] Habeeb JJ, Bogovic CN. Reduced friction lubricating oils containing functionalized carbon nanomaterials. US patent 8,435,931. May 7, 2013.
[17] McLaughlin MJ, Mathur N. Nanoparticle additives and lubricant formulations containing the nanoparticle additives. US patent 8,333,945. December 18, 2012.
[18] Coleman KS, Bailey SR, Fogden S, Green MLH. Functionalization of single-walled carbon nanotubes via the bingel reaction. J Am Chem Soc 2003;125:8722–8723.

[19] Guldi DM, Menna E, Maggini M, Marcaccio M, Paolucci D, Paolucci F, Campidelli S, Prato M, Rahman GM, Schergna S. Supramolecular hybrids of 60- fullerene and single-wall carbon nanotubes. Chem Eur J 2006;12:3975–3983.

[20] Ashcroft JM, Hartman KB, Mackeyev Y, Hofmann C, Pheasant S, Alemany LB, Wilson LJ. Functionalization of individual ultra-short single-walled carbon nanotubes. Nanotechnology 2006; 17:5033–5037.

[21] Chen CS, Chen XH, Xu LS, Yang Z, Li WH. Modification of multi-walled carbon nanotubes with fatty acid and their tribological properties as lubricant additive. Carbon 2005;43:1660–1666.

[22] Chen C-S, Chen X-H, Liu T-G, Yang D, Zhang G, Yi G-J. Chemical modification of carbon nanotubes and tribological properties as lubricant additive. Acta Chim Sin 2004;62 (14):1367–1372.

[23] Worsley KA, Moonoosawmy KR, Kruse P. Long-range periodicity in carbon nanotube sidewall functionalization. Nano Lett 2004;4 (8):1541–1546.

[24] Hu H, Zhao B, Hamon MA, Kamaras K, Itkis ME, Haddon RC. Sidewall functionalization of single-walled carbon nanotubes by addition of dichlorocarbene. J Am Chem Soc 2003;125:14893–14900.

[25] Sun Y-P, Huang W, Lin Y, Fu K, Kitaygorodskiy A, Riddle LA, Yu YJ, Carroll DL. Soluble dendron-functionalized carbon nanotubes: preparation, characterization, and properties. Chem Mater 2001; 13:2864–2869.

[26] Gao F, Lu Q, Komarneni S. Fast synthesis of cerium oxide nanoparticles and nanorods. J Nanosci Nanotechnol December 2006;6 (12):3812–3819.

[27] Sun S, Zeng H, Robinson DB, Raoux S, Rice PM, Wang SX, Li G. Monodisperse MFe2O4 (M is Fe, Co, or Mn) nanoparticles. J Am Chem Soc 2004;126:273–279.

[28] Wang H, Zhu J-J, Zhu J-M, Liao X-H, Xu S, Dinga T, Chen H-Y. Preparation of nanocrystalline ceria particles by sonochemical and microwave assisted heating methods. Phys Chem Chem Phys 2002; 4:3794–3799.

[29] Qu J, Blau PJ, Dai S, Luo H, Meyer HM III. Ionic liquids as novel lubricants and additives for diesel engine applications. Tribol Lett 2009;35:181–189.

[30] Kapur GS, Sarpal AS, Mazumdar SK, Jain SK, Srivastava SP, Bhatnagar AK. Structure–performance relationships of viscosity index improvers: I microstructural determination of olefin copolymers by NMR spectroscopy. Lubr Sci October 1995;8 (1):49–60.

[31] Gieselman MD, Preston AJ. Lubricating composition containing a functionalized carboxylic polymer. US patent 8,557,753. October 15, 2013.

[32] Srivastava SP. Detergents and dispersants. In: *Advances in Lubricant Additives and Tribology*. New Delhi: Tech Books International; 2009. p 462.

[33] Sarpal AS, Bansal V, Sastry MIS, Mukherjee S, Kapur GS. Molecular spectroscopic studies of the effect of base oils on additive–additive interactions. Lubr Sci November 2003;16 (1):29–45.

[34] Round FG. Additive interactions and their effect on the performance of a zinc dialkyl dithiophosphate. ASLE/ASME Conference; October 5–7, 1976; Boston.

[35] Round FG. Some factors effecting the decomposition of three commercial zinc organic phosphate. ASLE Trans 1976;18:78–89.

[36] Round FG. Some factors effecting the decomposition of three commercial zinc organic phosphate. ASLE Trans 1978;21:91–101.

[37] Inoue K, Watnabe H. Interaction of engine oil additives. ASLE Trans 1983;26 (2):189.

[38] Spikes HA. Additive-additive and additive-surface interaction in lubricants. Lubr Sci 1988;2 (1): 4–23.

[39] Ramakumar SSV, Rao AM, Srivastava SP. Studies on additive-additive interaction. Wear 1992; 156:101–120.

[40] Ramakumar SSV, Aggrawal N, Rao AM, Sarpal AS, Srivastava SP, Bhatnagar AK. Studies on additive–additive interactions: effects of dispersant and antioxidant additives on the synergistic combination of over based sulphonate and ZDDP. Lubr Sci October 1994;7–1:25–38.

[41] Recchuite AD, Newingham TD. Effect of zinc dithiophosphates on axial piston pump performance. 30th ASLE Meeting; 1975; Georgia. Preprint No.75AM-4B-1.

[42] Sarpal AS, Christopher J, Mukherjee S, Patel MB, Kapur GS. Study of additive-additive interactions in a lubricant system by NMR, ESCA, and thermal techniques. Lubr Sci May 2005;17 (3):319–345.

[43] Kapur GS, Chopra A, Ramakumar SSV, Sarpal AS. Molecular spectroscopic studies of ZDDP—PIBS interactions. Lubr Sci August 1998;10 (4):309–321.

[44] Sarpal AS, Christopher L, Mukherjee S, Patel MB, Kapur GS. Study of additive–additive interaction in a lubricant system by NMR, ESCA and thermal technique, Proceedings of the 13th LAWPSP Symposium; 2003; Mumbai, India, Vol. 13, No 13, AD.01.01-.15.

[45] Arai K, Yamada M, Asano S, Yoshizawa S Ohira H, Hoshino K, Ueda F, Altiyama K. Lubricant technology to enhance the durability of low friction performance of gasoline engine oils. SAE paper 952,533; 1995.

[46] Korcek S, Johnson MD, Jensen RK, McCollum C. Retention of fuel efficiency of engine oils. Proceedings of the 11th International Colloquium; 1998; Esslingan.

[47] Srivastava SP, Tyagi BR, Subhash C, Rao GJ, Ahluwalia JS. Improved gear oil compositions. Indian patent 148995. 1979.

[48] Griffith MG. Lubricant compositions containing amine phosphate. US patent 5,552,068. 1996.

[49] Olson DH. Relationship of engine bearing wear and oil rheology. SAE paper 872128; 1987.

[50] Srivastava SP, Jayprakash KC, Goel PK. Development in industrial gear oil additive technology. 4th International Seminar on Development of Fuels, Lubricants, Additives; March 7–10, 1983; Cairo: Misr Petroleum Co.

[51] Kapasa PH, Martin JM, Blanc C, Georges JM. Antiwear mechanism of ZDDP in the presence of calcium sulphonate detergent. J Tribol 1981;103 (4):486–492.

CHAPTER 6

LUBRICATION, FRICTION, AND WEAR

The purpose of lubricant is to reduce friction and wear between the two contacting surfaces in particular equipment under the operating conditions of temperature, pressure, load, speed, stress, and time duration. These conditions are called tribological parameters, and the entire lubrication system along with the lubricant and contacting surfaces form tribosystem. The word tribology coined in 1966 is derived from Greek word *Tribos* or *Tribein* meaning to rub. The science of tribology investigates the interaction of contacting metal surfaces in relative motion, which generates friction at the contact zone under various operating conditions. In lubricated surfaces, lubricant becomes another important part of the system. Thus, tribology becomes a multidisciplinary science involving materials, metallurgy, mechanical engineering, surface science, physics, and chemistry, and a close coordination between these different disciplines is required to understand the science of tribology. A proper lubricant for a system can be developed with the understanding of tribology and tribosystem. Such an understanding can lead to longer equipment life, lower wear, energy saving, and longer lubricant life. Several studies [1] in the United Kingdom, United States, and other European countries have shown that improper application and understanding of tribology lead to tremendous monetary loss. There are several advance books on tribology dealing with friction, wear, and lubrication [2–8] and may be referred for detailed understanding. This chapter deals with the basics of tribology—friction, wear, and lubrication—which is required to be understood by the lubricant developers and users.

FRICTION

Friction is a common phenomenon and is observed in day-to-day life. It is both desirable and undesirable in certain situations. For example, friction is desirable while walking with new leather shoes on a smooth surface or on inclined smooth surface. On the other hand, sliding down on snow with a ski is due to low friction between the ski and snow. High friction is also undesirable in all machineries, since it consumes more energy. However, in automotive vehicles, friction is desirable between the brake and disk to stop the vehicle. Leonard da Vinci was the first to

---

Developments in Lubricant Technology, First Edition. S. P. Srivastava.
© 2014 John Wiley & Sons, Inc. Published 2014 by John Wiley & Sons, Inc.

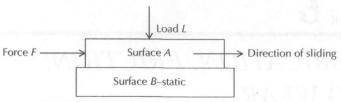

FIGURE 6.1 Frictional force and load relationship.

introduce the concept of coefficient of friction and defined it as the ratio of frictional force to normal load. G. Amontons, while studying two sliding nonlubricated surfaces, stated that frictional force is proportional to normal load and does not depend on the contact area.

Friction is now defined as resistance to motion. Consider that two loaded solid surfaces are in contact and a tangential force is gradually applied to move the upper surface (Fig. 6.1). The surface will move only after applying certain amount of force. This minimum force to move the upper surface is the resistive force parallel to the direction of motion and is defined as static frictional force. The tangential force required to continue the sliding movement of the surface is called kinetic or dynamic friction. Static friction is generally higher than the kinetic friction. The frictional resistance to motion will also depend on the applied load on the surface. Higher load will require higher force to move the surface.

Coefficient of friction is, therefore, defined as frictional force opposing motion divided by the perpendicular load applied to the surface.

$$\text{Coefficient of friction}\,(\mu) = \frac{\text{Frictional force opposing motion}\,(F)}{\text{Perpendicular load to the surface}\,(L)}$$

$$\text{Or frictional force}\,(F) = \mu L$$

The coefficient of friction μ between the two contacting surfaces is independent of load and area of contact. Whatever be the size of the solid surfaces or the load on them, the coefficient of friction remains constant. This relationship is valid only in dry or boundary friction conditions. With fluid film lubrication condition, it is not valid.

The solid surfaces are, however, not as smooth as depicted in the earlier diagram. The surfaces in fact are quite different from the bulk material and need to be understood properly. The surface is in contact with liquid or air and will have complex properties depending upon its interaction with the environment. First of all, the surface is never smooth, and no amount of polishing can produce a molecularly smooth surface. All surfaces have peaks or asperities and valleys. A typical roughness profile would appear like the one in Figure 6.2.

Surface profile can be traced by commercially available equipment, which moves a diamond stylus on the surface and traces the profile of asperities and valleys over the mean line. Now most of the fresh metal surfaces except the noble metals are reactive and form metal oxides in the presence of air. Presence of sulfur, chlorine, and nitrogen can further form sulfides, chlorides, and nitrides from a chemical reaction. There would then be physical absorption on the surface of various hydrocarbons

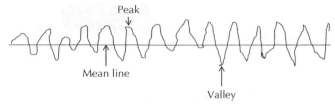

FIGURE 6.2   Typical roughness profile of surface.

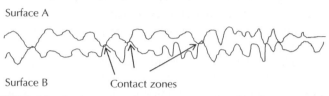

FIGURE 6.3   Contact zones between the two metal surfaces—asperities.

including lubricants, oxygen, and water vapors. Thus, the metal surface is quite complicated containing various layers of compounds depending upon the environment it has been subjected to. The characterization of these films is an important aspect of tribology and leads to the proper designing of the equipment and lubricants. These films can be characterized by several techniques. The absorbed and chemically reactive layers can be analyzed by techniques like Fourier transform infrared spectroscopy, Fourier transform nuclear magnetic resonance spectroscopy, X-ray photoelectron spectroscopy (XPS), and electron spectroscopy for chemical analysis (ESCA). Elemental analysis of the surface is carried out by X-ray fluorescence, electron probe micro analyzer, and Auger electron microscopy techniques. Metallurgical properties of the deformed layer can be determined by examining polished surface by a high-resolution optical and electron microscope. Crystalline structure of the surface can be investigated by X-ray diffraction analyzer.

INTERFACE BETWEEN THE TWO CONTACTING SURFACES

When the two metal surfaces having asperities and valleys come in contact with each other, actual contact will take place only at a few places on the asperities. As the load is increased, contact area increases by compressing the asperities. With still higher loads, surface deformation will take place. This deformation could be elastic (reversible) or plastic (permanent deformation). Thus, the area of contact between the two metal surfaces depends on load, surface roughness, and material properties and is independent of the geometric area. When the surfaces are sliding, the load shifts from one asperity to another in the direction of sliding, and the contact zone is progressively shifting. However, when plastic deformation takes place, load sharing will be disturbed due to the formation of strong junctions, and friction will increase even if the load is reduced. The contact zones between the two metal surfaces are shown in Figure 6.3.

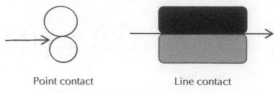

Point contact   Line contact

FIGURE 6.4   Contact geometry.

In different equipment, the contact zone will also vary; for example, between two balls, the contact is a point contact; in two rotating cylinder, it is a line contact. In a gear system, both sliding and rolling actions take place and the contact geometry will be different (Fig. 6.4).

Lubricants play a major role in defining and modifying the surface asperity interaction and friction arising out of such interaction. A thin film of lubricating oil between the contact zones can modify friction to a great extent. This thin film of oil has to be of right viscosity and other physicochemical characteristics to provide adequate protection to the moving surfaces.

FRICTION: TYPES

Different types of frictions have been identified in a tribosystem. They are as follows:

1. Solid friction
2. Fluid friction
3. Static friction
4. Dynamic or kinetic friction
5. Stick-slip friction
6. Sliding friction
7. Rolling friction
8. Boundary friction
9. Mixed friction

Solid friction occurs when two solid surfaces come in contact directly without a separating layer of lubricants. The fresh metal surfaces will be bound by a strong force of adhesion. This is also related to the free energy of the surfaces. However, the metal surfaces are not in their elemental form. These are reacted with atmospheric oxygen to form oxides. These oxide films play an important role in friction and wear behavior of surfaces. In solid-to-solid contacts, friction and wear are high. This condition must be avoided in equipment and engines.

Fluid friction is referred to the friction arising out of the lubricant film separating the two moving solid surfaces. Lubricant viscosity or rheological properties are responsible for this friction. Such friction occurs when the two surfaces are completely separated by the fluid under hydrodynamic or hydrostatic lubrication conditions.

Static friction as explained earlier corresponds to the force that must be overcome to initiate motion between the two static surfaces.

Dynamic or kinetic friction occurs under relative motion and corresponds to the force required to maintain the motion. This friction is usually lower than the static friction.

Stick-slip generally occurs in slow-moving machine tools on the slide/guide ways, where the static coefficient of friction is much larger than the kinetic friction. In such cases, larger force is applied to initiate the motion, but as the tool moves, it slips due to the much lower kinetic friction. Stick-slip causes chatter mark on the components and is undesirable. Specific lubricants are used to overcome stick-slip phenomenon in machine tools.

Pure sliding friction occurs in sliding motion such as in guide ways of machine tools. In roller bearings also, sliding takes place between the rolling element and cage.

Rolling friction takes place in rolling contacts such as steel or aluminum sheet rolling operations. Often, sliding and rolling are combined in several applications such as in gears and roller bearings. The ratio of sliding and rolling, however, varies in different systems. When gear teeth mesh with each other, both sliding and rolling motions take place. Similarly, in rolling bearing, there is rolling between the rolling elements and sliding between the cage and rolling elements.

Boundary friction occurs when the load between the two moving surfaces is very high, and lubricant film cannot carry the load. In such a situation, metal-to-metal contact takes place, and friction rises. Such higher loads can be carried only when the contacting surfaces are separated by another molecular layer of a substance that can influence the friction and wear characteristics. Extreme pressure (EP) additives present in the lubricant generally have an active element like sulfur or phosphorous or both, which react with the contacting surface asperities and generate a fresh layer of surface, capable of shearing under the load and thus carry the applied load. This is similar to the solid lubrication where the solid lubricants like graphite or molybdenum disulfide form a protective layer between the surfaces and carry the load. Lubricant and chemical additive thus play a greater role in influencing boundary friction.

Mixed friction occurs in equipment experiencing both boundary and fluid film lubrication. Such condition exists in equipment mainly operating under hydrodynamic condition (fluid film lubrication), but experience boundary condition during the starting and stopping of the equipment. Lubricating oils of such system would need certain amount of EP additives to take care of the situation.

WEAR

Friction and wear are two independent properties and have no relationship. One may think that higher friction will lead to higher wear and vice versa, but this is not true. Friction is the property of the material, while wear is the consequence of the load and the relative motion of the two contacting surfaces. Wear is desirable when new equipment is undergoing running-in. Wear in such a case progressively decreases, and smoother surfaces are created at the end of the process. Wear in most other cases is undesirable. Excessive wear can lead to the equipment failure. In wear, there is material redistribution or loss from the surfaces due to a variety of reasons. Some of the important routes through which wear takes place are described as follows:

1. Abrasion
2. Adhesion
3. Fatigue
4. Erosion
5. Corrosion
6. Fretting
7. Cavitation

Abrasive wear is related to the hardness of the surface. Abrasion takes place when a harder surface rubs over the softer layer and material is removed from the asperities. This happens in grinding and polishing processes where abrasive wear is desirable. In lubricated surfaces, contamination with abrasive matter such as silica from dust can cause abrasive wear by breaking surface peaks. Such abrasive particles can further generate metallic abrasive particles from the surface. Air filters and oil filters are helpful in removing these abrasive contaminants and protect moving surfaces.

Adhesion wear is complex in nature. The material transfer from surfaces arising out of interfacial adhesion is called adhesive wear. When two metal surfaces are in contact with each other, adhesion between the asperities can take place, and certain amount of force will be required to separate these surfaces. It has been found that when freshly cleaned metal surfaces are brought together in a vacuum chamber, strong adhesion takes place, and when separation of the surfaces is attempted, fracture can occur. This also then leads to material transfer from one surface to the other. The presence of metal oxide layer, or lubricant film, however, reduces adhesive friction considerably. In adhesion, both physical and chemical interactions take place, involving Van der Waals force or chemical bonds. The moving surfaces can break the spot joints bound by adhesion (sometimes called cold welding) and cause wear. The shearing of these micro spots is responsible for substantial amount of friction, especially when equipment is started after a long downtime. This adhesion wear can be substantially reduced by a thin film of lubricant or a molecular layer formed by boundary additives.

In several equipment, periodic severe loads are applied at the contact zones. This leads to the surface fatigue or fracture and ultimately damage takes place. There is a critical stress factor that needs to be applied so that the material undergoes fracture. At lower stress, a crack will develop slowly, and over a period of time when the crack is larger, the lower stress will be able to fracture the crack. On breaking these cracks, metal particles are also generated, and the surface will have pits. This is called pitting. While adhesive wear and abrasive wear take place slowly, fatigue wear takes place suddenly, and it therefore is difficult to monitor. Fatigue wear is related to the useful life of the equipment. Fatigue wear cannot be eliminated but can be controlled to a great extent by proper lubricants containing appropriate EP and antiwear additives.

Erosion wear takes place when high-pressure liquid droplets or liquid containing solid particles impinges on the solid surface. Erosion wear is more frequent in slurry pumps, hydraulic turbines, gas turbine blades, helicopter blades, and other solid-handling machines. Erosion wear is related to the kinetic energy or velocity and mass of the impinging particles and droplets. The angle of impact is also important.

Corrosive wear is also referred to as chemical wear and takes place when a chemical compound repeatedly reacts with the surface and removes material from the surface. The common interaction is that of oxygen and water vapors or chlorine (under sea environment). In internal combustion engine, sulfur is another factor in fuels, which on combustion gets converted to oxides and then to acid in contact with water. Dissolved gases like carbon dioxide can also cause corrosion of metal surfaces. Corrosive wear in internal combustion engine is controlled by using over-based detergent and dispersant, which continuously remove acids by neutralizing it with alkali base. Corrosive wear can increase with the increase in temperature, load, or rubbing speed in the contact zone.

Fretting wear arises out of vibrations. All machines are subjected to some kind of oscillation or vibrations during their operation. In vibration, the parts are subjected to oscillating load, and all sorts of previously discussed wear can take place. The most affected parts by fretting wear are nut and bolt fasteners, flexible couplings, axels, shafts, and bearing housing. Solid lubricants like molybdenum dioxide and graphite are effective in controlling fretting wear in nut and bolt fasteners.

All lubricants have some dissolved air in them (about 8%) but can suck in more air through leaks especially in high-pressure hydraulic systems. This excess air or gas gets dispersed in oil in the form of tiny bubbles or cavities. At low-pressure side, these bubbles or cavities will be released at the surface with great force, causing spherical shape wear. This is called cavitational wear. The extent of wear is related to the surface energy of the bubble and can be best controlled by not allowing ingress of air/gas into the system. Some amount of cavitation wear can be controlled by using lubricants having good air release value.

WEAR MECHANISM AND MEASUREMENTS

As discussed earlier, wear is a complex phenomenon and occurs due to a variety of unrelated reasons. Wear occurs due to contamination, chemical action, and mechanical stresses. Mechanical wear arises from the metal-to-metal contact and ultimately leads to material failure. Metallurgical parameters such as hardness and yield strength, and tribological parameters like contact geometry, load, and speed play important role in the wear of surfaces in relative motion. In engineering applications, metals are not used in their elemental form. Most materials are alloys, and their wear behavior may be quite different from pure metals.

Several attempts have been made to model friction and wear, and a large number of equations have been developed. Several reviews [9, 10] suggested that more than 300 equations have been developed using more than 100 variables to correlate wear and friction in both dry and lubricated surfaces. However, none of the models have found universal acceptance. The problem in wear and friction modeling arises from the fact that no solid surface is completely clean, homogeneous, and smooth as presumed. The clean new surface is also severely influenced by the atmospheric contaminations such as moisture, gases, and hydrocarbons. When the surface is not precisely characterized, the wear and friction arising out of the surface interaction become difficult to predict. Lubricant brings in additional complexity. The EP and antiwear additives present in them modify the surface and generate completely new reactive surfaces. With these complexities, the best way is to measure and evaluate wear and friction in actual equipment or in several simulating test rigs designed to measure wear and friction. There are several standard ASTM tests available to do this.

The antiwear and EP properties of molybdenum dialkyldithiophosphate, dibenzyl disulfide, molybdenum dialkyldithiocarbamate, zinc dialkyldithiophosphate, chlorinated paraffin wax, and triaryl phosphate were evaluated by four-ball friction and wear tests. This was followed by scanning electron microscopy (SEM), XPS, and X-ray photoelectron imaging analyses of the worn surfaces to determine the structure of the boundary lubrication film and the mechanism of the tribochemical reaction taking place during the friction process. The enhanced antiwear and load-carrying capacity of the additive-containing oils was attributed to the formation of a complex boundary lubrication film formed between the surfaces during the friction process as a result of the tribochemical reaction. The studies [11] indicated that the lubricating properties of the additives depend on their chemical nature and reactivity with metal surfaces.

Wear particles and wear debris can be monitored by several techniques. When wear is mild, the particle size of wear material is small (up to 10 μm), and the surfaces are smooth. This is normal wear and a desirable situation. In severe wear, particles of the order of 15–200 μm are generated. In lubricated surfaces, these wear debris enter lubricating oil and can be monitored through a condition monitoring system. This provides useful information about the nature of wear, and corrective action can be taken to protect equipment. For example, the number of particles present per milliliter of oil can be counted in different size ranges by using automatic particle counter. The oil can be filtered, and the separated particles can be examined in optical microscope for their shapes and sizes. Technique of ferrography can further separate magnetic particles and identify their origin and shape. The chemical nature of wear particles can be determined by analytical techniques like X-ray fluorescence, rotating disk emission spectrograph, and inductively coupled argon plasma (ICAP)-emission spectrograph. The ICAP technique, although expensive, is the most effective technique to find out the chemical nature of wear particles. In other techniques, the sizes of the particles play some role of interference. Based on this information, major equipment manufacturers fix rejection limit of the oil to avoid failures. Generally, higher amount of iron and silica indicate higher abrasive wear. Wear of other trace metals present in the surfaces will also be indicated through this analysis and provide warning signals. Advance analytical techniques such as SEM and ESCA/Auger electron microscopic methods are used for more specific and detailed information.

Quantitative estimation of wear after detection is quite complex. There is no direct correlation between various laboratory tests and actual field conditions. Wear is generally estimated by two well-known test methods: four-ball wear test under point contact and Falex block-on-ring under line contact conditions.

Four-ball wear test method is one of the most widely used methods for determining wear in laboratory. This test is also part of a large number of industrial oil and engine oil specification.

The four-ball wear test is covered in ASTM D-4172 method. In this method, three balls of 12.7 mm diameter are clamped together and covered with the test lubricant. A fourth 0.5-in. steel ball is then pressed with a force of 147 or 392 N into the cavity formed by the three balls. Thus, it becomes a three-point contact system. The top ball is rotated at 1200 rpm for 60 min at a temperature of 75°C. Wear appears in the form of scar on the lower three balls, and the scar diameter is measured and reported as average wear scar diameter in mm. The method can also be used under different operating conditions of load and speed to simulate various machine operating conditions [12–16]. For example, high load–low speed condition (30–60 kg and 600 rpm) represents sliding boundary condition. High load and high speed (1800–3000 rpm) represents elastohydro-dynamic condition.

Falex block-on-ring test and the Timken-type blocks-on-ring methods do not produce highly reproducible results but are used for several lubricants for comparative evaluation.

EXTREME PRESSURE TESTS

There are several standard tests to evaluate the load-carrying properties of lubricant under boundary conditions. Some of the important tests are as follows:

Four-ball EP test (ASTM D-2783)

Timken EP test (ASTM D-2782)

FZG load-carrying test (DIN 51364)

Oscillating SRV-friction machine test (DIN 51834 part 3)

In the four-ball test, the top steel ball is rotated at 1780 rpm over the three stationary balls for 10 s with increasing loads till the four balls weld together. This condition indicates the failure of lubricant EP film, and metal-to-metal contact takes place. The weld point is reported as the lowest load in kilogram when the rotating ball welds with the three stationary balls. Timken EP test is carried out according to ASTM D-2782 procedure with a test block and a rotating cylinder under line contact geometry. The cylinder cup is rotated at 800 rpm under increasing load condition (30 lbs and above). The minimum load that will rupture lubricant film and the maximum load that will not rupture the lubricant film are noted. The maximum load that will not rupture oil film is called Timken OK load.

FZG load-carrying test is a gear test and is generally used for evaluating gear oil, although the method is used in several antiwear hydraulic, turbine, and compressor oils as well. In this test, the gears are loaded at different stages up to 12 stages. The failure or pass stage is reported according to the damage on the gears.

LUBRICATION REGIMES

The interaction of *hills* and *valleys* of the two surfaces under specific conditions leads to different regimes of lubrication.

Following types of lubrication regimes have been identified as a function of friction:

1. Hydrodynamic
2. Quasi-hydrodynamic
3. Hydrostatic
4. Boundary
5. Elastohydrodynamic

These regimes are best represented by the classical Streibeck–Hersey [17, 18] curve depicting the change of coefficient of friction with ZN/P in a plain bearing, where Z is the viscosity of lubricant, N is the velocity, and P is the applied load. Originally, Streibeck in 1902 plotted coefficient of friction as a function of load, speed, and temperature. Hersey in 1914 brought out the present-day curve by plotting coefficient of friction with ZN/P. The curve is, however, popularly known after Streibeck (Fig. 6.5). The value of ZN/P will be high, when Z (viscosity) and speed (N) are high and P (load) is low. With the increase in load or decrease in viscosity and speed, ZN/P value can become low. This curve is able to explain various lubrication regimes except elastohydrodynamic regime based on friction and ZN/P value. The Streibeck curve is based on the start-up of a plain journal bearing, and the regions of boundary, mixed, and hydrodynamic lubrication can be clearly seen. At the start-up of the bearing, the oil film is very thin; friction is high due to metal-to-metal contact (point A). As the speed or revolution increases, hydrodynamic oil film is created. Friction is

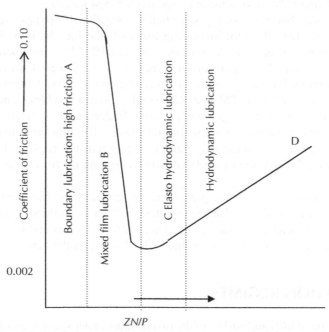

FIGURE 6.5   Typical lubrication regimes by Streibeck curve in plain bearing.

reduced and reaches a mixed film lubrication regime (point B). With further increase in speed (N) with constant viscosity (Z) and constant load (P), friction reaches its lowest value (point C). At this point C, full fluid lubrication is attained. With further increase in speed, lubricant internal friction is generated, and total friction starts increasing (point D). It is between point C and point D that the bearing is operating under hydrodynamic lubrication. The curve shows how, with the combination of suitable parameters like viscosity, speed/revolution, and load, full fluid lubrication and other lubrication conditions can be obtained. At lower values of ZN/P, there is essentially boundary lubrication condition.

The friction coefficients in pure hydrodynamic lubrication are low and are normally less than 0.01. In boundary lubrication, the friction coefficient is in the range 0.06–0.09. In mixed regime, the friction shall be between these two values. These regimes are further explained in the following text.

HYDRODYNAMIC OR FLUID FILM LUBRICATION

Figure 6.6 shows an example of hydrodynamic or fluid film lubrication, where the two surface asperities are completely separated by oil film of sufficient thickness. Usually under hydrodynamic lubrication, fluid film is at least three times thicker than the average asperity height.

The frictional resistance in these conditions arises mainly from the viscous shearing of the fluid, that is, its viscosity. Thus, under hydrodynamic lubrication, viscosity of the lubricant determines the friction, temperature rise, rate of flow of lubricant, and load-carrying capacity of the bearing. Other physicochemical properties of the lubricant and of the metallic surfaces are of secondary importance. A well-designed shaft bearing operating under the conditions of low-to-medium loads and high speeds with good supply of oil of appropriate viscosity will operate with fluid film lubrication.

HYDROSTATIC LUBRICATION

In a thrust or journal bearing, when the journal starts rotation, it is at the rest position and in contact with the bottom metal portion of the bearing (Fig. 6.7). This creates eccentricity between the centers of journal and bearing, and a convergent wedge is

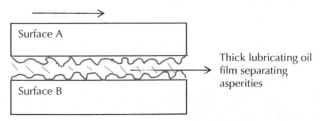

FIGURE 6.6   Hydrodynamic lubrication/fluid film lubrication.

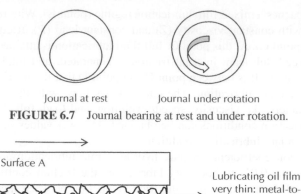

Journal at rest   Journal under rotation

FIGURE 6.7   Journal bearing at rest and under rotation.

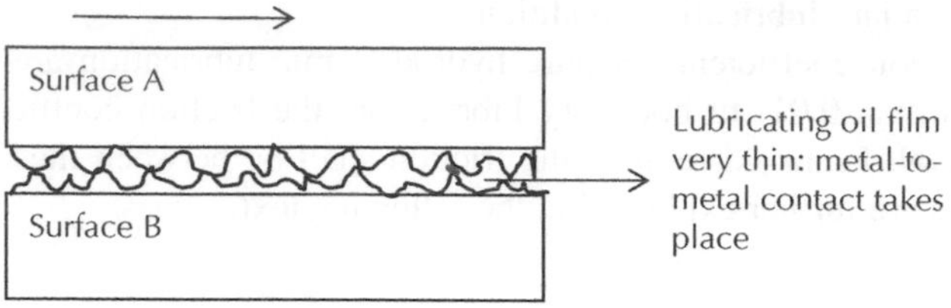

Surface A

Surface B

Lubricating oil film very thin: metal-to-metal contact takes place

FIGURE 6.8   Boundary lubrication—metal-to-metal contacts.

created. When the lubricant flows through this wedge, pressure is created at the wedge, and the velocity distribution across the film is such that the overall flow is constant. As the rotation of the journal is increased, due to higher pressure at the wedge, the journal lifts, and full fluid lubrication is attained. However, at the start-up and stop, there would be metal-to-metal contact, and friction will be high and wear will take place. The surfaces can, however, be separated by applying external pressure before the start-up. In heavily loaded bearing in steel rolling mills, the journal is lifted by applying pressure to reduce such wear at the start and stop cycle. Fluid film lubrication thus exists under conditions of high relative velocity of moving surfaces and moderate-to-high loads. This type of lubrication is known as hydrostatic lubrication.

BOUNDARY LUBRICATION

Fluid film thickness under boundary lubrication is very thin and starts increasing as it approaches mixed lubrication. When the loads are high, and speed and oil viscosity are low, boundary lubrication regime is reached (Fig. 6.8). During the period of starting and stopping, the velocity is too low, oil film is not capable of supporting the entire load, and the bearing, which normally operates with fluid film lubrication, will operate with *thin film* or boundary lubrication. Boundary lubrication exists also in a journal bearing if the bearing load becomes too high or if the viscosity of the lubricant is too low.

The lubricant properties, additive present in it, and the bearing surfaces play important roles in boundary lubrication. Certain animal and vegetable oils, esters, and compounds have better oiliness than mineral oils due to their polar nature. These polar molecules are asymmetrical and have a strong affinity for a metal surface, forming adsorbed layers at the rubbing metal surfaces. These layers can prevent a direct metal-to-metal contact in the bearing at moderate loads. However, under certain higher load and temperature, the boundary film can fail, so that direct metal-to-metal contact will take place. Initially, the surface asperities come in contact with each other, generating *hot spots*, where the temperature rises to a level that the welding of metals occurs. These points are then broken by the relative motion of the surfaces. Lubrication under this mode in many machines can be classified as either *boundary* or *fluid film*. But in actual practice, most of such equipment operates with a mixture of both at the same time.

Mixed lubrication is encountered in many gears, ball and roller bearings, seals, and even some conventional plain bearings. The additives in oil, specially EP and antiwear agents, play greater role in such lubrication regime. Viscosity attains secondary importance, but is important to protect the equipment under the fluid film lubrication mode. The difference between hydrodynamic and boundary lubrication lies in the fluid film thickness. When the oil film thickness is three times the composite roughness of the surfaces, fluid film lubrication takes place. Boundary lubrication is attained when fluid film thickness is equal or less than the composite roughness of the surfaces [19, 20]. Thus, in mixed lubrication regime, both oil viscosity and the presence of EP/antiwear additives attain importance. Different polar and saturated compounds react differently with organometallic EP additives [19, 20].

EP and antiwear additives contain active sulfur or/and phosphorous, which reacts with metal surface to form a completely new surface. This new surface can be sheared easily under applied loads. Thus, in lubrication, the two important parameters are surface roughness and oil film thickness. However, other parameters like surface hardness and operating conditions of the equipment play important role in managing friction, wear, and lubrication.

ELASTOHYDRODYNAMIC LUBRICATION

Elastohydrodynamic lubrication (EHD) takes place in concentrated contacts where the high Hertzian pressures cause high increase in oil viscosity and elastic deformation of metal surfaces. Under such conditions, a high-viscosity thin film of oil can carry larger loads than predictable. EHD can thus be considered a special form of hydrodynamic lubrication. The EHD theory takes into account both elastic deformation and pressure influence on viscosity. Friction in pure EHD contacts is governed by complex rheological behavior of thin films and contacts. It is however to be noted that EHD friction is normally higher than in the case of hydrodynamic situation, and typical EHD traction coefficients range from 0.03 to 0.08. In EHD contact, again mixed lubrication is possible when asperity contact occurs through films.

In many types of equipment, both rolling and sliding contacts take place (e.g., gears). In pure rolling contacts, the rolling velocities of both the cylinders are same. While in rolling/sliding mixed contacts, unequal velocities are observed. When the film thickness of the lubricants in such contacts was calculated based on hydrodynamic theory, it came to a negligible value, although equipment were running with very low wear, which was suggestive of adequate separating oil film. In this calculation, when the effect of pressure on oil viscosity and elastic deformation of surfaces was taken into consideration, the film thickness was found to be adequate to justify the operation of the equipment like gears. Film thickness in EHD is in the range 0.1–$1.0\,\mu m$ and is generally lower than found in hydrodynamic lubrication [21, 22].

Wear takes place in all the lubrication regimes with a varying degree of severity. In engines and loaded gear boxes, all the lubrication regimes are experienced during their operation. Cam, tappet, piston rings, and cylinder bore of IC engine undergo most severe operating parameters of temperature and pressure. In these areas, boundary and mixed lubrication regimes exist [23]. Bearings may experience hydrodynamic lubrication.

While designing a lubricant, these basic principles have to be carefully taken into consideration, and the selection of proper chemical additive, especially the EP, antiwear, lubricity, and friction modifiers, has to be based on the tribological operating conditions of the equipment. An interesting area of nanotribology is emerging to understand the subject of friction, wear, and lubrication [24] from a new perspective. Techniques like atomic force microscope are being used to understand the subject at nanolevel.

REFERENCES

[1] Peter Jost H. Economic impact of tribology. Proceedings of the 20th Meeting of the Mechanical Failure Preventive Group, NBS Special Publication, Volume 423; May 8–10, 1976; Gaithersburg.

[2] Bhushan B. *Introduction of Tribology*. New York: John Wiley & Sons; 2002.

[3] Dowson D, Hingginson GR. *Elastohydrodynamic Lubrication*. Oxford: Pergamon Press; 1997.

[4] Srivastava SP. *Advances in Lubricant Additive and Tribology*. New Delhi: Tech Books International; 2009.

[5] Bhushan B. *Modern Tribology Handbook*. Volumes 1 & 2, CRC Press; 2000.

[6] Neale MJ. *The Tribology Handbook*. 2nd ed. Boston: Butterworth-Heinemann; March 1996.

[7] Bruce RW. *Handbook of Lubrication and Tribology*. Volume 2, Boca Raton: CRC Press; 2012. Theory and design.

[8] Czichos H. Basic tribological parameters. In: Olson D, editors. *ASTM Handbook*. Volume 18, Friction, lubrication and wear technology. Metals Park: ASTM International; 1992.

[9] Hsu SM, Klaus EE, Cheng HS. A mechano-chemical descriptive model for wear under mixed lubrication condition. Wear 1988;120:307.

[10] Meng HC, Ludema KC. Wear model and predictive equations: their form and content. Wear of Material Conference; April 5–9, 1995; Boston.

[11] Unnikrishnan R, Christopher J, Jain MC, Martin V, Srivastava SP. Comparison of boundary lubrication films formed under four-ball friction test conditions with different AW/EP additives—an X-ray photoelectron spectroscopy study. Tribotest June 2003;9 (4):285–303.

[12] Hsu SM. Review of laboratory bench tests in assessing the performance of automotive crankcase oils. Lubr Eng 1981;37 (12):722.

[13] Claus EE, Bieber HE. Effect of some physical properties of lubricants on boundary lubrication. ASLE Trans 1964;7 (1).

[14] Brown ED. Friction and wear testing with modern four ball machine. Wear 1971;17:381.

[15] Miller AH. Considerations in interpreting four ball data. Wear 1973;23(1):121–127.

[16] Baldwin BA. Comparison of wear measurements between four ball and falex wear test machines. Wear 1984;93:233.

[17] Streibeck R. Characteristics of plain and roller bearings. Zeit des VDI 1902;46.

[18] Hersey MD. The laws of lubrication of horizontal journal bearing. J Wash Acad Sci 1914;4:542–552.

[19] Hsu SM. Boundary lubrication, current understanding. Tribol Lett 1997;3:1–11.

[20] Hsu SM. Boundary lubrication film formation and lubrication mechanism. Tribol Int 2005;38 (3):305–312.

[21] Dowson D, Hingginson GR. *Elastohydrodynamic Lubrication*. Oxford: Pergamon Press; 1977.

[22] Hamrock BJ, Dowson D. *Ball Bearing Lubrication: The Elastohydrodynamic of Elliptical Contacts*. New York: Wiley; 1981.

[23] Ludema KC. *Friction, Wear, Lubrication: A Test Book on Tribology*. Boca Raton: CRC Press; 1996.

[24] Bhushan B, Israelachvili JN, Landman U. Nano-tribology, friction, wear and lubrication at the atomic scale. Nature April 13, 1995;374:607–616.

PART II

INDUSTRIAL LUBRICANTS

CHAPTER 7

STEAM AND GAS TURBINE OILS

Three distinct types of turbines are operated by steam pressure, gas combustion, and water pressure to produce electricity in a power plant. The nuclear power plants utilize steam turbines to produce electricity where steam is produced by the radioactive material. Similarly, steam can also be produced by several other routes such as by the combustion of coal, naphtha, biomass, and gas. These turbines mounted on bearing are the prime movers for a variety of machines such as compressors, ship propellers, centrifugal pumps, and electric generators requiring rotary input power. The bearings of these turbines operating continuously round the clock require a very specific lubricant called turbine oil. The quality of oil thus assumes importance and should have long life without deteriorating its performance and characteristics over a period of time. Several turbines have seen a life of oil as long as 20 years or more with proper periodic maintenance, oil top-up, and cleanup. The unit size of a turbine could vary from 25 MW to 1300 MW (in a nuclear power plant). During the last 50 years, the design of turbines has undergone tremendous change, making them more compact. Steam pressures and temperatures have gone up; load to oil ratio (kW/l) has gone up, and consequently, oil temperature in the tank has also gone up. These conditions require further improvement in oil quality. The severe quality demands require more stable base oils with higher reserve of oxidation stability. Most turbine manufacturers have formulated their own specifications of turbine oil and require a qualification approval before the use of the oil. The most critical property of the oil is its high oxidation stability and low air release value. The oxidation stability can be achieved by the use of suitable antioxidant, but air release value cannot be improved by the known additives. Well-refined base oils with high viscosity index and lower level of polar compounds generally have low air release value. Silicon-based antifoam compounds tend to deteriorate air release property, and therefore, these compounds are avoided in turbine oil formulation. Synthetic base oils such as polyalphaolefins and phosphate esters are also used for specific-purpose turbines.

CLASSIFICATION OF TURBINE OILS

Turbine oils have been classified by ISO according to the application and severity of operations. There are generally two types of turbine oils, rust and oxidation inhibited and antiwear type. These two categories are again subdivided depending on the type of base

Developments in Lubricant Technology, First Edition. S. P. Srivastava.
© 2014 John Wiley & Sons, Inc. Published 2014 by John Wiley & Sons, Inc.

TABLE 7.1 Classification of lubricants for turbines: ISO 6743-5, family T

Application	Composition and properties	ISO symbol	Typical requirements
1. Steam turbine directly coupled or geared to the load, normal service	Highly refined petroleum oil with rust protection and oxidation stability	ISO-L-TSA	Power generation and their associated control systems, marine drives, where improved load-carrying capacity is not required for the gearing
2. Gas turbine, directly coupled or geared to the load, high load carrying		ISO-L-TGA	
3. Steam turbine directly coupled or geared to the load, high load-carrying capacity	Highly refined petroleum oil with rust protection, oxidation stability, and enhanced load-carrying capacity	ISO-L-TSE	Power generation and industrial drives and marine gear drives and their associated control system where the gearing requires improved load-carrying capacity
4. Gas turbine, directly coupled or geared to the load, high load carrying		ISO-L-TGE	
5. Gas turbine, directly coupled or geared to the load, higher-temperature service	Highly refined petroleum oil with rust protection and improved oxidation stability	ISO-L-TGB	Power generation and industrial drives and their associated control system where high temperature resistance is required due to hot spots

6. Synthetic lubricants, ISO-L

TSC—Synthetic steam turbine fluids with no specific fire-resistant properties (such as PAO)
TSD—Synthetic steam turbine fluids based on phosphate esters with fire-resistant properties
TGC—Synthetic gas turbine fluids with no specific fire-resistant properties (such as PAO)
TGD—Synthetic gas turbine fluids based on phosphate esters with fire-resistant properties
TCD—Synthetic fluids for control systems based on phosphate esters with fire-resistant properties

oil used such as mineral oil or synthetic oil (polyalphaolefins or phosphate esters). Phosphate ester base fluids are utilized for imparting fire-resistant properties to the oil. Another category of gas turbine oil has also been identified for operation at higher temperature requiring thermally stable oil. ISO has identified five classes of mineral oil-based turbine oils into ISO-L-TSA, ISO-L-TGA, ISO-L-TSE, ISO-L-TGE, and ISO-L-TGB categories. Another five classes have been identified for turbine oils based on synthetic oils (ISO-L-TSC, ISO-L-TSD, ISO-L-TGC, ISO-L-TGD, and ISO-L-TCD). The term TS stands for turbine steam, TG for turbine gas, and TC for turbine control. Table 7.1 provides details of ISO 6743-5 family T classification for turbine oils.

SPECIFICATIONS

Turbine oil specifications have been developed extensively by various standard organizations. In addition, turbine manufacturers have also developed specifications to protect their equipment.

The new second edition ISO 8068-2006 provides specifications for all the turbine oil categories described in ISO 6743-5:2006, while the original 1987 version described only the specification of TSA and TGA. New turbine technologies have emerged in recent years leading to the changes in lubricant requirements. For example, the development of single-shaft combined-cycle turbines has resulted in the use of a common lubrication system for both the gas and steam turbines. The lubricant should therefore meet the requirements for the lubrication of both gas and steam turbines. ISO-8068 does not specify the requirements for wind turbines. These are covered in ISO 12925-1. While power generation is the primary application for turbines, steam and gas turbines can also be used to drive rotating equipment, such as pumps and compressors. The lubrication systems of these driven loads can be common to that of the turbine. Turbine installations incorporate complex auxiliary systems requiring lubrication, including hydraulic systems, gearboxes, and couplings. Depending upon the design and configuration of the turbine and driven equipment, turbine lubricants can also be used in these auxiliary systems.

Some of the important standards are described below and should be considered for turbine oil development:

1. ISO 8068
2. JIS-2213
3. DIN 51515 parts 1 and 2
4. BS 489
5. Alstom HTGD 90 117 V 0001T
6. Solar ES 9-224
7. Siemens TLV 901304
8. GE GEK 107395A, GEK-46357E, and 46506D
9. Mitsubishi MS4-MA-CT, 001-003

A comparison of important properties in ISO, DIN, and Siemens specification is shown in Table 7.2. Original standard contains many other general test requirements, and readers are advised to refer to the complete specification.

DIN 51515 part 2 specifies turbine oils for high-temperature application. Viscosity grades (VG) 32 and 46 have been covered in this standard. The differences between part 1 and part 2 standards are that it is more severe in oxidation tests and copper corrosion test as compared to the part 1 standard. Part 2 specifies the following additional test:

1.	Cu corrosion 3 h at 125°C	2 max.
2.	TOST life	3500 h min.
3.	Rotary pressure vessel oxidation test (RPVOT) minutes	750 min.
4.	Modified RPVOT % time of normal RPVOT	85

DIN 51525 part 2 standard may be compared with Mitsubishi turbine oil standard (Table 7.3), where both specify RPVOT, modified RPVOT test, and TOST (ASTM D-943) tests, but the severity levels are slightly different.

TABLE 7.2 Comparison of important turbine oil properties

Properties	ISO-8068	DIN 51515 part 1 L-TD	Siemens TLV 901304
VG	32, 46, 68, 100	32, 46, 68, 100	32, 46
Viscosity index	80 min	—	95 min
Rust test D-665 A	—	No rust	0-B pass
Rust test D-665 B	Pass		
Demulsibility			
ASTM D 1401	30 min max.		20 min max.
Steam emulsion No IP-19 seconds	300, 300, 360, 360 max.	300 max.	300 max.
Foaming (s)			
Seq. 1	450/0 for 32 and 46 450/40 for 68 and 100	VG 100 only 450/0	400/450 max.
Seq. 2	100/0 for 32 and 46 100/10 for 68 and 100	100/0	
Seq. 3	450/0 for 32 and 46 450/40 for 68 and 100	450/0	
Air release value at 50°C, min	5/6/8	5/5/6, no limit for 100 grade	4 max.
Oxidation test			
ASTM D-943, TAN after 1000 h	—	2.0 max.	2.0 max. after 2500 h
Hours to reach 2.0 TAN (mg/KOH·g)	2000 h for 32, 46, 68 grades 1500 h for 100 grade	2000 h for 32, 46; 1500 h for 68 1000 h for 100 grade	
Oxidation IP-280			
Acidity (mg/KOH/g)	1.8 max.	—	
Sludge (%)	0.4 max.		—
RBOT ASTM D-2272	—	—	500 min
Modified RBOT N_2 blown	—	—	85% of the original unmodified test

British standard for turbine oil BS 489 relies on IP-280 oxidation test (total oxidation product (TOP) 0.7–0.8 and sludge 0.30–0.35%) besides other usual tests. IP-280 is a short-duration test and is carried out at higher temperature. There seems to be no correlation between various oxidation tests due to variation in test conditions such as temperature, duration, and catalysts [1].

Mitsubishi has specified additional modified oxidation tests based on ASTM D-943 and ASTM D-2272 to take care of higher-temperature application, and these along with other important properties are indicated in Table 7.3.

From these standards, it is clear that turbine oils need good-quality base oils with low air release value, appropriate dose of antirust, and carefully selected antioxidants. The base oils for turbine oils were earlier produced in the refineries by deeper extraction and hydrofinishing to obtain higher VI and improved stability. However, currently, base

TABLE 7.3 Key Mitsubishi turbine oils specification, MS 4-MA-CL 001, CL 002, 003

Tests	CL 001 steam turbine oils for low-temp application	CL 002 for high-temp (above 250°C) application	CL 003 for high-temp (above 250°C) application with EP properties
Oxidation stability ASTM D-943 TAN after 1000 h	0.4 max.	0.4 max.	0.4 max.
Oxidation stability ASTM D-943, hours to 2.0 TAN	2000 min.	4000 min.	4000 min.
Dry TOST (ASTM D-943) at 120°C, sludge (mg/kg)	100 at 25% RPVOT	100 at 25% RPVOT	100 at 50% RPVOT
Min. life hours	400 h to 25% RPVOT		700 h to 50% RPVOT
RBOT ASTM D-2272 (min)	220 min	700 min.	700 min.
Modified RPVOT, % of original RPVOT	85% min	85% min.	—
Total sulfur ppm max.	Report	1000	1000
Zinc ppm max.	60	10	10
FZG gear ring test D-5182, failure stage min.	—	—	9
Evaporation loss at 150°C, 22 h % max.	—	10	10

oils produced through hydroprocesses are quite suitable for formulating turbine oils since these are virtually free from aromatics, sulfur, and nitrogen compounds that are responsible for poor stability, water-separating characteristics, and air release value. Similarly, base oils produced through gas-to-liquid (GTL) technology are also suitable for turbine oils. The selection of antioxidant is most important since there are large numbers of chemical compounds capable of providing protection at different temperatures. For example, hindered phenol is good up to 90–100°C. Beyond this temperature, the chemical compound itself starts evaporating. Amine-based antioxidant provides higher-temperature protection. Phenothiazine-based antioxidants provide still higher-temperature protection and are used in synthetic gas turbine oils. A combination of more than one antioxidant is generally more useful in improving oxidation stability. Currently, long-life turbine oils are commercially available in the market providing greater than 10,000-h TOST life and more than 1,000-min rotating bomb oxidation test (RBOT) life. The advantages with such oil can only be derived if the lubrication system is properly maintained and oil kept clean. Such a long life can be achieved by using API group III or GTL base oils with high dosage of carefully selected and evaluated antioxidants. One of the GEK specifications, GEK-107395a, has specified very high oxidation stability for a VG 32 turbine oil to be used in a single-shaft STAG high-temperature turbine application. Key parameters of this specification are provided in Table 7.4.

This specification demands that the oil should be zinc-free, have moderate antiwear properties (FZG passed eight stages), and have exceptionally high thermal and oxidation stability. Such oils need to be formulated with high-quality base oils [2–5] such as API group II/III or GTL base oil utilizing antiwear (phosphate ester), metal deactivator (benzotriazole or its derivative), and antioxidant combination (phenolic and amine based), which are effective at high temperatures.

TABLE 7.4   Key properties of GEK-107395A turbine oil specification

Properties	Limits	Test method
ISO VG	VG 32	
Viscosity index	98 min.	D-2270
Foaming characteristics, sequences 1, 2, 3	50/0, 50/0, 50/0	D-892
Air release value (min)	5 max.	IP-313
Demulsibility (min)	30 max.	D-1401
Rust test	Pass	D-665 B
TOST life, hours to 2 TAN (mg KOH/g)	7000 min.	D-943
RBOT (min)	1000 min.	D-2272
Modified RBOT, percent time of unmodified test	85 min.	D-2272
FZG A/8.3/90	8 stages passed	D-5182
Zinc content	<5 ppm	D-4951
Evaporation loss, 149°C (wt.%)	6% max.	D-972
Autoignition temp (°C)	357 min.	E-659
Thermal stability, Cincinnati Milacron proc. A	Report	CM-A
Panel Coker test	Report	FTM 791A-3462
Volatility/oil thickening	Report	DIN 51356

PROPERTIES AND FUNCTIONS OF TURBINE OILS

Oil is circulated in the bearings of the turbine through a centralized lubrication system. The same oil is often used in the hydraulic control system of the turbine. In several systems, separate hydraulic oil is used if specific property such as fire resistance is required in the oil, which is not present in the conventional turbine oil. The following multiple functions are carried out by the turbine oils:

1. Reduce friction and wear between the contacting metal surfaces

2. Remove frictional heat from the system

3. Provide cushioning effect from vibration and shocks

4. Keep the system clean from contamination, wear debris, sludge, etc.

5. Protect the equipment from rusting and corrosion

Since the turbines in power plants run continuously, the oil must be capable of retaining these functions and properties over a long period of time. Some of the important properties of turbine oils are now described in the following pages.

VISCOSITY

The bearings of turbines run under hydrodynamic lubrication regime, that is, under full fluid lubrication system. Viscosity is therefore the most important property since it is responsible for maintaining oil film thickness. Optimum lubricant viscosity dissipates frictional heat generated at stress points and removes contaminants such as wear debris. A particular VG for the turbine is selected based on several operating parameters such as turbine speed, load, oil flow, and oil sump capacity. Steam turbines and gas turbines usually use ISO VG 32 and 46 oils. Geared turbine and those requiring antiwear properties generally require higher viscosity grades of ISO VG 68 and 100 grade oils. Each OEM recommends a particular grade of viscosity for the supplied turbine system based on the design and operating parameters.

RUST AND CORROSION PROTECTION

In steam turbines, ingress of water (through condensation of steam) in the lubrication system cannot be avoided, and therefore, all turbine oils contain an antirust additive that protects the equipment from rusting. This property is evaluated by ASTM D-665 A or D-665 B rust test for 24 h. Test A is conducted with distilled water, and test B is carried out with synthetic seawater. Test B is more severe and requires higher dosage of antirust additive. Turbines operating under marine environment need ASTM D-665 B rust test. Corrosion test is usually carried out by ASTM D-130 copper corrosion test at 100°C for 3 h. The oil must be noncorrosive to copper strip under specified test conditions.

WATER SEPARATION CHARACTERISTICS OR DEMULSIBILITY

The ingress of water or steam in turbine oil is unavoidable, and its continued presence can cause poor lubrication. The oil therefore should be capable of separating from water and steam quickly. This property depends upon the presence of surface-active or polar compounds in base oils. Highly refined or hydrotreated base oil has good water-separating characteristics. Demulsibility is evaluated by ASTM D-1401 test, and steam separation is evaluated by IP-19 test method.

AIR RELEASE

Air can get entrained into oil during its circulation in the system. This air, if not released from oil, can damage equipment due to cavitational phenomenon and also due to enhanced oil oxidation. This phenomenon has been discussed in detail [6]. It is the presence of polar surface-active compounds that are responsible for high air release values in base oils. API group II and III oils produced through hydroprocesses possess very low air release values due to very low or absence of polar aromatics, nitrogen, and sulfur compounds. Several common additives such as silicon-based antifoam compounds, detergents, pour point depressants, and ZDDP adversely affect air release property. Silicon antifoam compounds are most dangerous for air release value due to their concentration at the air–oil interface. These compounds are therefore not used in turbine oils, and instead, acrylate-based antifoam additives are used. The air release rate of lubricating compositions is significantly enhanced when the composition is formulated with one or more vinyl aromatic–olefin block copolymers [2] that form a micelle-like structure in the oil. Compositions having the specified copolymers retain less than 2.5% air after 1 min at 50°C when tested by ASTM

D-3427. There were no known chemical additives for base oils that can control air release value, but recent findings report the use of polymeric and ester-based products that can improve air release value. Air release value is measured by ASTM D-3427 or DIN 51381-1 or IP-313 test methods. Air release value of GTL base oil can be improved by the addition of a synthetic ester [7].

FOAM CONTROL

In a centralized oil circulation system, oil gets mixed with air and generates foam. Foam should not be confused with the air release property although both are generated by the mixing of air with oil. Air release is concerned with the release of finely dispersed air into the bulk oil, while foam is generated at the surface where the bigger air bubbles are surrounded and stabilized by oil film. These air bubbles called foam remain at the surface and collapse slowly. Excessive foam in turbine oils can promote oil oxidation and reduce the real flow of oil in the system. This can lead to reduced heat dissipation, higher wear, and poor lubrication. It is therefore important to control foam in turbine oil by using antifoam compound. As discussed earlier, the popular silicon-based antifoam compound cannot be used due to its adverse effect on air release property. Foaming characteristics are determined by ASTM D-892 test method in three sequences, and both foam formation and its stability are determined. The phenomenon of foaming and its control has been discussed in greater detail elsewhere [8].

ANTIWEAR PROPERTY

Turbines that are directly coupled or geared to the load require antiwear additives in addition to the rust and oxidation inhibitors. Most antiwear additives function by reaction at the metal surface to form a new surface and generally contain sulfur or phosphorous. Very high oxidation stability and good antiwear properties play opposite roles and require a very careful balancing of the oil formulation. Antiwear additive tends to lower oxidation stability, and the chemical structure of the two has to be such that both requirements are met. Many marine turbine applications require EP type of oils. U.S. military specification MIL-L-17331 specifies EP turbine oil. Antiwear property is measured by four-ball wear test and FZG four-square gear test machine. Usually, oil passing six to seven stages in FZG machine is considered suitable. Long-life EP turbine oil passing the ninth FZG stage and 7500-h TOST life has been reported based on suitable base oils, rust, and oxidation inhibitor along with a phosphate ester-based antiwear additive [9].

OXIDATION STABILITY

This is the most important property of turbine oils, and high oxidation stability means longer lubricant life. Base oils as produced in the refinery do not have adequate oxidation stability to support turbine oil performance. The more highly refined base oils are in fact poorer in oxidation stability due to the removal of natural antioxidants (certain sulfur and nitrogen compounds) present in the crude oil. This property is therefore obtained by the incorporation of an antioxidant molecule that functions by interaction with the free radicals produced during the process of hydrocarbon oxidation. The base oil and antioxidant interaction has been extensively investigated [10–14] and reported. Different base oils respond differently to antioxidants and need to be investigated thoroughly before arriving at the turbine oil composition [1, 15, 16,]. Oxidation stability of turbine oils is evaluated by a long-duration ASTM D-943 (also known as TOST) test method for 1000 h or more. Oils having TOST life of 1000 h (TAN after the test 2.0 mg/KOH/g) were considered acceptable earlier. However, currently, oils having long TOST life of 5,000–10,000 h are available. It is, however, not clear about the specific advantage obtained by the use of such oils, but these have high antioxidant reserve and are supposed to provide trouble-free long service in turbines.

In the early 1970s, the short-duration (164 h) IP-280 test also known as CIGRE test was developed [3, 4] as an alternative to the long-duration ASTM D-943 test for assessing oxidation stability of turbine oils. This test is also part of British and ISO standard for turbine oils. IP-280 test is carried out at 120°C for 164 h in the presence of soluble iron and copper catalyst. After the test, the amount of acid and sludge formed is measured, and from these values, TOP is calculated. TOP values below 1.0 are regarded as acceptable. On the other hand, D-943 test is carried out at 95°C in the presence of solid copper and iron catalysts. General correlation between the two tests is not found [5]. Oils showing good results in D-943 test could fail in IP-280 test. Both the tests are antioxidant specific. Hindered phenols in adequate amount would show good result in D-943 test at 95°C but would sublime at 120°C in the IP-280 test and show poor result. The IP-280 test, therefore, require high-temperature antioxidants. It is, however, possible to design turbine oil by using complex mixtures of antioxidants, which will show up good results in both ASTM and IP tests.

Another oxidation test used frequently for monitoring turbine oil quality during service is ASTM D-2272 test also known as RBOT. The test is conducted under the pressure of oxygen in a bomb and is most suited to monitor the residual life of turbine oil or as a quality control tool to quickly assess the oil quality with respect to the antioxidant presence. There is again no correlation between various oxidation tests due to the obvious differences in test parameters [5]. Mitsubishi specification for heavy-duty turbine oils requires RPVOT (ASTM D-2272) retention value in addition to the ASTM D-943 test.

GAS TURBINE OILS

Gas turbines are compact machines that can be started and stopped quickly. These are useful in industry or in an urban power generation plant. In the combustion zone where the gas is injected, temperature of the order of 800–900°C is attained. The exhaust gas temperature of the order of 400–500°C may also reach. These turbines have small oil sump capacity, and due to higher temperature, the oil undergoes faster deterioration. This requires faster oil replacement as compared with the steam turbine oils. Generally, normal rust- and oxidation-inhibited steam turbine oil is used in gas turbines with quick drain intervals. For longer drain, synthetic turbine oils based on polyalphaolefin can be used. Mineral oil-based turbine oil may last only about a year in land gas turbine, while a synthetic oil may be used up to 4 or 5 years. The actual drain interval is decided by oil condition. Aviation gas turbines operate under more severe low- and high-temperature conditions. Aviation turbine oils are low-viscosity VG 10 or 22 oils with neopolyester base oil. These ester-based oils are capable of operating under wide low- and high-temperature conditions, exhibit high thermal/oxidation stability, and possess good antiwear and load-carrying properties. Specific chemical additives are used in aviation synthetic gas turbine oils [17, 18] to meet various civil and military specifications. Steam and gas turbine oils have been formulated [19] using API group II or III base oils with an antioxidant alkylated diphenylamine; substituted hydrocarbyl monosulfide (n-dodecyl 2-hydroxyethyl sulfide, 1-(tert-dodecylthio)-2-propanol); a dispersant based on polyetheramine; borated/nonborated succinimide dispersant; Mannich reaction product of a dialkylamine, an aldehyde, and a hydrocarbyl-substituted phenol; and other usual additives such as a rust inhibitor, a metal deactivator, a demulsifier, and an antifoam agent.

Turbine oils are thus high-performance and long-life lubricants containing rust inhibitor and oxidation inhibitors. For higher antioxidation properties, a small dose of metal deactivator is incorporated, which synergizes with the antioxidant properties. Mineral oil-based products may use API group II or group III base oils [20]. Most of the major additive manufacturers provide additive recommendations for formulating these oils. Total additive dosage does not exceed 0.5%. In antiwear-type oil, additional antiwear additive is added. Phosphate ester-based products are preferred since these do not affect other properties of turbine oils to a considerable extent [21]. Polyalphaolefin-based turbine oils generally utilize a similar additive combination. Care is required in selecting additives with respect to their solubility in PAO, since these have limited solubility for certain solid antioxidant and metal deactivator. If the PAO base oil is intended for higher-temperature application, the selection of antioxidant needs special consideration, since hindered phenols do not provide protection at temperatures higher than 100°C. Several commercially available amine-based products are useful as high-temperature antioxidants. A combination of amine-based and phenolic antioxidants provides [22] improved performance.

Ester-based aviation turbine oils constitute a different category and need to be formulated with greater care [17, 18] using special high-temperature antioxidants such as phenothiazine. All oils may also contain a nonsilicon copolymeric foam inhibitor in ppm dosage. Silicon-based antifoam additive tends to increase air release values and is therefore not used in turbine oils. Improved thermal, oxidative stability, antirust, and multifunctional properties have been reported by an amine-functionalized [23] polymeric additive suitable for turbine oils, hydraulic fluids, and greases. Lubricating oils for wind turbine gearbox [24] require special formulations and have been formulated with perfluoropolyether (PFPE) base oils. Phosphate esters have been used in steam turbine governor hydraulic system as a fire-resistant fluid [25, 26].

During manufacture, special care is also required to avoid contamination of these oils to control in-service deterioration of oil properties. After blending operation, the oil is generally filtered and filled in clean barrels to remove and control any external contamination. Composition with improved cleanliness for lubrication of steam and gas turbine systems has been reported [27].

REFERENCES

[1] Srivastava SP. *Advances in Lubricant Additives and Tribology*. New Delhi: Tech Books International; 2009. p 365–368.

[2] Deckman DE, Baillargeon DJ, Horodysky AG. Lubricant air release rates. US patent 8,389,451. March 5, 2013.

[3] Wilson ACM. Problems encountered with turbine lubricants and associated systems. Lubr Eng 1976;32 (2):59–65.

[4] Murray DW, MacDonald JM, White AM, Wright PG. The effect of base stock composition on lubricant oxidation performance. Pet Rev February 1982:36–40.

[5] Jayprakash KC, Srivastava SP, Anand KS, Goel PK. Oxidation stability of steam turbine oils and laboratory method of evaluation. Lubr Eng February 1984;40 (2):89–95.

[6] Basu B, Chand S, Jayprakash KC, Srivastava SP, Goel PK. Air entrainment phenomenon in mineral lubricating oils. 39th ASLE Annual Meeting; 84-AM-4 D-3; May 7–10, 1984; Chicago.

[7] Poirier M-A. Method for improving the air release rate of GTL base stock lubricants using synthetic ester, and composition. US patent 7,910,530. March 22, 2011.

[8] Srivastava SP. Anti foam additives. In: *Advances in Lubricant Additive and Tribology*. New Delhi: Tech Books International; 2009. p 227–236.

[9] Saxena D, Mooken RT, Pandey LM, Gupta A, Mishra AK, Srivastava SP, Bhatnagar AK. Long life anti-wear turbine oils. Petrotech 1999; New Delhi; 1999. p 203–208.

[10] Okazaki ME, Militante SE. Performance advantages of turbine oils formulated with group II base oils. In: Herguth WR, Warne TM, editors. *Turbine Lubrication in the 21st Century*. West Conshohocken: American Society for Testing and Materials; 2001. ASTM Special Technical Publication, 1407.

[11] Schwager BP, Hardy BJ, Aguiilar GA. Improved response of turbine oils based on group II hydrocracked base oils compared with those based on solvent refined oils. In: Herguth WR, Warne TM, editors. *Turbine Lubrication in the 21st Century*. West Conshohocken: American Society for Testing and Materials; 2001. ASTM Special Technical Publication, 1407.

[12] Irvine DJ. Performance advantages of turbine oils formulated with group II and group III base stocks. In: Herguth WR, Warne TM, editors. *Turbine Lubrication in the 21st Century*. West Conshohocken: American Society for Testing and Materials; 2001. ASTM Special Technical Publication, 1407.

[13] Gatto VJ, Grina MA. Effects of base oil type, oxidation test conditions and phenolic antioxidant structure on the detection and magnitude of hindered phenol/diphenylamine synergism. Lubr Eng January 1999;55:11–20.

[14] Yano A, Watanabe S, Miyazaki Y, Tsuchiya M, Yamamoto Y. Study on sludge formation during the oxidation process of turbine oils. Tribol Trans 2004;47:111–122.

[15] Mooken RT, Saxena D, Basu B, Satpathy S, Srivastava SP, Bhatnagar AK. Dependence of oxidation stability of steam turbine oils on base stock composition. 52nd STLE Annual Meeting, Kansas City, May 18–22, 1997. Lubr Eng October 1997: 19–24.

[16] Srivastava SP. *Modern Lubricant Technology*. Dehradun: Technology Publications; 2007. p 181–189.

[17] Yaffe R. Synthetic aircraft turbine oil. US patent 4,096,078. June 20, 1978.

[18] Yaffe R, Reinhard RR. Synthetic aircraft turbine oil. US patent 4,157,971. June 12, 1979. US patent 4,188,298. February 12, 1980.

[19] Kramer D, Lok B, Krug R. The evolution of base oil technology. In: Herguth WR, Warne TM, editors. *Turbine Lubrication in the 21st Century*. West Conshohocken: American Society for Testing and Materials; 2001. ASTM Special Technical Publication, 1407.

[20] Simmons GF, Glavatskih S, Müller M, Byheden Å, Prakash B. Extending performance limits of turbine oils. Tribol Int 2014;69:52–60.

[21] Yoshiharu B. Lubricating oil composition. US patent 2004,0053,794 A1. March 18, 2004.

[22] Tatsumi Y, Umehara K, Iino S. Antioxidant composition and lubricating oil composition containing same. US patent 2013,0184,190. July 18, 2013.

[23] Crawley SL, Sivik MR, Butke BJ. Lubricating composition containing a carboxylic functionalised polymer. US patent 2013,0203,638 A1. August 8, 2013.

[24] Boccaletti G, Riganti F, Jungk M. Method for lubricating wind turbine gearbox. US patent 2011,0067,957 A1. March 24, 2011.

[25] Phillips WD. The use of triaryl phosphates as fire resistant lubricants for steam turbines. Lubr Eng 1986;43:228–235.

[26] Phillips WD. Turbine lubricating oils and hydraulic fluids. In: Totten GE, editor. *ASTM Fuels and Lubricants Handbook: Technology, Properties, Performance, and Testing.* West Conshohocken: American Society for Testing and Materials; 2003.

[27] Butke BJ, Barber AR. Composition with improved cleanliness for lubrication of steam and gas turbine systems. US patent 2013,0281,332. October 24, 2013.

CHAPTER 8

HYDRAULIC FLUIDS

Hydraulic fluids constitute a wide variety of fluids based on mineral oil, vegetable oils, synthetic oils, emulsions, microemulsions, aqueous solutions, and water–glycol mixtures. This is the reason that these are preferred to be referred as fluids and not oils. Among the industrial oils, these are also the largest group of products utilized across all the industries to transmit power through pressure and flow hydrostatic. Pascal in 1650 discovered that if pressure is exerted on a confined liquid, it can be transmitted in all directions with equal force and equal area, and that is now known as the Pascal law. This law formed the basis of a hydraulic system, which has the capability of transmitting and transforming force and energy in any direction. The example of an automobile hydraulic jack is one of the fine examples of transforming and transmitting force, where a smaller force can be utilized to lift larger and heavy object such as a vehicle.

Figure 8.1 shows how power is transmitted through the hydraulic fluid to lift heavyweight automobiles by applying only a fraction of force on piston A. For example, if 10 lbs of weight is applied on piston A with 1-square-inch area, it will exert a pressure of 10 psi on the hydraulic fluid in all directions. Thus, if there is a piston B of 10-square-inch area on the other end of the system, it will experience a pressure of $10 \text{ psi} \times 10 = 100 \text{ psi}$ and can support a load of 100 lbs. This means that by manipulating piston area in hydraulic systems, varying amount of forces can be generated. Braking system of the cars is another simple example of hydraulic system where power is transmitted through a fluid (brake fluid).

All industrial hydraulic systems consist of a pump supplying oil under pressure to a cylinder (hydraulic motor), which is responsible for mechanical movement. In addition, there will be pressure control valves, filters, seals, fluid tank, etc. In hydraulic system, pump is the main equipment that needs proper lubrication. These pumps could be of various designs such as gear pump (150–250 bar), sliding vane pump (200–300 bar), and radial or axial piston pumps (up to 450 bar). Thus, pumps are selected according to the pressure requirement of the system. The pressure generated by the pumps is then required to be converted to linear motion by the use of hydraulic motor or cylinder. These hydraulic cylinders could be single acting or double acting. In double-acting cylinders, the fluid flow direction can be changed to move the piston in different directions. There are different types of valves such as flow control valves, relief valves, and servo control valves to carry out precise operations. These valves could be operated manually, mechanically, electrically,

Developments in Lubricant Technology, First Edition. S. P. Srivastava.
© 2014 John Wiley & Sons, Inc. Published 2014 by John Wiley & Sons, Inc.

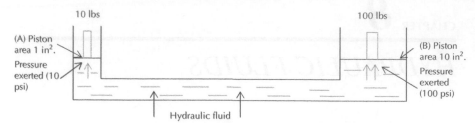

FIGURE 8.1 Example of power transmission in hydraulic jack.

hydraulically, or pneumatically. Different circuits are designed to operate complicated machines automatically [1]. Thus, a hydraulic system essentially consists of
a hydraulic pump, a fluid reservoir, a pressure control valve to regulate pressure, a
fluid directional control valve, and an actuator to convert hydraulic power into
mechanical motion. The hydraulic fluid has to lubricate all the moving parts coming
in contact with each other and also be compatible with various seals and gaskets
of different elastomers present in the system [2]. The oil should also be capable of
removing heat from the hot spots to the sump and protect the metallic parts from
rusting and corrosion. During the compression and decompression processes, air and
moisture gets into the system. The oil must be capable of separating quickly from
water and air. Therefore, appropriate properties are to be provided in the hydraulic
fluids [3, 4] to carry out these functions over long duration of time.

CLASSIFICATION OF HYDRAULIC FLUIDS

There are large numbers of hydraulic fluids available to the industry depending upon
the service requirements. The following categories have been identified:

1. Straight mineral oils without additives (rarely used now)
2. Rust- and oxidation-inhibited (R&O) oils
3. Antiwear hydraulic oils
4. Heavy-duty antiwear oils
5. Antiwear oils with improved viscosity index (VI)
6. Antiwear oils with improved thermal stability
7. Antiwear oils with stick–slip behavior
8. Antiwear oils with dispersant/detergent
9. Synthetic fluids without fire-resistant properties (polyalphaolefin (PAO) and
   ester based)
10. Synthetic oils with fire-resistant properties based on phosphate esters
11. Oil-in-water emulsions
12. Water-in-oil emulsions
13. Water–polymer solutions
14. Aqueous chemical solutions

15. Water–glycol mixtures
16. Biodegradable fluids based on vegetable oils
17. Polyglycol-based fluids
18. Food grade fluids (FDA, USDA H1 and H2, based on white oil, PAO)
19. Universal tractor transmission oils (UTTO and STOU)
20. Aviation and military hydraulic fluids

All these categories of oils are commercially available for different applications from major lubricant manufacturers. Hydraulic fluids thus constitute much diversified products for different applications. OEM recommendations must be kept in mind while recommending a fluid for the system. The development of hydraulic fluids has been discussed in greater detail [5–7] in several research papers. And several guidelines have also been reported [8] for the selection and test methods for hydraulic fluids.

ISO has classified mineral oil and non-fire-resistant synthetic-based fluids into seven categories, and details along with their applications are provided in Table 8.1.

TABLE 8.1   Classification of hydraulic fluids: ISO 6743/4, category H, hydrostatic–hydraulic systems

Category ISO-L[a]	Composition and characteristics	Field of application and operating temperature (°C)
HH	Noninhibited mineral oils	Hydraulic systems without specific requirements (generally not used now). −10 to 90°C
HL	Refined mineral oils with improved antirust and antioxidant properties	Hydrostatic drive systems with high thermal stress; need for good water separation. −10 to 90°C
HM	Oils of HL type (i.e., R&O) with improved antiwear properties	General hydraulic systems that include highly loaded components; need for good water separation. −20 to 90°C
HR	Oils of HL type with additives to improve viscosity–temperature behavior	Enlarged range of operating temperatures compared with HL oils. −35 to 120°C
HV	Oils of HM type with viscosity modifiers to improve viscosity–temperature behavior	Hydrostatic power units in construction and marine equipment. −35 to 120°C
HS	Synthetic fluids with no specific inflammability characteristics and no specific fire-resistant properties	Specific applications in hydrostatic systems with special properties. −35 to 120°C
HG	Oil of HM type with additives to improve stick–slip behavior	Machines with combined hydraulic-way systems where vibrations or intermittent sliding (stick–slip) at low speed must be minimum. −35 to 120°C

[a] Category L: Lubricant, industrial oils, and related products.

TABLE 8.2   ISO 6743/4, classification of fire-resistant hydraulic fluids

Category	Composition	Applications and operating temperatures
Water-containing fire-resistant hydraulic fluids		
HFAE	Oil-in-water emulsion, mineral oil, or synthetic ester, water content >80%	Power transmissions, about 300 bar, high working pressures, powered roof support
HFAS	Mineral oil-free aqueous synthetic chemical solutions, water content >80%	Hydrostatic drives, about 160 bar, low working pressures 5 to <55°C
HFB	Water-in-oil emulsions, mineral oil content about 60%	In British mining industry, not approved in Germany, 5–60°C
HFC	Water–polymer solutions, water content >35%	Hydrostatic drives, industry and mining applications, −20 to 60°C
Water-free, synthetic fire-resistant hydraulic fluids		
HFDR	Water-free synthetic fluids, consisting of phosphate esters, not soluble in water	Lubrication and control of turbines, industrial hydraulics −20 to 150°C; hydrostatic applications 10–70°C;
HFDU	Water-free synthetic fluids of other compositions (e.g., carboxylic acid esters)	hydrostatic drives, industrial hydraulic systems, −35 to <90°C

TABLE 8.3   Water-free, rapidly biodegradable hydraulic fluids

VDMA 24568 and ISO/CD 15380-E		
HEPG	Polyalkylene glycols soluble in water	Hydrostatic drives, e.g., locks, water hydraulics, −30 to <90°C
HETG	Triglycerides (vegetable oils) not soluble in water	Hydrostatic drives, e.g., mobile hydraulic system, −20 to <70°C
HEES	Synthetic esters not soluble in water	Hydrostatic drives, mobile, and industrial systems, −30 to <90°C
HEPR	PAOs and/or related hydrocarbons, not soluble in water	Hydrostatic drives, mobile, and industrial systems, −35 to <80°C

The same ISO standard, ISO 6743/4, has further classified fire-resistant fluids into six categories, and details are provided in Table 8.2. Biodegradable and water-free fluids have been described in Table 8.3 according to VDMA 24568 and ISO/CD 15380-E standard.

SPECIFICATION OF HYDRAULIC FLUIDS

Each category of the aforementioned fluids is defined by specifications. Table 8.4 describes various specifications for mineral oil- and vegetable oil-based fluids.

R&O oil requirements are simple, and these require API group I or group II base oils with minimum amount of rust inhibitor and antioxidant to meet the ASTM

TABLE 8.4   Specifications for different category of hydraulic fluids

R&O fluids	ISO 11158-HL; DIN 51524 part1; Denison HF 1; U.S. Steel 126; Cincinnati Milacron (CM) P-38, P-54, P-55, P-57; SAE MS 1004-HL; ASTM D-6158-HL
General antiwear fluids	ISO 11158-HM, DIN 51524 part 2, Denison HF 2, U.S. Steel 127, ASTM D-6158-HM
Premium antiwear fluids	ISO 6743/4, Denison HF 0, CM P-68, P-69, P-70, SAE MS 1004-HM
High VI premium antiwear fluids	ISO 11158-HV, DIN 51524 part 3, Poclain Z00-032-42, Vickers and Hitachi standards, SAE MS 1004-HV, ASTM D-6158-HV
Zinc-free fluids	GM-LH-02/03/04/06-1-00
Biofluids	Denison TP 30560 HF 6, VDMA 2456, vegetable oil base

D-665 A rust test for 24 h and D-943 oxidation test for 1000 h. Antiwear type of oils may need additional antiwear additives based on secondary zinc dialkyldithiophosphate (ZDDP). The presence of such antiwear additives may change the requirement of antioxidant since ZDDP itself acts as antioxidant. However, ZDDP suffers from poor hydrolytic stability, and oils requiring higher protection in this property may need special antiwear additives based on other chemistry such as phosphorous- or sulfur–phosphorous-containing additives. High VI premium antiwear hydraulic oils shall contain an additional amount of shear-stable viscosity modifier at a dosage to meet the VI requirements of the specification. There are large numbers of specifications of hydraulic oils, but the essential elements of these are presented in Table 8.5 for R&O and antiwear type of oils. Different OEMs also have their specifications with additional requirement, which must be met to obtain their approvals. Actual standards may be referred for complete information.

One of the comprehensive specifications of hydraulic fluids has been developed by Denison (HF 1, HF 2, and HF 0). HF 1 specifies R&O types of oil and is generally in line with the ISO standard. HF 2 specifies antiwear types of oils and incorporates Denison vane pump wear test, CM thermal stability test, Denison filterability procedure, and hydrolytic stability tests. Denison HF 0 is more severe standard as compared to HF 2 and involves an additional Denison axial piston pump wear test. These limits put several limitations on the selection of base oils and antiwear/antioxidant additives. It is, however, possible to formulate a universal hydraulic fluid meeting most of the standards described with carefully selected antiwear and antioxidant additives. Denison HF 3 and HF 4 are water-based fluids containing 37–45% and 40–45% water, respectively. Denison TP 30560 also mentions HF 5 and HF 6 fluids. HF 6 is a biofluid. DIN 51524 part 3 describes high VI oils. Key properties of this specification are provided in Table 8.6.

The standard also calls for behavior toward rubber seals in addition to other normal physicochemical tests.

Zinc-free oil described by GM limits zinc level to 10 ppm maximum and includes a thermal stability test according to ASTM D-2070. Antiwear properties are evaluated by Vickers vane pump 35VQ25 test and FZG (10 stages failed) test. CM standards heavily rely on the thermal stability test originally developed by them (CM procedure A).

TABLE 8.5 Key properties of R&O and antiwear hydraulic oil specifications

Properties	ISO 6743/4	DIN 51524 part 1	DIN 51524 part 2
Viscosity grades (VG)	10, 15, 22, 32, 46, 68, 100, 150	10, 22, 32, 46, 68, 100	10, 22, 32, 46, 68, 100
Pour point (°C)	−30, −21, −18, −15, −12, −12, −12, −12	−30, −21, −18, −15, −12, −12	−30, −21, −18, −15, −12, −12
Copper corrosion 100°C, 3 h	2 max.	2 max.	2 max.
Rust test, procedure A	Pass	Class 0	Class 0
Air release 50°C, max. min	5, 5, 5, 5, 10, 10, 14, 14	5, 5, 5, 10, 10, 14	5, 5, 5, 10, 10, 14
Water separation, time to 3-ml emulsion	30, 30, 30, 30, 30, 40, at 82°C-30, 30	30, 40, 40, 40, 60, 60	30, 40, 40, 40, 60, 60
Oxidation stability; 1000 h			
TAN (mg KOH/g)	2 max. only for the last six grades	2 max.	2 max.
Insoluble sludge	Report		
Wear protection FZG, fail stage	10 stages for the last five grades only	—	10 stages for VG 32–100
Vane pump wear Ring wt loss (mg, max.) Vane wt loss (mg, max.)	Report for VG 22–68	—	For VG 32–100 120 30
Elastomer compatibility ISO 6072 test for 1000 h	Report	—	—
Elastomer compatibility	—	% change in vol, 0–18 for VG 10; 0–15 for VG 22; 0–12 for VG 32, 46; 0–10 for VG 68, 100	
DIN 53538 part1, after 7 days at 100°C		Change in shore hardness, 0 to −10 for VG 10; 0 to −8 for VG 22; 0 to −7 for VG 32, 46; 0 to −6 for VG 68, 100	

TABLE 8.6 Key properties of DIN 51524 part 3 standard for HVLP oils

Properties	Limits				
ISO VG	15	32	46	68	100
Kinematic viscosity at 0 and 100°C	To be specified by supplier				
VI	140 min (15–68 grades)		120 min for 100 grade		
Pour point (°C, max.)	−39	−30	−27	−24	−21
Air release value 50°C, max. min	5	5	10	10	14
Demulsibility 54°C max. min	30	40	40	50	50
Oxidation D-943 1000 h neutralization no. mg/KOH/g, max.	2 for all grades				
FZG gear stage min.	10 stages for VG 32–100				
Vane pump test DIN51389/2 wt loss Ring Vane	120 mg max. for VG 32–100 30 mg max. for VG 32–100				
Shear stability, 250 Bosch injector cycles	To be specified by the supplier				

SAE has also developed comprehensive standard for hydraulic fluids. Details of this standard are provided in Table 8.7.

In addition to the aforementioned tests, SAE standard also specifies flash point, pour point, foaming characteristics, demulsibility, copper and steel corrosion, and rubber seal compatibility tests. Readers may refer to the complete standard for reference. SAE standard and DIN 51524 are the most suitable standards for formulating R&O, antiwear, and high VI with antiwear-type hydraulic oils, which would meet most of the industry requirements. Specific OEM requirements are, however, covered

TABLE 8.7   Key properties in SAE MS 1004 hydraulic oil specification

Properties	HL-R&O type	HM antiwear type	HV antiwear and high VI type
ISO VG	10,15,22,32,46, 68,100	10,15,22,32,46, 68,100	10,15,22,32,46, 68,100
Density, TAN and ash	To be specified by supplier	To be specified by supplier	To be specified by supplier
Air release value minutes max.	5 up to 32 grades and 10 for VG 46–100	5 up to 32 grades and 10 for VG 46–100	5 up to 32 grades and 10 for VG 46–100
1000-h D-943 TOST, TAN max.	2 mg KOH/g	2 mg KOH/g	2 mg KOH/g
Cleanliness as received ISO 4406	19/16/13	19/16/13	19/16/13
Thermal stability ASTM D-2070			
Acid no. change %	±50 max.	±50 max.	±50 max.
Viscosity change % max. 40/100°C	5		5
Sludge mg/100 ml max.	25	25	25
Cu rod color max.	5	5	5
Cu weight loss mg max.	10	10	1
Steel rod color max.	No discoloration	No discoloration	No discoloration
FZG A/8.3/90	No requirement	11 stages failed VG 46–100	11 stages failed VG 46–100
Hydrolytic stability D-2619			
Cu wt loss mg/cm3 max.	0.2	0.2	0.2
Acidity of water layer, mg KOH, max.	4	4	4
Vickers 35VQ25 pump test			
Ring wear mg max.	No requirement	120	120
Vane wear mg max		30	30
Denison piston pump test P-46	No smearing, scoring, scratching, corrosion	No smearing, scoring, scratching, corrosion	No smearing, scoring, scratching, corrosion

in their respective standards and have to be dealt with separately. ASTM D-6158 standard for HL-, HM-, and HV-type fluids is also on the line of SAE standard with minor deviations.

HYDRAULIC OIL PERFORMANCE TESTS

The performance of antiwear premium-quality hydraulic oils is carried out in actual vane and axial piston pump tests. These tests are in addition to the physicochemical and rig tests to judge the field performance. Conventional antiwear hydraulic fluids are evaluated [9–11] by ASTM D-2882 procedure using Vickers V104 or V105 vane pumps. The new-generation heavy-duty fluids are evaluated under more severe operating conditions by employing Vickers 35VQ25 vane pump and Abex Denison P-46 axial piston pump and HP T5D-42, HP T6-20 vane pumps. Silent features of these tests conditions are described in Table 8.8. These tests are based on Vickers and Denison vane pumps and Denison axial piston pumps. Vickers vane pump tests have been adopted by ASTM, IP, and DIN.

TABLE 8.8 Various hydraulic fluid pump tests and their test conditions[a]

Test	Speed (RPM)	Time (h)	Temp. (°C)	Pressure (bar)	Filter (μm)	Pass/fail criteria
ASTM D-2882/ IP-281 Vickers 104C/105C Vane pump	1200	100	65.5 or 79.4	136	25	Ring and vane weight loss 500 mg max.
DIN 51389/2 Vickers 104C vane pump	1400	25	Depends on viscosity	140	25	Ring and vane weight loss Ring 120 mg max. Vanes 30 mg max.
Vickers HP 35VQ25 Vane pump	2400	150[a]	93.3	204	10	Ring and vane weight loss Ring 75 mg max. Vanes 15 mg max.
Denison HP T6-20 vane pump	1800	300 300 with 1% water	80±5	250	6	Visual assessment
Denison HP T5D-42 vane pump	2400	60 h at 71.1 40 h at 98.9		170	10	Vane contour increase 0.015 in. max.
Denison HP P-46 Axial piston pump	2400	60 h at 71.1 40 h at 98.9		340	10	Appearance of bronze on shoes, wear plate, and port plate

[a] 3 × 50-h tests using the same oil but a new cartridge in each test. If one fails, two additional tests must be run.

TABLE 8.9   NAS and ISO cleanliness categories

ISO 4406 codes	NAS 1638	Applications
8/5	00	Superclean oil
9/6	0	Superclean oil
10/7	1	Superclean oil
11/8	2	Superclean oil
13/10	3–4	Aviation hydraulic oils
15/12	4–6	Oils for high-pressure servo mechanism
16/13	7–8	Industrial hydraulic oils
18/15	8–10	Mobile medium-pressure hydraulic systems
19/16	9–11	General-purpose hydraulic oils
21/18	12	—

HYDRAULIC OIL CLEANLINESS

Modern industrial hydraulic systems contain numerically controlled systems with fine tolerances and micronic filters. This requires that the oil cleanliness must be maintained throughout the operation. Aircraft hydraulic oil requires superclean oil, and cleanliness standards have been developed as the National Aerospace Standards (NAS). Clean hydraulic oils can be produced by filtering the blended oil through appropriate filters and then filling in precleaned containers. This requires elaborate filtration system with container/drum cleaning facilities in the blending plant. NAS 1638 describes various oil cleanliness categories based on particle size distribution. ISO 4406 standard has also categorized cleanliness levels based on particles greater than $10\,\mu m/ml$ of oil. Table 8.9 describes NAS 1638 and ISO 4406 standard along with typical applications of hydraulic fluids. NAS 1638 is more comprehensive and is based on the range of particles present in $100\,ml$ of oil. This is often determined by automatic particle counter or by microscopic technique.

ISO 4406 has several other categories, but we have only described the corresponding categories to the NAS 1683 classes. Table 8.10 provides details of NAS 1638 cleanliness classes along with particle size distribution.

The strict control on particle size in hydraulic oils during service requires that oils be filtered continuously to maintain cleanliness level. Oil in hydraulic systems is generally filled through a series of filters. Electrostatic cleaner is also used off-line for achieving maximum cleanliness level. The service life of hydraulic oils is greatly improved by effective oil filtration since this removes most of the solid contaminants.

HYDRAULIC FLUID PROPERTIES

In addition to the normal physical properties such as viscosity, pour point, and flash point, hydraulic oils are required to possess the following important characteristics to provide good performance:

1. Thermal stability
2. Oxidation stability

TABLE 8.10 NAS 1638 cleanliness class and particle size in μm, no. of particles/100 ml

Class	5–15 μm	15–25 μm	25–50 μm	50–100 μm	>100 μm
00	125	22	4	1	0
0	250	44	8	2	0
1	500	89	16	3	1
2	1,000	178	32	6	1
3	2,000	356	63	11	2
4	4,000	712	126	22	4
5	8,000	1,425	253	45	8
6	16,000	2,850	508	90	16
7	32,000	5,700	1,012	180	32
8	64,000	11,400	2,025	360	64
9	128,000	22,800	4,050	720	128
10	256,000	45,600	8,100	1140	256
11	512,000	91,200	16,200	2680	512
12	1,024,000	182,400	32,400	5760	1024

3. Rust and corrosion protection
4. Good water-separating characteristics (demulsibility)
5. Hydrolytic stability
6. Low air release value
7. Filterability
8. Good seal swell characteristics (elastomer compatibility)
9. Antiwear protection
10. Oil cleanliness

R&O oils require the first four properties, while antiwear and premium antiwear oils require all these properties. Steam turbine oils thus make good R&O hydraulic oils although their cost may be slightly higher.

Thermal and Oxidation Stability

Hydraulic systems offer an ideal condition for thermal and oxidative oil degradation due to the presence of air/oxygen, higher temperatures, water, and metals. Oil oxidation can generate harmful acids and sludge leading to system failure. These properties are measured by ASTM D-943 (TOST), ASTM D-2272 (RBOT), and CM heat tests. Longer TOST and RBOT life and good heat test result will ensure good oil stability leading to longer oil life. Right choice of base oil and antioxidant additive ensures good stability of oils.

Rust and Corrosion Protection

Oil must be capable of protecting metal surfaces from rusting and corrosion. Corrosion protection is measured by ASTM D-130 copper corrosion test for 3 h at 100°C. Antirust properties are measured by ASTM D-665 A and B procedures for 24 h in the presence of either distilled water or synthetic sea water (B procedure). Copper corrosion can be prevented by the use of a metal deactivator, and rusting is prevented by a long-chain polar compound forming a protective layer on the metal surface.

Demulsibility or Water-Separating Characteristics

In hydraulic fluid, the ingress of moisture in the system cannot be prevented, and therefore, the oils must be capable of separating out from water quickly. This property is assessed by the ASTM D-1401 test. Good oils should be able to separate out from water in 10–30 min.

In addition to these characteristics, antiwear oils must have the following properties as well.

Hydrolytic Stability

Antiwear additives usually contain sulfur and phosphorous and are inherently unstable compounds. ZDDP is the most common antiwear additive for hydraulic oils, and it undergoes hydrolysis in the presence of water and high temperature. On hydrolysis, these compounds generate acidic and insoluble material, which can choke fine filters and lead to system malfunctioning. It is therefore necessary to stabilize these molecules or use an additive that is not easily hydrolyzable. This property is measured by the beverage bottle test ASTM D-2619. The requirements of this test are stipulated in Denison HF 2 specification.

Air Release Value

This property has been discussed in the previous chapter on turbine oils and is also important for hydraulic oils. The ingress of air through seals is common, and in the pumps, air may get compressed. This air in low-pressure areas can be released with a great force and may cause cavitation wear of the metal surfaces. The oils are therefore required to possess low air release value or should be able to release air quickly. Air release value is measured by IP-313 or DIN 51381 test methods. Low air release value of 5–10 min ensures good protection.

Filterability

Oil filtration is important since all hydraulic systems have built-in filters to keep the oil clean. The filter pore size depends upon the application. It has been observed that the oil filtration is also influenced by the presence of water and Denison HF 0 specification demands that the filtration time with water should not be more than twice the value of neat oil. In Denison test, 75 ml of oil is filtered through 1.2-μm membrane under 65 cm of vacuum suction. The time to filter oil is recorded. The filtration is repeated with added water. It is necessary that [12] the aged oil should also be capable of quick filtration for smoother operation. The additive used should be stable in presence of water or should have good hydrolytic stability to pass filtration test. Ashless hydraulic oil additives provide improved performance in filtration test.

Elastomer Compatibility

In hydraulic system, several seals and elastomers are present, and oil comes in contact with these materials. If the oil shrinks these elastomers, oil can leak out. Highly paraffinic oil can do that. Modern API group II or group III oils are highly paraffinic and so are PAOs and GTL base oils. To avoid oil leakage through seals, it is necessary that the oils slightly swell the elastomers. Aromatic compounds are used as seal sell additives in such paraffinic base oils. This property can be evaluated by DIN 53538 test for 7 days, and limits are specified in DIN 51524 part 2 standard. ISO 6072 specifies a 1000-h test.

Antiwear Protection

In a hydraulic system, pump is the main equipment that generates fluid pressure and requires wear protection. Pumps could be of different designs such as gear, vane, and radial or axial piston types. These will require different kind of evaluation procedures due to inherent design differences. Antiwear protection to the gear pumps can be evaluated by shell four-ball wear test and FZG gear failure load test. Antiwear protection to the vane and piston pumps are evaluated by running an actual pump test as indicated in Table 8.8. ZDDP provides the most effective means of antiwear protection in hydraulic oils. There are different types of ZDDP (with primary or secondary alcohols or aryl ZDDP) available with varying degrees of protection. Aryl ZDDPs have higher thermal stability as compared to alkyl ZDDP, but secondary alkyl ZDDP offers better antiwear protection. For higher level of protection, ZDDP has been combined with other antiwear additives such as phosphate ester. Ashless antiwear additives based on sulfur and phosphorous chemistry are now available, which are thermally stable and also provide good antiwear protection in low dosage.

HYDRAULIC FLUID TECHNOLOGY

Industry has wide choice of fluids to be used in hydraulic system depending on the type of pump and operating parameter. Under moderate conditions, R&O fluids can be used. These fluids also contain a pour depressant and antifoam additive to control low-temperature operability and foaming. With the increase in severity, normal antiwear fluids are used. These fluids shall contain additional antiwear additives in the conventional R&O oils. ZDDP also acts as an antioxidant, and therefore, part of the ashless antioxidant can be taken out of the R&O oil. As the severity increases further, premium antiwear oils are used. These could be based on S–P ashless antiwear additive [13, 14] or stabilized ZDDP or a combination of ZDDP with phosphate ester to achieve higher level of wear protection. When the hydraulic system is operated in an environment where wide variation in temperatures are encountered such as in heavy earth-moving equipment, high-VI hydraulic fluids are used [15, 16]. These types of fluids can also be energy efficient [17]. These oils contain additional highly shear-stable polymeric VI improver. Several PMA-based products are commercially available for this application. Other classes of polymers can also be considered after complete evaluation. Biodegradable fluids are used in areas where soil contamination with oil is considered hazardous such as in forest and agricultural application. Hydraulic oil cleanliness level is quite important for modern systems with servo-controlled mechanism having fine clearances. Cleanliness level is maintained right from the manufacturing stage where the oils are filtered through series of finer filters. The oils, however, have to be regularly cleaned in the system during their operation. The use of such filtration also improves the life of oil due to the removal of all solid contaminants.

FIRE-RESISTANT HYDRAULIC FLUIDS

Fire-resistant hydraulic fluids (FRHFs) are used as a safety measure in those applications where danger exists of explosive firing of combustible fluid escaped from broken hydraulic fluid pressure lines and striking at hot surfaces. Fire-resistant fluid can burn but is more resistant to ignition and, if ignited, is less capable to propagate fire as compared to non-fire-resistant mineral oil-based fluid. There are a large number of chemical compounds and their mixtures that can be used as FRHF, but the following six categories have been identified in ISO 6743/4 and DIN 51502 standards (Table 8.11):

1. Oil-in-water emulsions
2. Synthetic aqueous solutions
3. Water-in-oil emulsions
4. Water–polymer (glycol) solutions
5. Phosphate esters
6. Other synthetic fluids

TABLE 8.11   Classification of fire-resistant hydraulic fluids and biodegradable fluids

ISO 6743/4	Composition	Applications and operating temperatures
Water-containing fire-resistant hydraulic fluids		
HFAE	Oil-in-water emulsion, mineral oil, or synthetic ester, water content >80%	Power transmissions, about 300 bar, high working pressures, powered roof support
HFAS	Mineral oil-free aqueous synthetic chemical solutions, water content > 80%	Hydrostatic drives, about 160 bars, low working pressures 5 to <55°C
HFB	Water-in-oil emulsions, mineral oil content about 60%	In British mining industry, not approved in Germany, 5–60°C
HFC	Water–polymer solutions, water content >35%	Hydrostatic drives, industry and mining applications, −20 to 60°C
Water-free, synthetic, fire-resistant hydraulic fluids		
HFDR	Water-free synthetic fluids, consisting of phosphate esters, not soluble in water	Lubrication and control of turbines, industrial hydraulics −20 to 150°C; hydrostatic applications 10–70°C.
HFDU	Water-free synthetic fluids of other compositions (e.g., carboxylic acid esters)	Hydrostatic drives, industrial hydraulic systems, −35 to <90°C

HFAE: OIL-IN-WATER EMULSIONS

In these fluids, 20% mineral oil is emulsified in 80% water with the use of emulsifier. These are low-cost materials and can be used in the temperature range of 5–55°C. When subjected to flame or high heat, water is converted to steam and serves as a blanket to prevent burning of petroleum oil. Emulsions have inherent instability, and care is required to keep them stable during storage and applications. Water hardness is important to keep the emulsion stable. Usually, a concentrate is marketed by oil companies and is diluted at the site to obtain the emulsion of appropriate concentration.

HFAS: AQUEOUS CHEMICAL SOLUTIONS

About 1–5% chemical solution in water is used as hydraulic fluid. These chemicals are generally anticorrosion, antirust, lubricity, and antiwear compounds. The use of HFA-type fluids is very limited in coal mining industry in certain countries.

There are other classes of hydraulic fluids under this category and are described by the following designations:

HFAES: Semisynthetic with emulsified oil.

HFAM: Microemulsions; these are transparent emulsion with droplet size less than 0.1 μm.

HFAT: Thickened microemulsions.

HFB: INVERT EMULSION (WATER-IN-OIL)

These fluids are water-in-oil emulsions, where 40–45% water is emulsified in about 60% oil with the help of emulsifiers and stabilizers. In these fluids, oil is the continuous phase and is limited to maximum 60%. Due to this, invert emulsions have limited fire-resistant properties. The presence of water also limits their application temperature range to 5–60°C. Development of an advanced oil-in-water emulsion hydraulic fluid and its application as an alternative to mineral hydraulic oil in a high-fire-risk environment have been reported [18].

Both HFA- and HFB-type emulsions require specific emulsifier or surface-active compounds having different hydrophilic–lipophilic balance (HLB) values to support the dispersed phase.

HFC: WATER–GLYCOL SOLUTIONS

These are generally water–glycol solutions with some amount of polymer thickener to provide adequate viscosity to the fluid. Usually, 35–60% water is used. The presence of glycol lowers the freezing point of the solution, and water–glycol fluids can be used in the temperature range of −10 to +60 °C. Water–glycol solutions remain quite stable during storage and also possess good antiwear properties. These fluids can be used in high-pressure hydraulic pumps.

TABLE 8.12   Standard tests for fire-resistant properties

Tests	Procedure and observations
Spray ignition, NCB-570/1970, Appl. A	Sustained combustion is observed when flame is introduced into atomized fluid spray
Wick ignition, NCB-570/1970, Appl. B	An asbestos wick soaked in fluid reservoir is ignited and observed for continued burning
Hot surface test, factory mutual res. corp.	High-pressure fluid spray is directed onto steel channel at 704°C and observed for ignition
Hot manifold test, aeronautical material spec 3150C	Fluid is dripped into a tube at 704°C and observed for burning
Spark ignition, British min. of aviation spec. DTD 5526	Atomized fluid spray is exposed to high-tension electric arc and observed for ignition
Compression ignition, U.S. Navy test	MIL-H-19457C fluid is injected into an engine as fuel. Combustion should not occur at a compression ratio of <42:1

HFD: NONAQUEOUS FIRE-RESISTANT FLUIDS

These are generally synthetic phosphate ester-based products although other synthetic oils can also be used. Mixtures of mineral oil and phosphate esters have also been used as FRHF. These fluids are thermally stable and can be used in the temperature range of −20 to +150°C. The presence of phosphorous in the molecule imparts anti-wear properties, and no extra additive is required for this purpose to pass hydraulic pump [19] wear tests. Physicochemical properties of phosphate esters have been described in the technical brochures of PE manufacturing companies [20, 21] as well as in some papers [22]. PE fluids have been successfully used in most of the commercially available hydraulic pumps [23–25]. Phosphate esters have high ignition temperatures (500–600°C) and will not support combustion if ignited [26]. Phosphate ester fluids find applications in aircraft hydraulics [27], electrohydraulic furnaces, and door control of aluminum smelting plants [28, 29].

Fire-resistant properties of fluids are measured by several specialized tests (Table 8.12). There is no single test to characterize resistance to ignition and flame propagation. Properties like flash point, fire point, and autoignition temperature do not fully assess the safety aspect of the fluid. Tests described in Table 8.12 are used to evaluate fire-resistant properties of hydraulic fluids. Phosphate esters, however, degrade in the presence of water and high temperatures [30, 31] during the service, and their performance is required to be monitored [32]. These also require specific techniques to recondition used fluid [33, 34]. Ion exchange treatment has been used to reclaim degraded used fluids [35].

AIRCRAFT AND MILITARY HYDRAULIC FLUIDS

Both civil and military aircrafts use hydraulic systems for flight controls and landing gears. The oils used in these systems are quite sophisticated and are required to operate at high pressures in the range of 1000–3000 psi and temperatures from −60 to 120°C. The precise controls and extremely low clearances in these systems demand lubricants of very high quality and cleanliness levels. Earlier mineral oil-based fluids with high VI and antiwear properties were used. The advanced modern aircrafts now use either PAO-based fluids or phosphate ester-based fire-resistant hydraulic oils [36–39]. Some of the specifications issued for aircraft hydraulic systems and other military applications are provided in Table 8.13.

MIL-H-5606H is a low-pour naphthenic mineral oil-based hydraulic fluid containing VI improver and antiwear additive. Due to fire hazards, these fluids were replaced with a phosphate ester-based product (SAE Publication AS 1241). These fluids, however, had compatibility problems with mineral oils, and PAO-based hydraulic fluids were developed (MIL-PRF-83282) to resolve this issue. PAO-based fluid has high-temperature-range operability as compared to mineral oils. This product was further improved to meet the low-temperature operability (MIL-PRF-87257). Advanced and completely nonflammable hydraulic fluids based on chlorotrifluoroethylene have been developed, but these are quite costly and may find application [40] in the future (MIL-H-53119). These hydraulic oils contain antiwear, VI improver [41], antioxidants, and antirust additives and are produced under a clean environment so that the oil is free from particulate matter.

The cleanliness property is quite important, since the hydraulic systems operate with very fine clearances for precise controls. In order to achieve this, the oils after blending are filtered through series of filters to achieve desired cleanliness level. These cleanliness levels have been defined and classified as NAS levels/ISO 4406 categories. In this system, particles are counted in 100 ml of fluid either by a microscope or by an automatic particle size analyzer in the range of 5–15 µ, 15–25 µ, 25–50 µ, 50–100 µ, and above 100 µ. Depending upon the permissible limits of particles in each range, ISO and NAS define the cleanliness categories. For aviation application, NAS 3 to 4 cleanliness levels are required. A more detailed and comprehensive discussion on synthetic lubricants and fire-resistant fluids can be found in classical work by Gunderson–Hart and Shubkin [42–44].

TABLE 8.13 Military hydraulic fluid specifications

Specification	Designation	Base fluid
MIL-H-5606H	NATO Code H-515 hydraulic fluid	Mineral oil −54 to 135°C
DEF STAN 91-48/2	OM-15, hydraulic fluid	Mineral oil
MIL-PRF-83282D	NATO Code H-537, OX-19, hydraulic fluid	Synthetic (PAO) −40 to 204°C
MIL-PRF-87257B	NATO Code H-538 hydraulic fluid	Synthetic (PAO) −54 to 200°C
MIL-PRF-46170B	Contains additional Ba rust inhibitor	Synthetic PAO, rust-inhibited version of MIL-PRF-83282D
MIL-H-27601B	Fire-resistant high temp., flight vehicles	Synthetic PAO
MIL-H-83306	Fire-resistant fluid	Phosphate esters
MIL-H-53119	For nonflammable armored vehicles	Chlorotrifluoroethylene

BIODEGRADABLE AND ENVIRONMENT-FRIENDLY HYDRAULIC FLUIDS

The use of biodegradable products is increasing due to environmental considerations, especially in agricultural and forest applications. Such products can be formulated with the use of polyglycols, vegetable oils, synthetic esters, and PAOs [45–47]. VDMA guideline 24568 has specified minimum requirements for the first three categories of oils. ISO/CD 15380-E has additionally included PAO-based products. Application and operating range of these oils are provided in Table 8.14. Key properties of these fluids are compared in Table 8.15. These are specialized products and are used only in those applications where fluid biodegradability [48] is of prime importance. The main difference in these fluids is the oxidation test. In vegetable oil- and ester-based fluids,

TABLE 8.14   Water-free biodegradable hydraulic fluid categories and their applications

VDMA 24568	Composition	Applications
HEPG	Polyalkylene glycols soluble in water	Hydrostatic drives, e.g., locks, water hydraulics, −30 to <90°C
HETG	Triglycerides (vegetable oils) not soluble in water	Hydrostatic drives, e.g., mobile hydraulic system −20 to <70°C
HEES	Synthetic esters not soluble in water	Hydrostatic drives, mobile, and industrial systems −30 to <90°C
ISO/CD 15380-E	Composition	Application
HEPR	PAOs and/or related hydrocarbons, not soluble in water	Hydrostatic drives, mobile, and industrial systems −35 to <80°C

TABLE 8.15   Comparison of key properties of VDMA fluid specification

Properties	HEPG fluids	HETG fluids	HEES fluids
ISO VG	22, 32, 46, 68	22, 32, 46, 68	22, 32, 46, 68
Pour point	−21, −18, −15, −12	Report	−21, −18, −15, −12
Aging test, TAN after 1000 h, DIN 51587	2 max.	—	—
Baadar oxidation test at 95°C, 72 h, viscosity increase at 40°C, DIN 51554 part 3	—	20% max.	20% max.
Air release value	5, 5, 10, 10	7, 7, 10, 10	7, 7, 10, 10
Demulsibility	No requirement	Report	Report
FZG fail stage	10 for 46 and 68 grades	10 for all grades	10 for all grades
Pump wear test DIN 51389 part 3			
Ring (mg, max.)	120	120	120
Vane (mg, max.)	30	30	30

Baadar test is used, since the usual long-duration 1000-h TOST does not yield good result. There are also minor differences in other tests. Full details can be seen in the original standard.

Thus, the biodegradable ester-based hydraulic fluids have certain limitations as compared to the mineral-based products, especially with respect to the oxidation life, water separation characteristics, and air release value. They, however, have improved lubricity characteristics and biodegradability. Such products would thus be used in applications where it is mandatory to use biodegradable products.

Hydraulic fluids are one of the most interesting groups of products, ranging from simple R&O-type fluids to antiwear, superclean, highly sophisticated synthetic fluids for aircrafts, and all of such products have been clearly defined by several governing specifications. Several OEMs have also defined the products with specific requirements to protect their equipment. The development of each of these fluids is a real challenge to the formulators and chemical additive developers.

REFERENCES

[1] Totten GE, De Negri VJ, editors. *Handbook of Hydraulic Fluid Technology*. 2nd ed. Boca Raton: CRC Press; 2012.

[2] Totten GE, Kinker BG, editors. Fluid viscosity and viscosity classification. In: *Handbook of Hydraulic Fluid Technology*. New York: Marcel Dekker; 2000.

[3] Totten GE, Webster GM. Review of the testing methods hydraulic fluid flammability; 1996. SAE Paper 932436 (see also ASTM STP 1284, 42-60).

[4] Totten GE, Reichel J. *Fire resistance of industrial fluids*. ASTM Special Technical Publication 1284; 1996.

[5] Papay AG, Harstick CS. Petroleum based industrial hydraulic oils-present and future developments. Lubr Eng 1975;31:6–15.

[6] Grover KG, Perez RJ. The evolution of petroleum based hydraulic fluids. Lubr Eng 1990;46 (1):15–20.

[7] Wamback WE. Hydraulic systems and fluids. Lubr Eng 1983;39:483–486.

[8] Radhakrishnan M. *Hydraulic Fluids—A Guide to Selection, Test Methods and Use*. New York: ASME Press; 2003.

[9] Soul DM. Development and evaluation of anti-wear additive system: performance International symposium. Institute of Petroleum, UK; October 1978. p 133–148.

[10] Jayne GG, Daffy P. Hydraulic fluids—a pump durability test. 6th International Colloquium; January 12–19, 1988; Esslingen, Germany, 18-6.1-11.

[11] Perez JM, Hansen RC, Klaus EE. Comparative evaluation of several hydraulic fluids in operational equipment- a full scale pump stand test and the four ball wear tester, part II, phosphate esters, glycols and mineral oils. Lubr Eng 1990;46:249–255.

[12] Saxena D, Mooken RT, Srivastava SP, Bhatnagar AK. An accelerated aging for anti-wear hydraulic oils. Lubr Eng October 1992;49 (10):801–809.

[13] Humblin PC. New ashless industrial oil package. In: Srivastava SP, editor. Proceedings of International Symposium on Fuels and Lubricants (ISFL); New Delhi. New Delhi: Tata McGraw Hill; 1997. p 113–122.

[14] Okada M, Yamashita M. Development of extended life hydraulic fluids. Lubr Eng 1987; 43 (6):459–466.

[15] Cocks R, Hutchinson P, Neveu CD. Conserving energy with high VI hydraulic oils, ISFL-2004, New Delhi, tech. session II p-25; 2004.

[16] Neveu C, Schweder R. Computer aided hydraulic oil formulation. In: Srivastava SP, editor. Proceedings of the International Symposium on Fuels and Lubricants; New Delhi; 2000. New Delhi: Allied Publisher. p 101–106.

[17] Ajay Kumar I, Singh D, Saxena V, Martin G, Sharma K, Raje NR, Srivastava SP, Bhatnagar AK. Frictional characteristics of hydraulic fluids in Vickers vane pump test. Proceedings of the International Symposium on Fuels and Lubricants; 2000; New Delhi. New Delhi: Allied Publisher. p 119–124.

[18] Young KJ, Kennedy A. Development of an advanced oil-in-water emulsion hydraulic fluids, and its application as an alternative to mineral hydraulic oil in a high fire risk environment. Lubr Eng 1993;49:873.

[19] Perez JM, Hansen RC, Klaus EE. Comparative evaluation of several hydraulic fluids in operational equipment. A full-scale pump stand test and four ball wear tester: part 2, phosphate esters, glycols and mineral oils. Lubr Eng 1989;46 (4):249–255.

[20] Reolube HYD Fire-Resistant Fluids', 'Reolube Turbofluids' Chemtura Corp. technical data sheets.

[21] 'Fyrquel', 'Fyrquel EHC' Supresta Corp. Technical data sheets.

[22] Gamrath HR, Hatton RE, Weesner WE. Chemical and physical properties of alkyl aryl phosphates. Ind Eng Chem 1954;46:208–212.

[23] Castleton VW. Practical considerations for fire resistant fluids. Lubr Eng 1998;54 (2):11–17.

[24] Phillips WD. Comparison of fire-resistant hydraulic fluids for hazardous industrial environments (Pts 1 & 2). J Syn Lub 1997 14,3,211–235 and 1998;14 (4):303–330.

[25] Phillips WD. 'Fire-Resistant' hydraulic fluids—an accurate description? STLE Annual Meeting; 2010; Las Vegas.

[26] Khan MM. Spray flammability of hydraulic fluids, fire resistance of industrial fluids. ASTM Special Technical Publication 284. Philadelphia: American Society for Testing Material; 1996.

[27] Snyder CE, Gschwender LJ. Fire resistant hydraulic fluids. Lubr Eng 1995;51 (7):549–553.

[28] Sherman JV, Mateew MM. Property and performance requirements of fire-resistant hydraulic fluids from a global perspective, STLE Annual Meeting; 2004; Toronto.

[29] Burkhardt E. An examination of the fire properties of various hydraulic fluids. STLE Annual Meeting; 2008; Cleveland.

[30] Brown KJ, Billings LM, Austin EM. Effect of water on properties of fire-resistant EHC fluids. STLE Annual Meeting; 2004; Toronto.

[31] Phillips WD. The high temperature degradation of hydraulic oils and fluids. J Syn Lubr 2006;23 (1):39–69.

[32] Shade WN. Field experience with degraded synthetic phosphate ester lubricants. Lubr Eng 1987; 43:176–182.

[33] Phillips WD. The conditioning of phosphate ester fluids in turbine applications. Lubr Eng 1983; 39:766–780.

[34] Wooton D, Livingstone G. Phosphate ester fluids—a review of the degradation and treatment chemistries with relationship to the electro hydraulic control system performance. STLE Annual Meeting; 2007; Philadelphia.

[35] Duchowski JD. Ion exchange treatment of phosphate esters in industrial hydraulic systems: lessons learned from power generation applications. STLE Annual Meeting; 2010; Las Vegas.

[36] Snyder CE Jr, Gschwender L, Campbell WB. Development and mechanical evaluation of non-flammable aerospace (−54 to 135°C) hydraulic fluids. Lubr Eng 1982;38:41–51.

[37] Snyder CE Jr, Gschwender L. A survey of fire resistant hydraulic fluids. STLE 50th Annual Meeting; Chicago; 1995. p 4. Preprint N.95 AM-4D-1.

[38] Snyder CE Jr, Gschwender L. Nonflammable hydraulic system development for aerospace. J Syn Lubr 1984;1:188–200.

[39] Gschwender L, Snyder CE Jr, Driscoll GL. Alkyl benzenes, candidate-high temperature hydraulic fluids. Lubr Eng 1990;46:377–381.

[40] Snyder CE Jr, Gschwender L. Trends toward synthetic fluids and lubricants in aerospace. In: Rudnick LR, editor. Synthetic mineral oils and bio-based lubricants, Chemistry and technology. Boca Rotan: CRC/Taylor & Francis; 2006.

[41] Snyder CE Jr, Gschwender L, Paciorek K, Kratzer R, Nakahara J. Development of shear stable viscosity index improvers for use in hydrogenated PAO based fluids. Lubr Eng 1986;42:547–557.

[42] Gunderson RC, Hart AW. Synthetic Lubricants. New York: Reinhold; 1962.

[43] Shubkin RL. Synthetic lubricants. In: Lappin GR, Sauer JD, editors. Alphaolefins Application Handbook. New York: Marcel Dekker; 1993.

[44] Hombek R, Buczek M, Marolewski T. Next generation of fire resistant fluids. Proceedings of the 48th National Conference on Fluid Power; 2000; Milwaukee. Paper 100-1.10.

[45] Carpenter JF. Biodegradability and toxicity of polyalphaolefins base stocks. Proceedings of the 9th International Colloquium Ecology and Economic Aspects of Tribology, Esslingan, Volume 1; 1994. p 4.6.1–4.6.6.

[46] Carpenter JF. Biodegradability and toxicity of polyalphaolefins base stocks. J Syn Lubr 1995; 12:13–20.

[47] Carpenter JF. The biodegradability of polyalphaolefins base stocks. Jpn J Tribol 1994;39:573–577.

[48] Harvey D. Biodegradable fluids and lubricants. Ind Lubr Tribol 1996;48:17–26.

CHAPTER 9

COMPRESSOR, VACUUM PUMP, AND REFRIGERATION OILS

Compressor is a machine that increases the pressure of a gas by transferring energy to it and in doing so reduces the volume and increases density and temperature of gas. Similarly, if the compressed gas pressure is released suddenly (increasing volume and reducing density), cooling effect is observed. This is the principle used in preparing solid ice with pressurized carbon dioxide gas. When the gas is required to be compressed to a high pressure, it is to be carried out in several stages with intercooler in between to bring down the gas temperature. For example, if air is compressed from an initial pressure of 14.7 psi at a temperature of 15.6°C to a pressure of 100 psi in a single stage under adiabatic condition (i.e., no heat leaves or enters the system), discharge air temperature could reach 247°C. However, if this compression is carried out in two stages with intercooling, lower air discharge temperature of 114°C would be obtained. This is based on the assumption that all the mechanical work done is converted into heat in the gas, and there is no heat loss from the system. Actual temperatures could be lower since heat is lost through the cylinder walls.

Compressors are of different designs and handle a variety of gases including air, oxygen, nitrogen, hydrogen, ammonia, carbon dioxide, refrigerant, hydrocarbon, and other process gases. The choice of lubricant, therefore, varies according to both compressor design and gas handled by it. Compressors can be classified either as dynamic compressors or positive-displacement compressors in the first place.

DYNAMIC OR TURBO COMPRESSORS

Dynamic compressors could be centrifugal or axial flow type and are used for handling large-volume process gas with medium pressures. These are also referred to as turbo compressors and supply compressed gas continuously. In a multistage version, several rotors are mounted on the same shaft. In centrifugal compressors, radial vanes of the rotor provide high radial velocity to the gas. In axial flow, the gas flow is parallel to the axis of the rotors whose angled blades provide acceleration and drive it past a ring of stationary blades. In such compressors, the gas usually does not

Developments in Lubricant Technology, First Edition. S. P. Srivastava.
© 2014 John Wiley & Sons, Inc. Published 2014 by John Wiley & Sons, Inc.

come in contact with the lubricant. The shaft of the compressor is mounted on the bearings that need lubrication, and the oil is selected based on the bearing design and its operating parameters. Sometimes, gases could leak through the seal into the bearings, and oil compatibility with the gas becomes important. The reverse leakage of oil into gas is avoided due to the positive pressure of gases. Hydrocarbon gases if leaked into oil system can reduce oil viscosity and create lubrication problems. These parameters are taken into consideration while selecting oil for such compressors.

POSITIVE-DISPLACEMENT COMPRESSORS

There are different types of positive-displacement compressor designs in which successive volumes of gas are trapped in a closed space, compressed, and delivered at higher pressures. Thus, the gases are delivered intermittently in pulses. To obtain continuous supply, compressed gases can be routed through a storage tank. Different types of positive-displacement compressors are described below:

1. Reciprocating
2. Membrane compressor without crank mechanism
3. Reciprocating piston and swash plate compressor with crank
4. Rotary vane and rolling piston with one shaft
5. Rotary screw with two shafts
6. Lobe or root-type compressors
7. Liquid-ring-type compressors

RECIPROCATING COMPRESSORS

Conventional reciprocating compressors are similar to internal combustion engine in that these have pistons, piston rings, cylinders, valves, connecting rods, and crankshafts to be lubricated. Due to this similarity, these compressors were earlier lubricated by engine oil. It was soon realized that compressors do not need engine oils containing detergents and dispersants, since fuel combustion process is absent. Pistons, however, could be single acting or double acting. In single-acting compressor, air is compressed only on one side of the piston, while in the double acting type, there are two compression strikes per revolution. The piston layouts in reciprocating compressors are several such as single cylinder, twin parallel, Vee arrangement, and double-acting crosshead configurations. There are oil-free compressors using graphite piston/rings that do not require lubrication in the compression zone. In multistage compressors, air compressed in the first stage is sent to the next-stage cylinder for further compression. Air is, however, cooled after every stage to avoid excessive temperature rise. The intercoolers could be water- or air-cooled heat exchangers. These could also be just finned pipes, depending on the operating conditions.

ROTARY COMPRESSORS

Sliding Vane Compressors

In these compressors, an eccentric cylindrical rotor is housed in the cylindrical housing. The rotor is slotted to take sliding vanes. The vanes are made of steel, PTFE, or some synthetic resins. When the rotor moves, vanes slide out of the slots due to centrifugal force. The volume of the pressure chamber varies from high to low in each revolution, leading to the compression of gases. Gas or air enters at a point where the space is expanding and is delivered near the full contraction point. Multistage units with intercooler are employed for higher delivery pressures. Rotary piston compressors also work on the same principle. Such compressors are compact in design and provide vibration-free continuous gas flow. In rotary vane compressors, lubricant is required to provide a seal between the sliding vanes and cylinder wall.

Screw Compressors

Screw compressors are both dry screw type and oil flooded type. In dry-screw-type machine, the two screws do not come in contact, but are geared to each other. These two screws counterrotate on two separate shafts in a single housing. One screw has a helical convex cross section, and the other has a helical concave cross section. When the two shafts rotate, volume progressively becomes smaller and gas gets compressed. In oil-flooded screw compressors, the two screws come in contact with each other and do not require timing gears. Air or gas enters at the *nonmeshing* point and is compressed by the contraction of the inner lobe space. The oil along with the compressed gas is delivered from the outlet, and oil separation step is necessary in this design. Oil does the role of both sealing and cooling functions in oil-flooded machines. Screw compressors are of compact design and provide continuous vibration-free air/gas.

Lobe or Root Compressors

These have two symmetrical lobed rotors (∞ shaped) in housing and connected by external gears. The counterrotating lobes do not come in contact with each other and therefore do not require internal lubrication. Only the gears and bearing need lubrication. These compressors thus provide large-volume and oil-free gas at moderate pressures.

Liquid-Ring Compressors

In these compressors, the rotor has fixed vanes and runs on a hollow shaft. The elliptical casing contains water, which forms a trapped *ring* moving in and out of the intervane spaces, thus creating two compression cycles per revolution. There are holes between the roots of the vanes that sweep past ports in the hollow shaft for the entry and exit of air.

CLASSIFICATION AND SPECIFICATIONS OF COMPRESSOR OILS

There are two well-known classifications of compressor oils, one by DIN 51506 and the other by ISO 6743-3A (1987) family D. DIN 51506 classifies reciprocating compressor oils into five categories: VB, VBL, VC, VCL, and VDL with different performance levels. L stands for additive-treated oils. Oils meeting VB and VBL standards can be used up to a maximum compressed air temperature of 140°C. VC and VCL oils can be used up to 160–180°C in compressors with storage tanks and up to 220°C in compressors on moving equipment for brakes, signals, etc. VDL grades are highest-performance oils and can be safely used up to 220°C in all types of reciprocating compressors.

ISO 6743-3A (1987) family D classifies compressors into six categories, but has not defined the performance characteristics. ISO/DP 6521 provides performance standard for reciprocating compressor oils L-DAA and L-DAB and for rotary screw compressors L-DAH and L-DAG. ISO reciprocating compressor specification has been based on DIN 51506 standard. In the absence of complete specifications for all types of compressors, OEM recommendation and their oil standards are to be followed. For reciprocating compressors, DIN 51506 standard is followed widely. DIN 51506 (Table 9.1, Table 9.2, and Table 9.3) stipulates that lubricants proposed for use in reciprocating compressors should pass certain laboratory tests that simulate the operational conditions encountered in the discharge systems of these compressors. In this specification, several levels of performance are defined, as determined by the compressor application and the air discharge temperature. The table shows that the most severe category VDL grades relate to discharge temperatures up to 220°C. Each category has a range of viscosity grades (VG) (e.g., VDL 22, 32, 46, 68, 100, and 150). The main feature of the standard is the inclusion of an oxidation test known as Pneurop oxidation test or in short known as POT. These stipulations include the maximum level of evaporation loss and carbon residue formation. ISO has also issued a draft standard (ISO/DP6521) for ISO L-DAA, DAB, DAH, and DAG categories. These are enclosed in Table 9.4 and Table 9.5.

SAE has classified synthetic compressor oils based on esters and polyalpholefins. Six categories of ester-based (DEA, DEB, DEC, DEG, DEH, and DEJ) and six categories of polyalphaolefin (PAO)-based products (DPA, DPB, DPC, DPG, DPH,

TABLE 9.1 Air compressor lubricant standard DIN 51506

	Compressed air temperature (°C max.)	
Oil classification	For compressors on moving equipment for brakes, signals, and tippers	For compressors with storage tanks and pipe network systems
VDL	220	220
VC and VCL	220	160[a]
VB and VBL	140	140

[a]Some types of compressors up to 180°C with VCL or engine oils.

TABLE 9.2 Air compressor lubricant standard DIN 51506: VE, VBL, VC, and VCL grades

Lube oil group	VB and VBL									VC and VCL				
ISO viscosity class (DIN 51519)	ISO VG 22	ISO VG 32	ISO VG 46	ISO VG 68	ISO VG 100	ISO VG 150	ISO VG 220	ISO VG 320	ISO VG 460	ISO VG 32	ISO VG 46	ISO VG 68	ISO VG 100	ISO VG 150
Kinematic viscosity min. (DIN 51561) At 40°C (mm²/s, max.)	19.8–24.2	28.8–35.2	41.4–50.6	61.2–74.8	90–110	135–165	198–242	288–352	414–506	28.8–35.2	41.4–50.6	61.2–74.8	90–110	135–165
At 100°C (mm²/s, min.)	4.3	5.4	6.6	8.8	11	15	19	23	30	5.4	6.6	8.8	11	15
Flash point (°C, (COC) min.)	175		195		205	210	255		255	175	195		205	210
Pour point (°C, max.)	−9					−3	0			−9				−3
Ash (% m/m, max. (DIN 51675))	VB and VC: 0.02 oxide ash VBL, VCL sulfated ash to be specified by the supplier													
Water-soluble acids	Neutral													
TAN (mg KOH/g, max.)	VB and VC: 0.15 VBL, VCL to be specified by the supplier													
Water (% mass)	0.1 max.													

(Continued)

TABLE 9.2 (Continued)

Lube oil group	VB and VBL			VC and VCL
% mass CRC max. after air aging (DIN 51352 part 1)	2.0	2.5	1.5	2.0
% mass CRC max. of 20% distillation residue (DIN 51356)	Not required		0.3	0.75

Note: Grades VB and VC are mineral oils; grade VDL contains additives to increase aging resistance. Grades VBL and VCL are additive-treated oils.

TABLE 9.3   Air compressor lubricant standard DIN 51506: VDL grades

Lube oil group	VDL				
ISO VG	ISO VG 32	ISO VG 46	ISO VG 68	ISO VG 100	ISO VG 150
Kinematic viscosity (DIN 51561) at 40°C (mm²/s max.)	28.8–35.2	41.4–50.6	61.2–74.8	90–110	135–165
At 100° (mm²/s, min.)	5.4	6.6	8.8	11	15
Flash point (°C, (COC) min. (ISO 2592))	175	195		205	210
Pour point (°C max. (ISO 3016))	–9				–3
Ash (% m/m, max. (DIN 51575))	Sulfated ash to be specified by the supplier				
Water-soluble acids (DIN 51558)	Neutral				
TAN (mg KOH/g max. (DIN 51558 part 1))	To be specified by the supplier				
Water, % mass (ISO 3733)	0.1 max.				
% Mass CRC max. after air aging (DIN 51352 part 1)	Not required				
% Mass CRC max. after air/ Fe$_2$/O$_3$ aging (DIN 51352 part 2)	2.5		3.0		
% Mass CRC max. of 20% distillation residue (DIN 51356)	0.3				0.6
Kinematic viscosity at 40°C max. of 20% distillation residue mm²/s (DIN 51356)	Maximum of five times the value of the new oil				

TABLE 9.4   Air compressor lubricant standard ISO/DP 6521 Draft 1983 mineral oil-based lubricants for reciprocating compressors

Category	ISO-L-DAA					ISO L-DAB					Test method
VG	32	46	68	100	150	32	46	68	100	150	ISO 3448
Kinematic viscosity At 40°C (mm²/s, +10%)	32	46	68	100	150	32	46	68	100	150	ISO 3104
At 100°C, mm²/s	To be stated					To be stated					
Pour point (°C, max.)	–9					–9					ISO 3106
Copper corrosion (max.)	1b					1b					ISO 2160
Rust	No rust					No rust					ISO/DP 7120A

(Continued)

TABLE 9.4 (Continued)

Category	ISO-L-DAA	ISO-L-DAB		Test method
Emulsion characteristics Temperature (°C)	No requirement	54	82	ISO/DP 6614
Time (min) to 3 ml emulsion (max.)		30	60	
Oxidation stability after aging at 200°C	15	No requirement		ISO/DP 6617 part 1
Evaporation loss (%, max.)				
Increase in Conradson carbon residue (%, max.)	1.5	2.0		
After aging at 200°C	Not applicable	20		ISO/DP 6617 part 2
Evaporation loss, %, max.				
Increase in Conradson carbon residue (%, max.)		2.5	3.0	ISO/DP 6616 with ISO/DP 6615 and ISO 3104
Distillation residue (20% vol) Conradson carbon Residue, %, max.	Not applicable	0.3	0.6	
Ratio of viscosity of residue to that of new oil (max.)		5		

TABLE 9.5 Air Compressor Lubricant Standard ISO/DP 6521 Draft 1983-mineral oil-based lubricants for rotary screw compressors

Category	ISO-L-DAH					ISO-L-DAG					Test method
VG	32	46	68	100	150	32	46	68	100	150	ISO 3448
Kinematic viscosity at 40°C (mm²/s, +10%)	32	46	68	100	150	32	46	68	100	150	ISO 3104 (IP-71)
Pour point[a] (°C, max.)	−9					−9					ISO 3106 (IP-15)
Copper corrosion (max.)	1b					1b					ISO 2160 (IP-154)
Rust	No rust					No rust					ISO/DP 7120A (IP-135A)
Emulsion characteristics[b]											
Temperature, °C time (minutes) to 3 ml emulsion max.	54		82			54			82		ISO/DP 6614 (ASTM D-1401)
	30					30					

(*Continued*)

TABLE 9.5   (Continued)

Category	ISO-L-DAH	ISO-L-DAG	Test method
Foaming characteristics sequence I			ISO/DP 6247 (IP-146)
Tendency (ml, max.)	300	300	
Stability (ml, max.)	Nil	Nil	
Oxidation stability	To be decided	To be decided	To be established
Evaporation loss (%, mass) Increase in viscosity (%) Increase in acidity (%) Sludge (wt.%)			

[a]When VG32 or VG46 oils are used in cold climate, pour points lower than –9°C are required.

[b]Only in those applications where condensation of atmospheric moisture is a problem. Where this does not apply, oils with dispersant additives, which tend to have poor water-separating properties, may be used satisfactorily.

TABLE 9.6   Key performance characteristics of SAE MS 1003-2 synthetic compressor oil specification

Properties	Ester-based oils		PAO-based oils		
	DEA, DEB, DEC	DEG, DEH, DEJ	DPA, DPB	DPC	DPG, DPH, DPJ
ISO Vis, grade	VG 32 to VG 150		VG 32 to VG 100		
Demulsibility D-1401	—	40-37-3	—	—	40-40-0
D-2711 Water in oil, 5 h Emulsion after centrifuge Total free water	Report		Report		≤1% ≤2 ml ≥60 ml
Oxidation D943 Hrs to 2 TAN	1000, 1500, 2000	1000, 1500, 2000	2000, 3000	4000	2000, 3000, 4000
Autoignition temp. (°C)	380 min.		No requirement		
Four-ball wear scar dia. mm at 40 kg		≤0.4		≤0.4	
CCR % of base oil	0.05		—	—	—
Thermal stability ASTM D-2070 Comparative IR Change in TAN Viscosity change Sludge (mg/100 ml) Cu rod color Cu wt. loss mg Steel rod color (CM)	Report 0.15 ≤5% ≤25 ≤5 ≤10 1 max.		Report 0.15 ≤5% ≤25 ≤5 ≤10 1 max.		

and DPJ) have been identified in SAE MS 1003-2 specification (2004). The specification is characterized by ASTM D-943 oxidation test and a thermal stability test according to ASTM D-2070 test procedure. Behavior toward rubber seals has also been specified. TOST life (to 2 TAN) of 1000–2000h for ester-based products and 2000–4000h for PAO-based products has been specified (Table 9.6).

Major OEMs have their own specifications of compressor oils requiring specific tests on antiwear performance, cleanliness levels, seal compatibility, and autoignition temperature.

FUNCTIONS OF COMPRESSOR OILS

Compressor lubrication was not properly understood till 1970s, and these were lubricated by oils developed for other applications such as engine oils, hydraulic oils, and turbine oils. There were several cases of compressor fires in high-pressure compressors in Europe. Systematic investigation by European committee of manufacturers of compressors and vacuum pump (Pneurop) during the 1960s led to the conclusion that fires and explosions in reciprocating compressors are most likely to occur when oil-soaked carbonaceous deposits are present and sufficient heat is available to cause spontaneous ignition of this deposit. Carbon deposits are also responsible for reduced heat transfer leading to enhanced air temperature rise. Thus, it became necessary to control deposit formation in reciprocating compressors. A new test method (DIN 51352 part 2) was developed to evaluate deposit-forming tendencies of compressor oils and incorporated in DIN 51506 specification and subsequently in ISO standards. This test is also known as Pneurop oxidation test or in short POT. To meet this test requirement, it is necessary to use a high-temperature antioxidant and carefully selected base oil. This understanding of compressor oil solved the frequent fire and explosion hazard in reciprocating compressors.

LUBRICATION OF RECIPROCATING COMPRESSORS

Conditions in reciprocating-type piston compressors are to some extent similar to the internal combustion engine. Oil has to lubricate bearings, pistons, and piston rings. The major difference is, of course, the absence of fuel and its combustion products in a compressor in the path of oil. The other contaminants such as moisture, airborne dirt, and oil degradation products are broadly similar. In certain high-pressure compressors, the temperatures could, however, be higher, and compressor oils are required to be protected from thermal and oxidative degradations. Problems in reciprocating compressors are encountered due to lacquer formation on rings leading to loss of efficiency and carbonaceous deposits on valves, coolers, and pulsation dampers leading to fires and explosions. Oils are thus required to provide the following functions in a compressor:

Provide liquid seal

Remove heat of compression and act as a coolant

Provide rust protection

Protect moving metallic components from wear

Resist oxidation in the presence of high-pressure air and high temperatures

Minimize deposit buildup at the discharge valves and other places

Keep oil and air separators clean

COMPRESSOR OIL PROPERTIES

Viscosity is the most important primary property of the oil to keep the moving parts apart, and yet it should be fluid enough to minimize viscous drag. Oil viscosity is determined by the compressor design and its operating parameters of temperature and pressure. A high-pressure unit requires high viscosity although high-viscosity oils tend to form more carbon deposits, provide higher drag, but minimize blow by losses. The final viscosity choice is a compromise between these factors. Low-viscosity oils on the other hand are more energy efficient and also tend to form lower deposits. Earlier, when low-viscosity-index oils (naphthenic) were available, these were preferred in compressor oils due to their low deposit-forming properties. Currently, high-viscosity-index oils of API group I or II types of base oils are preferred, and their properties are suitably modified by the use of chemical additives. The DIN 51506 and ISO standards also provide restriction on the choice of base oils. These standards demand that a 20% distilled residue of the oil should possess not more than 0.3–0.6% carbon residue in DIN VDL and ISO DAB types of oils.

Oxidation and Carbon Deposit Formation

This is the most important property of reciprocating compressor oil since the oil in these compressors is subjected to severe oxidative and vaporizing conditions in the air discharge system leading to the deposit formation. Thin film of oil or oil droplets come in direct contact with pressurized air in the presence of metal surfaces at high temperature. The oxidation of oil leads to viscosity increase and ultimately producing tarry residue and oil coke. In reciprocating compressors, some oil is also carried forward along with the air to the discharge system. Bigger oil droplets tend to settle down at the hot walls of exhaust valve and continue to oxidize in an exothermic manner, thus providing conditions that favor spontaneous ignition. This condition can eventually lead to compressor fires and explosion. This aspect has been thoroughly investigated [1–4] by several authors and concluded that the absence of carbonaceous deposits will reduce the risk of compressor fires and explosions considerably in reciprocating compressors. Thus, the reciprocating compressors can be lubricated by an oil meeting DIN 51506 VDL specification. However, in high-pressure multistage reciprocating compressors, mineral oil-based lubricants are not adequate to lubricate the last stages of the unit where temperatures and pressures are quite high. In such cases, the last one or two stages are lubricated by a fire-resistant phosphate ester-based fluid. Phosphate esters have natural antiwear properties also, and therefore, no additional antiwear additive is required in PE fluids.

The problem of deposit formation in reciprocating compressors needs specific oil formulation to address two main tests provided in DIN 51506 specification. These are carbon residue of oil after aging at 200°C in the presence of iron oxide as catalyst and viscosity and carbon residue of the 20% residue after distillation. In order to meet these two requirements, it is necessary to select base oils having narrow boiling range (so that heavy oils are avoided) and not to blend the products with very low and very high viscosities. This can control carbon residue of the distilled residue. The oxidation test requirements at 200°C can be met by selecting high-temperature antioxidants based on amines such as alkylated diphenylamine and others [5, 6]. Mixture of several antioxidants would be useful in meeting oxidation test. Most hindered phenol-based antioxidants do not provide adequate protection at high temperatures. Compressor oils requiring antiwear performance may contain additional antiwear additive based on phosphate ester and others. It is necessary that the antiwear additive is also fairly stable [7].

LUBRICATION OF ROTARY COMPRESSORS

The sliding vane and screw-type rotary compressors also offer highly oxidizing atmosphere for oil, where it is sprayed into the hot compressed air chamber. Rotary compressor oils should have high oxidation stability in addition to good water-separating and antirust characteristics. The normal rust- and oxidation-inhibited oils of turbine and hydraulic type usually do not provide long service life since these are formulated with antioxidants such as phenolic type that only provide protection at lower temperatures. Several OEMs also demand antiwear protection and can be provided by incorporating an antiwear additive such as ZDDP or phosphate ester. Synthetic oils based on PAO or ester provides longer life in these compressors. Rotary piston compressors are lubricated by direct oil injection or by a total loss system, and the lubrication is similar to the reciprocating piston compressors. Sufficient quantities of oil are injected so that the air outlet temperature does not exceed 110°C. DIN 51506 VCL- or VDL-type oils are sufficient to lubricate rotary piston compressors.

Screw compressors are also lubricated by oil injection. The oil also lubricates bearings and the rotating rotors of the screw. Oil quantity is adjusted so that the exhaust air temperature does not exceed about 100°C. In rotary compressors, lubricant is repeatedly heated, cooled, and vigorously mixed with hot and wet air. The oil must therefore be quite stable under these operating conditions. Synthetic oils based on PAO or hydrotreated mineral oils of high viscosity index have been used with advantage in these compressors, providing longer oil life as compared to mineral oil [8–10]. Some of the OEM requires that the oil should have good antiwear/EP properties as evaluated by four-ball wear tester and FZG machine. Two-stage hydrotreated mineral base oils have been found to be effective compressor oils similar to synthetic oils [11].

LUBRICATION OF DYNAMIC OR TURBO COMPRESSORS

Oil usually does not come in contact with compressed air in these compressors since only the bearings on which the rotor is mounted is required to be lubricated. Usually, a simple rust- and oxidation-inhibited oil is sufficient for these applications. The machines on other than air service, however, would need specific oil to suit the requirement. For example, when hydrocarbon gases are compressed, these gases could leak into the oil and make it thinner. For such application, oil-insoluble glycol fluids can be used, or the higher-viscosity mineral oil-based products would be required to be changed or reclaimed frequently due to dilution by gases. Special oil may be needed for corrosive or reactive gases such as hydrogen sulphide and sulfur dioxide etc.

SYNTHETIC COMPRESSOR OILS

Synthetic oils are preferred when mineral oil-based products cannot meet performance characteristics and compact high-performance compressors offer ideal environment for their applications. As the operating temperature and pressure increase in the compressors, life of mineral oil is reduced. The use of synthetic oils increases the life of oil considerably, sometimes 10-fold. The following types of synthetics are used in compressors:

PAOs

Organic diesters and polyolesters

Phosphate esters which are also fire resistant

Polyalkylene glycols (PAGs)

Chlorofluorocarbons (CFCs), mainly for oxygen compressors

PAO-Based Compressor Oil

PAOs are saturated branched paraffinic hydrocarbons prepared from the oligomerization of α-olefin (usually α-decene or a mixture of other α-olefins) and are, therefore, chemically and thermally stable molecules. These have low pour point, high viscosity index, good oxidation stability and hydrolytic stability. Compressor oils can be formulated with PAOs using similar additive approach used for formulating mineral oil-based lubricants. However, care should be taken to ensure adequate solubility of certain additives such as benzotriazole and other high-molecular-weight antioxidants in PAOs. The paraffinic nature of PAO limits solubility of polar additives and its rubber seal swell characteristics. Both these properties can, however, be improved by the incorporation of 10–20% of diester or polyolesters. PAO when combined with a rust inhibitor, antioxidant, metal deactivator, and antiwear additive provides good long-life compressor oil [12, 13]. There has been intensive activity in developing synthetic compressor oils for different applications [14–20], and large number of papers have been published on this subject.

Organic Esters (Diesters and Polyolesters)

Esters are characterized by their high-temperature stability (especially polyolesters) and high solvency power. This property is utilized in formulating high-performance reciprocating and rotary compressor oils. These oils minimize the buildup of carbon deposits on discharge valves and hot pistons and reduce oil filter blocking.

SAE standard for synthetic compressor oil covers PAO- and ester-based lubricants. Key performance characteristics are provided in Table 9.6. Both R&O and antiwear-type oils have been specified. Oils are also characterized by high oxidation stability and thermal stability.

Phosphate Ester-Based Compressor Oils

These are trialkyl/tri-aryl phosphate esters having excellent fire-resistant, antiwear, antioxidant, and thermal stability properties and find applications where high temperatures and pressures in the compressor pose fire and explosion hazards. Similar phosphate esters are also used in fire-resistant hydraulic fluids. The technology to formulate both hydraulic fluids and compressor oils based on phosphate esters is similar except that the compressor oils may need higher antioxidant level to support higher-temperature operations. Due to their fire-resistant properties, these are used in high-pressure multistage compressors [21, 22]. High phosphorous content in the oil may create disposal problems, and special disposal measures have to be adopted.

Polyalkylene Glycols

PAGs are an interesting class of compounds that could be water soluble or oil soluble depending on the molecular structure and are thus used in a variety of applications ranging from coolant to metalworking fluids and hydraulic and compressor oils. These are prepared by the reaction of ethylene oxide or propylene oxide with alcohols or organic acids. A mixture of EO and PO in varying ratios can be used to produce PAG of desired characteristics. Higher ratio of ethylene oxide leads to water solubility. The strong C–O bond imparts typical solubility behavior different from mineral oil and is hygroscopic. The viscosity of PAGs can be adjusted during their production, and any desired viscosity can be obtained. High-molecular-weight and high-viscosity PAGs are used in the hydrocarbon compressors, such as in the multistage high-pressure ethylene compressors (used for the production of polyethylene).

Chlorofluorocarbons for Oxygen Compressors

Hydrocarbon-based lubricant cannot be used in oxygen compressor cylinders due to explosion hazards. Water or water–glycerine has been used for lubricating these compressors. High-pressure oxygen compressors are lubricated by CFC-based lubricant. Mineral oils can, however, be used for separate compressor drives, where it does not come in contact with the compressed oxygen.

VACUUM PUMP OILS

There are many industrial applications of vacuum such as in production of X-ray tube, television picture tube, solid-state power devices, integrated circuits, and lamps that involves thin film, plasma, evaporation, vacuum baking, and low-pressure chemical decomposition by pyrolysis. Any space with lower pressure than atmospheric pressure is called vacuum. Vacuum is created by taking out air or gas molecules from that space. If the compressor inlet is connected to that space, it will start taking out gas from the space. Thus, vacuum pumps are in fact compressors where the inlet is connected to the space where vacuum is required to be created. Therefore, good compressor oils can be used to lubricate vacuum pumps. DIN 51506 VDL grade lubricants make good vacuum pump oils for moderate vacuum. However, when the vacuum is very high, two requirements become important.

1. Oils must have high initial boiling point and should have narrow boiling range.
2. Oil vapor pressure should be very low or oil volatility should be low
3. Good oxidation and thermal stability
4. Compatibility with seal/elastomers

If these requirements are not met, high vacuum cannot be created, since the oil itself starts evaporating and losing low-boiling fractions. For high vacuum, the following synthetic oils are preferred:

1. PAOs
2. Polyolesters and phenyl esters
3. Silicone fluids

Any additive used in vacuum pump oils should also have high boiling point. For example, low-boiling phenolic antioxidant or additive containing low-viscosity mineral oils as diluent should not be used for blending vacuum pump oils.

REFRIGERATION COMPRESSOR OILS

Refrigeration is the heat removal process from a material that is required to be cooled. The most widely used method of heat removal is by vapor compression cycle, which is based on the fact that a gas that is at a temperature below its critical temperature can be liquefied by applying suitable pressure and on removing the pressure, the liquid can again be converted to gas or vapor. The heat of vaporization is utilized for cooling. Refrigerants having suitable properties are ammonia, carbon dioxide, propane, isobutene, and low-molecular-weight chloro- and/or fluorocarbons. Ammonia and chloro-/fluorocarbons are the most common refrigerants used in the industry. A refrigeration system is a closed circuit in which a refrigerating fluid is circulated (Fig. 9.1).

A simple refrigeration circuit consists of four elements linked together (compressor, condenser, evaporator, heat exchanger). The compressor circulates the fluid

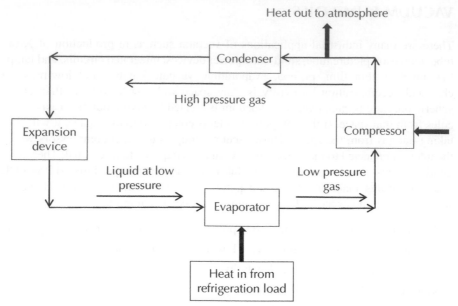

FIGURE 9.1 Typical refrigeration system.

vapor from the evaporator to the condenser. Gas is compressed and transmitted to the condenser. In the condenser, compressed gas is liquefied, and the heat of condensation is transmitted to outside environment. The expansion valve maintains the difference of pressure between the condenser and the evaporator. In the evaporator, the cooled liquid evaporates, absorbing heat of vaporization from outside, and thus provides a cooling effect.

Oil that lubricates the compressor comes in contact with the refrigerant and is thus carried away by the refrigerant in which it is present in varied proportions. The oil in turn returns to the compressor through the evaporator. The refrigeration compressor oil is thus required to be soluble in the refrigerant at all temperatures of the evaporator, which is the coldest part of the system.

REFRIGERANTS

Refrigerants form an important part of the compressor, and the choice of lubricants depends on them. The earliest known refrigerant is ethyl ether, which was used in 1748. In 1869, ammonia found commercial application. Following these developments, soon, carbon dioxide, sulfur dioxide, methyl chloride, ethyl chloride, and isobutene were utilized for refrigeration. In the early twentieth century, CFCs were found to have excellent properties for refrigeration applications. These became so popular that worldwide large quantities of CFCs were manufactured.

Some of the important CFCs, hydrochlorofluorocarbons (HCFCs), and hydro-fluorocarbons (HFCs) are as follows:

R12 $\longrightarrow$ Dichlorodifluoromethane

R22 $\longrightarrow$ Chlorodifluoromethane

R123 $\longrightarrow$ Dichlorotrifluoroethane

R134a $\longrightarrow$ 1,1,1,2-Tetrafluoroethane

R125 $\longrightarrow$ 1,1,1,2,2-Pentafluoroethane

R32 $\longrightarrow$ Difluoromethane

R143a $\longrightarrow$ 1,1,1-Trifluoroethane

R152a $\longrightarrow$ 1,1-Difluoroethane

CFCs (R12) have the highest potential to deplete ozone layer, and HFC has the least potential out of these products. R134a has been favored for refrigeration following Montreal protocol. However, the natural refrigerants like CO_2, hydrocarbons, and ammonia have zero ozone-depleting potential and would be favored for large industrial applications. Even R22 (HCFC) is about 20 times better in ozone-depleting property. In the current scenario, R22, R134a, and ammonia are the favored refrigerants. Various refrigerants are described in American Society of Heating, Refrigerating, and Air-Conditioning Engineers (ASHRAE) 34-1992 and DIN 8960.

REQUIREMENT OF REFRIGERATION OILS

The performance of the lubricant that is in intimate contact with the refrigerant under varying conditions of pressure and temperature plays a greater role in the choice of the lubricant. These lubricants thus need some important properties in addition to the normal properties required for the general compressors since the oil is subjected to both low and high temperatures.

Refrigeration compressor oil must have the following important properties [23, 24]:

1. Low pour
2. Low floc point (indicative of low wax content) with refrigerant
3. High dielectric strength
4. Compatibility with different refrigerants
5. High chemical and thermal stability
6. Adequate viscosity
7. Act as a coolant to remove heat (have good thermal conductivity)
8. Act as an oil seal
9. Provide energy efficiency

The lubrication of compressors for refrigeration and air-conditioning generally proceeds through hydrodynamic lubrication depending upon lubricant pressure, velocity, and viscosity. However, during the compressor start-up, shutdown, and overloading conditions, boundary lubrication regime may be encountered. Refrigeration compressor lubricants are based on either mineral oil or mineral oil synthetic blends or pure synthetic oils.

MINERAL OILS

Refrigeration oils are mainly required for closed compressor units (hermetically sealed) and are known to provide longer service life with less maintenance. Highly refined mineral oils similar to white oils have been commonly used for refrigeration applications. These lubricants may vary in their physical properties, chemical structure, and degree of refining, all of which influence performance in refrigeration applications. Linear or straight-chain paraffins are generally undesirable due to their precipitation as wax crystals at low temperatures, which cause clogging of valves and lines resulting in reduced heat transfer. Branched-chain paraffins have good viscosity retention at high temperature and better low-temperature fluidity. Naphthenes or cycloparaffins have good low-temperature characteristics. All of these components are saturated hydrocarbons that have good chemical stability, higher aniline point, low specific gravity, low refractive index, and higher molecular weight.

Mineral Oil: Synthetic Blends

Synthetic oils are sometimes blended with mineral oils to improve their overall performance. Mixtures of alkyl benzene and mineral oil blends have been successfully used in refrigeration oils. These alkyl benzenes are obtained from linear alkyl benzene production plants as by-products and have very low pour points. These are also fully miscible with mineral oils.

SYNTHETIC OILS

Mineral oil-based products have certain limitations such as low solubility in R13, HFC, CO_2, and R502 refrigerants. Alkyl benzenes have been found to possess improved performance in solubility characteristics. Solubility data of large number of synthetic oils have been characterized and reported [25, 26] and is useful in selecting appropriate lubricant.

The use of synthetic oils for refrigeration was first proposed in 1929. These were considered for their ability to solve problems experienced with the use of mineral oils such as wax separations, low miscibility with some refrigerants, and carbonization of valves in reciprocating compressors. Alkyl benzenes, polyvinyl ether, PAG, PAOs, and polyolesters have been utilized for several applications with benefits. Synthetic lubricants offer a wide range of the following desirable properties for any refrigeration system:

- Better viscosity/temperature characteristics resulting in improved hydrodynamic lubrication of compressor bearings

- Good low-temperature properties
- Improved lubricity in the presence of refrigerants
- Improved stability in the presence of refrigerants at high temperature
- Reduction in power requirements for cold temperature starting by providing lower cold-temperature viscosity while maintaining the viscosity required at higher operating temperature
- Better sealing in compression chamber leading to improved efficiency
- Additional energy savings through improved heat transfer in refrigeration system

SPECIFICATIONS OF REFRIGERATION OILS

DIN 51503: I describe the minimum requirements of refrigeration oils. Oils have been classified into the following six categories:

1. KAA—oils not soluble in ammonia
2. KAB—oils soluble in ammonia
3. KB—oils for carbon dioxide
4. KC—oils for chlorinated and fluorinated hydrocarbons (CFC, HCFC)
5. KD—oils for fluorinated hydrocarbons (FC, HFC)
6. KE—oils for hydrocarbon refrigeration like propane and isobutene

KAA oils are generally naphthenic mineral oil, mineral oils with alkyl benzene, and PAOs. Some hydrogenated mineral oils are also used in this application.

KAB oils are usually PAGs with low water content level.

KB oils for CO_2 compressors are polyolesters and PAGs. PAGs have limited solubility in carbon dioxide, and PAOs are not miscible.

KC oils for CFC and HCFC are low-pour naphthenic mineral oils and alkyl benzene blends.

KD oils for fully fluorinated hydrocarbons (HFC and FC) are generally polyolesters and PAG with low hygroscopic nature.

KE oils for hydrocarbons such as propane and isobutene are mineral oils, PAOs, and alkyl benzene, which are completely miscible. Polyolesters and PAG can also be used with low water content in them.

Refrigeration oils require special testing facilities to evaluate their compatibility with the refrigerant. Following environmental regulation, R134a (CH_2FCF_3) is becoming the choice refrigerant for vehicle air-conditioning. For such applications, PAO or polyglycol- [27] or polyolester-based oils are recommended. Refrigerant miscibility is tested by DIN 51514 and refrigerant compatibility with R134a by ASHRAE 97/1999 sealed tube test methods. DIN 51503-2 and ASHRAE 34-1992 describe criteria for evaluating used refrigeration oils.

Multipurpose compressor oil for refrigeration, hydraulic, metalworking, and heat treating applications has been based on a specific base oil with certain Ca, Cp, and Cn values. In order to obtain sufficient thermal and oxidation stability in conventional rotary gas compressor oils, a large amount of antioxidants are added, which sometimes lead to the formation of sludge. This sludge may adhere to the bearing of the rotary gas compressor and cause clogging and damage of the bearing. A specific antioxidant based on p-branched-chain-alkylphenyl-alpha-naphthylamine has been suggested, which solves these problems [28].

REFERENCES

[1] Evans EM, Hughes A. Lubrication of air compressors. Proceedings of the International Mechanical Engineering; September 1969–1970; London, 184, pt. 3R paper 10, 93–105.

[2] Smith AC, Thomas A. Fire and explosion hazards. J Inst Prod Eng August 1958;37(8):482.

[3] Thoenes HW. Safety aspect for selection and testing of air compressor lubricants. Lubr Eng 1975;25:409–411.

[4] Hampson DFG. Reducing the risks of fire in large reciprocating compressors systems. Wear 1975;34:399–407.

[5] Srivastava SP, Rao AM, Goel PK. Developments in air compressor lubricating oils. 4th International Seminar on Fuel, Lubricants and Additives; March 7–10, 1983; Cairo: Misr Petroleum Co.

[6] Srivastava SP, Rao AM, Rao GJ, Ahluwalia JS. Improved air compressor oil. Indian patent 151316. 1979.

[7] Senapati SK, Jaiswal AK, Mooken RT, Mishra AK, Srivastava SP, Bhatnagar AK. A new quality compressor oil with anti-wear performance. In: Srivastava SP, editor. Proceedings of the 2nd International Symposium on Fuels and Lubricant (ISFL), New Delhi; March 10–12, 2000. New Delhi: Allied Publisher. p 143–148.

[8] Mills AJ, Tempest MA, Thomas AA. Performance testing of rotary compressor fluids in Europe. Lubr Eng 1986;42 (5):278–286.

[9] Sugiura KM, Nakano H. Laboratory evaluation and field performance of oil flooded rotary compressor oils. Lubr Eng 1982;38:510–518.

[10] Miller JW. New Synthetic food grade rotary screw compressor lubricants. Lubr Eng 1984;40:433–436.

[11] Cohen SC. Development of synthetic compressor oil based on two stages hydrotreated petroleum base stocks. Lubr Eng 1988;44:230–238.

[12] Bell NJ, Nagakari M, Smith PWR, Bell NJ, Nagakari M, Smith PWR. Lubricating oil composition. US patent 2007,0129,268. June 2007.

[13] Negoro M, Kawata K. Lubricant composition and triazine containing compounds. US patent 2006,0068,997. March 2006.

[14] Arbocus G, Weber H. Synthetic compressor lubricants- state of the art. Lubr Eng 1978;34:372–374.

[15] Miller JW. Synthetic HVI compressor lubricants. J Syn Lubr 1989;6:107–122.

[16] Mathews PHD. The lubrication of reciprocating compressors. J Syn Lubr 1989;5:271–290.

[17] Glen D. Short, development of synthetic lubricants for extended life in rotary-screw compressors. Lubr Eng 1984;40:463–470.

[18] Miller JW. Synthetic and HVI compressor lubricants, J Syn Lubr 1989, 6, 2, 107–122.

[19] Miller JW. Synthetic lubricants and their industrial applications, J Syn Lubr 1984;1 (2):136–152.

[20] Legeron JP, Beslin L. Rotary vane compressors – some technical aspects of long life lubricants. J Syn Lubr January 1999;6(4):299–309.

[21] Landsdown AR. *High Temperature Lubrication*. London: Mechanical Engineering Publication.

[22] Chhatwal VK, Rao AM, Jayprakash KC, Srivastava SP. Synthetic phosphate esters for reciprocating compressors. Proceedings of the 7th LAWPSP Symposium; 1990; IIT Mumbai, India. Paper No. A 7.1–A 7.6.

[23] Lubricants in refrigeration systems. In: *ASHRAE Handbook, Refrigeration System and Application*. 1990.

[24] Kruse HH, Schroeder M. Fundamentals of lubrication in refrigeration system and heat pumps. ASHRAE J 1984;26:5–9.

[25] Sanvordenker, KS, Larime MW. A review of synthetic oils for refrigeration use. Symposium on Lubricants and Refrigerants, ASHRAE Annual Meeting; June 22–29, 1972; Nassau.

[26] Glen D. Short, synthetic lubricants and their refrigeration applications. Lubr Eng 1989;46 (4):239–247.

[27] Sundaresan SG. Status report on polyalkenyl glycol lubricants for use with HFC-134a in refrigeration compressors. Proceedings of the ASHRAE/Purdue CFC Conference; 1990; Purdue. p 138–144.

[28] Tagawa K, Shimomura Y, Sawada K, Takigawa K, Yoshida T, Mitsumoto S, Akiyama E, Shibata J, Suda S, Yokota H, Hata H, Hoshino H, Nakao H, Konishi S. Compressor oil composition, US patent 8,299,006. October 30, 2012.

PART III

GEAR OILS AND TRANSMISSION FLUIDS

CHAPTER 10

INDUSTRIAL AND AUTOMOTIVE GEAR OILS

Gears are important part of industrial and automotive equipment and offer most challenging environment for lubricants. The different designs of gears and their operating and tribological parameters (such as load and contact geometry) make each system unique, requiring a carefully formulated lubricant. In this chapter, both industrial gear oils and automotive gear oils are discussed together, although the lubricants used by each category are different. These oils do have separate standards and performance requirements due to the differences in operating parameters. The basic function of any gear is to transmit power and control speed. Large enclosed industrial gears have an oil sump, and oil is circulated through a centralized system. Industrial gears found in almost all industrial sectors are of two types, enclosed gears and open gears. Open gears are slow moving and are generally lubricated by a high-viscosity product having adhesive properties. Automotive gears are mobile on the vehicles and have to operate under a different set of conditions with variable load, speed, and torque. Gear lubrication was always a challenge ever since the discovery of gears by Aristotle in 384 BC. Lewis in 1886 first designed a gear tooth, and this was followed by the development of bevel and bevel-spiral gears. Initial gears were lubricated by vegetable oils or animal fats, but as the loads and speed increased, these were no longer suitable for lubrication. Need for a load-carrying additive was felt, and gears were the first to use an extreme pressure (EP) additive in lubricant based on lead soap and active sulfur. With the discovery of hypoid gears in 1925 for automobiles, which can carry high load quickly, lubrication again became a problem. In hypoid gear, the drive shaft is lowered, which can carry high loads and operate quickly. This also changes the vehicle profile and its center of gravity. When gears operate, gear teeth mesh with each other, and both sliding and rolling contacts take place. The degree of sliding, rolling, and contact geometry varies in different gears. This calls for specific lubrication requirements.

TYPES OF GEARS

The following broad categories of gears operate in the industry [1]. These gears have constant gear ratio.

Developments in Lubricant Technology, First Edition. S. P. Srivastava.
© 2014 John Wiley & Sons, Inc. Published 2014 by John Wiley & Sons, Inc.

1. *Spur Gears.* These have straight tooth, cylindrical shape, and parallel shafts with linear contact over the width of meshing teeth. In such gears, usually only one or two teeth mesh at a time, and therefore, their load-carrying capabilities are limited. Noise label is also relatively higher. The direction of sliding is at 90° to the line of contact.

2. *Helical Gears.* Like spur gears, these also have parallel shafts and cylindrical shape, but the teeth are either helical or curved. This shape of tooth increases contact area and is thus capable of carrying higher loads with lesser noise. The tooth shape is also responsible for generating axial thrust load, which requires specific bearing design. The thrust load can be eliminated in double helical gears or herringbone gears. In such gears, the direction of sliding is not at the right angle to the line of contact, and thus, some sliding takes place at the line of contact.

3. *Bevel Gears.* These gears are cut from a cone and their axes are at right angle to each other. Gear teeth mesh in a linear manner. The direction of sliding in these gears is at right angle to the contact line. According to the tooth profile, these are further subdivided into the following two classes.

    3a. *Spur-Bevel Gears.* Bevel gears having straight teeth like spur gears are called spur-bevel gears.

    3b. *Spiral-Bevel Gears.* Bevel gears having helical or curved teeth are described as spiral-bevel gears. These are capable of carrying much higher loads due to larger contact area.

4. *Worm Gears.* Worm gears are reduction gears and consist of a worm gear and a screw-type gear. Gear shafts are at right angle to each other. Worm gears experience very high sliding. In addition to sliding, rotation introduces high rate of side sliding. These conditions are responsible for high frictional losses and consequently higher operating temperatures.

5. *Hypoid Gears.* Hypoid gears are used in automotive vehicles for carrying higher loads at high speed and torque. These are actually modified spiral-bevel gears with offset axis. The axis of the pinion is below the axis of the ring gear. This arrangement allows greater contact area and sliding, resulting in smoother operation. The use of hypoid gears permitted the lowering of chassis to the ground and streamlined vehicle design. Engines with hypoid gears could now apply higher horsepower to the driveline. However, the demand on lubricant is much higher, and specially formulated oils are required to meet their requirements.

These are the basic gear system, and several other combinations and variations are available to meet the performance. In transmission systems, such as manual or stepless mechanical variable (constant variable transmission (CVT)) and torque converters, gears of variable ratio are utilized.

Gears operate under all lubrication regimes, that is, hydrodynamic, boundary, and elastohydrodynamic regions, depending upon the design and operating parameters [2–7]. Lubricants for hydrodynamic gears such as hydraulic clutch coupling and torque converters require good thermal and oxidation stability, but do not need antiwear and EP additives. However, heavily loaded industrial gears and automotive hypoid gears do need EP and antiwear additives in addition to other additives. Sliding and rolling contacts in hypoid gears also require high lubricant viscosity at the operating temperature. In the operation of a gear system, gear tooth engagement and disengagement take place continuously, making it a vibrating system, where the dynamic load distribution differs from the static load. In such a vibrating system, high lubricant viscosity is responsible for reducing the noise level. At the point of gear tooth engagement, lubrication conditions are favorable due to rolling movement. While at the disengagement point, high sliding action is present and metal-to-metal contact can take place. Thus, in each cycle of gear engagement and disengagement, lubrication film is regenerated and breaks down. In gear lubrication, sliding and rolling speed and contact pressures, loads at the tooth contact, and the rotational speed of the gear system are important parameters. This is the unique condition in a gear system and requires careful lubricant formulation generally different from other industrial equipment. Inadequate lubrication and gear tooth metallurgy can lead to gear failures such as tooth fracture, pitting, scuffing, and excessive wear. Detailed discussions on gear systems and their tribological aspects can be found in several publications [8–10].

Industrial gear units are of two types: enclosed and open. Open gear units are slow, with a pitch line speed below 150 m/min and adhesive oil is used to lubricate gear unit. Oil is not circulated in these open units, and it is a total loss or once-through operation. The enclosed gear units are more popular in industry covering a wide variety of loads and speeds. In enclosed gears, lubrication is through a centralized system with properly formulated EP gear oil.

TYPES OF GEAR OILS

The lubricant requirement of various gear systems described earlier is different depending on the operating gear speed, load factor, transmitted power, operating temperature, and method of application. Oil recommendation is therefore made after considering these factors and also OEM suggestions. Industrial gear oils are divided into three categories:

Rust- and Oxidation-Inhibited Oils

These are mineral oil-based products containing antioxidant, rust inhibitor, and anti-foam additive and are used as circulating oil to lubricate lightly loaded gears, not requiring EP/antiwear properties. Such oils also lubricate the bearing of the system and are characterized by oxidation test, rust test, foam test, and water separation tests. These oils are, for example, covered in DIN 51517 part 2 CL lubricating oil. Rust- and oxidation-inhibited turbine and hydraulic oils can also be used in gears if EP/antiwear properties are not required.

Compounded Gear Oils

Compounded oils are generally used in worm gear system (reduction gear) and are also based on mineral oil, antioxidant, fatty oil (to provide oil resistance to wiping and rubbing action), and antifoam additive system. Usually, 3–10% of fatty oil or synthetic fatty oil is used. Addition of antirust additives may not be necessary since the fatty material in addition to the oiliness/antifriction property also provides rust protection by forming a film on metal surfaces.

EP Gear Oils

EP gear oils are mostly formulated from mineral oils, but specific oils requiring high thermal stability and longer life may be formulated with synthetic oils (mainly polyalphaolefins (PAOs), polyalkylene glycol (PAG), and esters). Such oils are formulated with antioxidant, antirust, anticorrosion, antiwear, EP, metal deactivator [11], and antifoam additives. The EP property is improved by the incorporation of benzotriazole [12] or its derivatives to a mixture of sulfurized alkylene monomer and dihydrocarbyl phosphonate [13]. Antiwear turbine and hydraulic oils can also lubricate gears where the loads are moderate.

INDUSTRIAL GEAR OIL SPECIFICATIONS

Table 10.1 provides key performance properties of some of the important EP industrial gear oil specifications. There are, however, several other national and OEM standards for these lubricants. Most of the OEMs accept AISE (U.S. Steel) 224 or ISO 12925 standard, but some have their own test requirement. Flender is one of the major European OEMs of gears and has its own gear oil specification. The basic requirement in this specification is DIN 51517 part 3 with several additional tests on elastomer compatibility, foaming, FAG FE-8 roller bearing wear, and FZG. AGMA has issued detailed guidelines and specifications of gear oils [14–17].

Other standards on industrial gear oils that are sometimes referred by OEMs are as follows:

1. MAG Cincinnati machine gear lubricants: P-63, P-76, P-77, P-74, P-59, P-35, P-34, and P-78. These standards only specify Timken OK load 45 lbs as an EP test and rely more on CM thermal stability test. P-39 specification is for worm gears.

2. German steel industry SEB-181226—CLP-type gear oil standard.

3. GM-LR-104 gear oil standard.

Comparison of the above standards indicates that oils meeting AISE U.S. Steel 224 specification would generally satisfy other standards. Most of the oil developers, therefore, target this specification. Additive packages from additive manufacturers are also targeted to this standard. Specific Flender requirements can be met with supplement additives. Flender's requirements are provided in Table 10.2. They specify both mineral and synthetic oils. These requirements are such that a carefully selected base oil blend is required to meet elastomer test in combination with high-EP and antifoam properties.

The most important properties of industrial gear oils are its water separation characteristics, oxidation stability, and EP/antiwear characteristics. All modern industrial and automotive gear oils are now based on sulfur–phosphorous (S–P) additive chemistry with different chemical compounds and S–P ratios. The additive dosage in

TABLE 10.1   Key performance properties of EP industrial gear oil specifications

Property	ISO 12925-1, Cat. CKC	DIN 51517 part 3-CLP grades	AISE (USS)-224 Lead free EP	AGMA 9005-E02	David Brown S 1.53 101 EP type
ISO VG	32–1500	32–1500	68 and above	32 to >3200	Various grades
Rust test D-665 A and B	Pass	Pass, A method	Pass	Pass, B	Pass, A
Demulsibility D2711, free water Emulsion max. ml Water-in-oil max. ml.	80 ml min 1 2	**DIN ISO 6614** 30 min 60 min for VG 680 and above	80 ml min 1 2	D 2711—B 80 ml min 1 up to VG 320 and 4 for higher 2	60–50 1 2
Oxidation stability D-2893 at 95°C Vis. increase 100°C max. Precipitation no. max.	6 0.1	6 —	At 121°C 6	At 121°C 6 up to 320 grade 8 for 320 10 for 460 15 for 680 Report for higher —	10
FZG,A/8.3/90 Fail stage min.	12	12	11 pass	10 up to 46 grade 12 up to 150 and >12 for higher	11 up to 46 grade 12 up to 320 and >12 for higher grades
FAG FE-8 bearing wear, roller wear mg Cage wear	—	30 max. report	—	—	—
Timken OK load lbs, min.	—	—	60	—	—
Four-ball weld load kg Load wear index	—	—	250 45	—	—
Four-ball wear, 20 kg, 1800 rpm, 1 h, mm max.	—	—	0.35	—	—

TABLE 10.2 Additional requirements of Flender above DIN 51517 part 3

Characteristics	Method	Mineral oil	Synthetic oil
FZG gray staining test	FVA 54/7	≥10	≥10
FZG A/16.6/90		>12	>12
FZG A/8.3/90		>12	>12
FAG FE-8 roller bearing wear	DIN 51819-3	<30 mg	<30 mg
Cage wear		Report	Report
Compatibility with Mader and Loctile material	Mader and Loctile	Pass	Pass
Compatibility with elastomeric static, 72 NBR, 902, 168 hr.100°C and with 75FkH-585	DIN 53521		1008-h test
Shore hardness A		±5	±5
Volume swell		+5/−2	+5/−2
Tensile strength %		−50 to +20	−50 to +20
Elongation break %		−60 to +20	−60 to +20
Dynamic test, 72 NBR-902, 1000 h at 80°C and with 75 Flh-585	DIN 3761	To pass if limits exceed in static test	Pass
Flender foam test, original oil with 2% and 4% impurity of running-in-oil	Flender	15/10%	15/10%

industrial gear oils is, however, less than the automotive oils, since the latter requires higher level of antiwear/EP and thermal stability properties. Multipurpose automotive and industrial gear oil additive packages are also available from additive manufacturers with difference in treat level to meet the desired standard. These packages are based on EP/antiwear components such as sulfurized hydrocarbon (isobutene), phosphite, phosphate, phosphate ester, or thiophosphate. In addition, these contain antioxidant, antirust, and antifoam additives. The performance of the package depends on the chemical nature and balance between the active sulfur and phosphorous compounds as well as other additives added for oxidation stability, corrosion protection, friction modification, thermal stability, etc. Some specific additives may contain compounds based on nitrogen, boron, and molybdenum for energy efficiency.

GEAR FAILURES

Gear failures take place due to a variety of reasons, and some of the common ones are described. Improper gear metallurgy and lubrication can lead to several failures such as high wear, pitting, tooth fracture, scuffing, and scoring. All these gear failures can be adequately addressed by utilizing good gear metallurgy and properly formulated lubricant with balanced EP and antiwear characteristics. Slow and continuous wear in gear system is normal and leads to polishing, which will not affect the life of gears. Heavy abrasive or destructive wear can change the shape of tooth, resulting in noisy operation and reduced life of gears. Corrosive wear can also take place in the presence of acidic material and water.

Scoring takes place due to the failure of lubricating oil film, where metal-to-metal contact takes place due to heavy loads. This leads to removal of metal by tearing of welded spots. In certain cases, thin flakes of metal are removed from gear teeth. This generally takes place with softer steels or bronzes during the initial period of operation. Flaking is a surface fatigue phenomenon caused due to the sliding and rolling action. Pitting may also take place due to surface fatigue along the pitch line, but this could be corrected in due course of time when the high spots are smoothened. However, if the pitting increases in size, the tooth shape may be destroyed, and eventually, failure takes place due to fatigue breakage. When a larger particle of metal flakes out from the tooth surface, it is called spalling. This occurs in hardened steel at the top edges of the gear teeth. Severe pitting, spalling, or heavy abrasive wear can remove enough metal to weaken the gear tooth leading to its breaking point.

Ridging occurs on tooth surface of case-hardened hypoid pinions and bronze worm gears. It may appear as diagonal ridges in the direction of sliding. This is caused by higher loads and improper lubrication and if not controlled may lead to complete gear failure.

Plastic flow of metals can take place due to excessive loads/impact load and sliding action leading to the formation of fins at the edge of teeth and wavelike formation on the metal surface at right angle to the direction of sliding.

Gear failures can take place due to poor design, surface defects, overloads, or misalignment and unsuitable metallurgy. These conditions generate stresses beyond the tolerance limit of the metals, leading to the fatigue breakage.

PROPERTIES

Gear oils must have the following important properties to provide satisfactory performance.

Viscosity/Fluidity

Viscosity is the basic property of lubricating oil on which other properties are built. Under hydrodynamic lubrication conditions, this is the single important property of the lubricant, providing adequate oil film thickness. Viscosity changes with temperature, and proper oil film thickness at the operating temperature needs to be provided to protect the equipment at both ambient and operating temperatures. Higher-viscosity oils are capable of carrying higher loads. On the other hand, very-high-viscosity oils have poor fluidity, and such oils may cause oil starvation especially in colder climatic conditions. Proper oil viscosity also helps in removing heat, wear debris, and contaminants entering into the oil system.

Antiwear and EP Properties

As discussed earlier, gears operate under all lubrication regimes. Under boundary lubrication regimes, when the loads are higher, metal-to-metal contact can take place. The surfaces are then required to be protected by an EP additive. Gears also encounter severe conditions such as high loads at a varying speed, shock loads, vibrations, high temperatures, cold starts in winter, etc. These conditions require a well-balanced EP and antiwear properties in the gear oil that maintain a protective film between metal surfaces. Four-ball, Timken, and FZG tests are carried out to assess the EP/antiwear properties of industrial gear oils. Automotive gear oils are, however, assessed by a set of rig test described in the following pages.

Thermal and Oxidation Stabilities

Oxidation of oil at higher temperatures can lead to the formation of sludge and deposits. Viscosity will also increase due to degradation. Sludge and deposits clog oil passages, change heat transfer rates, and deplete additives' contents. Poor oxidation may thus reduce oil and equipment life. ASTM D-2893 method is used to evaluate oxidation stability of industrial gear oils. Automotive gear oils are evaluated by CRC L 60 method.

Demulsibility (Water Separation Characteristics)

Ingress of water in most industrial equipment is inevitable. This can cause poor lubrication and corrosion. Gear oils are, therefore, required to separate from water quickly. This property is known as demulsibility and is evaluated by ASTM D-1401 and ASTM-2711 tests. Specific additives can also be used to improve water-separating property of gear oils.

Foam Control and Air Entrainment

Foaming and air entrainment are caused by air ingress when gears run at high speed. This can cause wear, enhanced oxidation, and poor lubrication. A high amount of air in oil also reduces real oil flow, resulting in thinner oil film. Foaming characteristics are determined by ASTM D-892 foam test and IP-313 for evaluating air release properties.

Rust and Corrosion Protection

Ferrous corrosion or rusting is a common problem in the industry since the equipment do come in contact with water or moisture. Most industrial gear oil specifications require a pass rust test with ASTM D-665 A or B procedure. Oil should also not be corrosive to copper or brass. This property is evaluated by ASTM D-130 copper strip corrosion test.

TOXICITY AND SAFETY

Earlier gear oils contained lead as an antiwear additive, which is a known toxic material. However, lead has now been removed from the lubricant formulation. Modern gear oils are based on sulfur and phosphorous compounds. These compounds have unpleasant odor, and due care is required to be taken that the oil or its vapors are not inhaled. Proper ventilation is also required in the oil cellars.

GEAR OIL FORMULATION

Lightly loaded gears can be lubricated by simple base oil blends of desired viscosity with any antioxidant and a rust inhibitor. Moderate loads can be carried by compounded gear oil containing 3–10% fatty material and are recommended for worm gear. As the load further increases, need for EP/antiwear additives arises. The industrial gear oil technology underwent many changes in the past 70 years. In the early 1930s, gear oils used a combination of Pb–S compounds (lead naphthenate and a sulfur source in the form of a disulfide or sulfurized fat) and chlorinated paraffins to obtain EP/antiwear properties and were quite popular till the early 1970s. The ratio of lead to sulfur is important to realize adequate EP/antiwear properties. Lead and sulfur compound combination provided good antiwear and EP protection and has been used extensively in industrial gear oils for several years. These combinations, however, had poor water separation characteristics and only average oil life due to poor thermo-oxidative resistance. In the 1960s, more stable S–P-based additives were developed to achieve better antiwear, EP, and antioxidant properties. Modern S–P industrial gear oil contains antioxidant, EP, antiwear, antirust, metal deactivator, antifoam, and demulsifier additives. Oxidation stability can be imparted by the use of antioxidants such as hindered phenols, aromatic amines, and dithiocarbamates or by some EP additives themselves such as sulfurized olefin or sulfurized fat. EP and antiwear properties are achieved by the use of sulfurized olefins, sulfurized esters, thiophosphates, phosphates, phosphites, etc., as well as their combinations. Rust inhibition can be obtained with amines or amine salts, fatty amides, oxazolines, imidazolines, phosphoric acid esters, and carboxylic acid esters. Copper corrosion is usually prevented by the addition of metal deactivators or sulfur scavengers such as thiadiazoles, triazoles, and other heterocyclic compounds of nitrogen and sulfur. Demulsibility can be enhanced by the use of a wide range of demulsifiers that include polyethylene glycols, mixed polyethylene/propylene glycols, ethoxylated phenols, ethoxylated amines, carboxylic acids, and trialkyl phosphates. Foam control can be greatly improved with methyl silicone polymers or acrylate copolymers.

Currently, S–P-based additive package formulation is complex, and a suitable combination of additives may only satisfy both industrial and automotive gear oil requirements. Specific oils may also contain friction modifiers based on molybdenum [18] to impart energy efficiency to the oil. Slow-moving heavily loaded gears have also utilized solid lubricants such as molybdenum disulfide and graphite powders dispersed in high-viscosity oils [19, 20]. Compounds based on titanium and boron have also been reported to provide energy efficiency [21, 22]. A typical additive package for gear oil consists [23] of antiwear additive, antioxidant, rust inhibitor, metal deactivator, demulsifier, and an antifoam additive. The composition has less than 3.5% phosphorous, less than 1.7% ppm nitrogen, less than 1000 ppm sulfur, less than 100 ppm metals, and a TAN of less than 1.

Biodegradable API GL-4 quality gear oil has been developed [24] based on chemically modified nonedible vegetable oils. The oil contains antioxidant, EP additive, antifoaming agent, pour point depressant, corrosion inhibitor, and a detergent–dispersant additive.

Both automotive and industrial EP gear oils have to meet the following requirements:

1. EP/antiwear/antiscoring properties to prevent wear, pitting, spelling, scoring, tooth fracture, etc.
2. Protection against oxidation and thermal degradation to resist high temperature and consequent longer life.
3. Antirust and anticorrosion properties.
4. Water separation characteristics or demulsibility.
5. Nonfoaming characteristics.

Appropriate chemical additives as discussed earlier are selected to meet this requirement. Severity level, however, differs in industrial and automotive gear oil, and therefore, treat level of additives differs in these two classes of lubricants. Gear oil technology has been extensively reviewed in several publications [25–30]. There are several common commercial EP gear oil additive packages for automotive and industrial gear oils with slight difference in treat level. All of these oils use base oil in combination with antioxidant, antiwear, EP, corrosion inhibitor, and antifoam additives. Some oils may additionally contain friction modifiers and viscosity index improvers to provide improvement with respect to energy-saving and wide temperature operating capabilities.

Energy-Efficient Gear Oil

Energy efficiency of gear oils can be improved by the incorporation of a friction modifier that works under EP mode. Coefficient of friction in EP mode is reported to be in the range of 0.1–0.2, while friction modifiers can reduce [31] it to the level of 0.01–0.05. In gear oils, molybdenum-based additives have found to be more effective friction modifiers. Molybdenum dithiocarbamates and sulfurized oxymolybdenum organic phosphorous dithioate have been specifically useful. A commercial EP industrial gear oil based on 2.5% S/P additive and with an additional 0.4% of molybdenum compound (molybdenum dithiocarbamate) has been evaluated [32] in Falex No. 1 machine, and results showed a considerable reduction in wear rate, wear scar diameter, friction, and final temperature rise. In an actual gearbox [18], such oil led to about 37% reduction in frictional energy loss, and gear efficiency was improved by 1.8%. On an average, electrical power saving of the order of 3.5% was achieved. Finely dispersed MoS_2 and graphite [19, 20] have also been used extensively to achieve energy efficiency, but these have to be kept in dispersion by mechanical agitation or by some other means. Borated gear oils have also been reported to be energy efficient [33].

Synthetic Gear Oils

There are many situations where very long oil life is required such as in sealed-for-life gearbox. For such applications, synthetic gear oils based on PAOs are used to provide greater thermo-oxidative stability and longer life. The active additive content generally remains similar to the one used with mineral oils. However, the solubility factor needs to be considered, since certain additives like metal deactivator may have limited solubility in PAOs. Additive packages contain about 40–60% mineral oils as solvent; therefore, in synthetic gear oil additive packages, mineral oil is replaced with compatible synthetic oil. To improve the seal swell characteristics of PAO base oils, 10–15% synthetic ester may be added. Biodegradable gear oils can also be produced by using synthetic esters and vegetable oil with less toxic additive packages. Esters and vegetable oils have natural low-friction properties and thus require lower dosage to obtain antiwear properties.

PAGs also provide low friction and high natural antiwear and EP properties due to their polar nature, and industrial gear oils of various viscosities can be formulated. PAG behavior toward elastomers is, however, different as compared to mineral oils or PAOs, and care has to be taken to use compatible seal materials when using PAG-based gear oils. Again, the cost of PAOs and PAGs is much higher, but their service life is also higher than the mineral oils, and their application depends on the cost–benefit analysis. The gearbox life can also be extended with the use of synthetic gear oils [34].

Automotive Gear Oils

Gears in automotive vehicles are required to transmit power from the engine to the drive gears. The gearbox and drive axle utilize differential gears to transmit power to the wheels. Differentials contain several components, which are required to be lubricated by single oil. The ring and pinion gears operate under EP and sliding contact. The bearings operate under rolling motion where lubricant film strength is important, and limited-slip clutches require specific friction modifiers for proper operation. Gearboxes that change gear ratio are manual transmission with synchronization. This transmission can also be automatic transmission or CVT. In such cases, an automatic transmission fluid is used, which has to satisfy the lubrication requirements of gears, torque converter, retarders, wet brake, and wet clutches. The transmission fluid thus becomes more complicated and utilizes larger number of additive components to meet the lubrication requirement of various components. These transmission fluids are discussed in the next chapter.

Performance Specifications of Automotive Gear Lubricants   Gear oils are used in the manual transmission and transaxle of buses, trucks, and passenger cars. U.S. military was the first to come out with automotive hypoid gear oil specification and then went on issuing a series of gear oil standards. These standards have been followed worldwide, and API GL-5 standard has also been derived from U.S. MIL standards. These are as follows:

1. MIL-L-2-105A in the 1940s
2. MIL-L-2-105B in 1946
3. MIL-L-2105 in 1950
4. MIL-L-2105A in 1959
5. MIL-L-2105B in 1962
6. MIL-L-2105C in 1976

TABLE 10.3 SAE J306 automotive gear viscosity classification

Grades	70W	75W	80W	85W	80	85	90	140	190	250
Viscosity at 100°C, min., mm²/s	4.1	4.1	7.0	11.0	7.0	11.0	13.5	24.0	32.2	41.0
Maximum, mm²/s	No requirement					11.0	13.5	24.0 41.0 41.0		No requirement
Viscosity of 150,000 mPa.s, max. temp °C	−55	−40	−26	−12			No requirement			
20-h KRL shear (CRC L-45, T-93), KV 100 after shear, mm²/s	4.1	4.1	7.0	11.0	7.0	11.0	13.5	24.0	32.2	41.0

7. MIL-L-2105D in 1987

8. MIL-PRF-2105E in 1995

9. SAE 2360 in 2001

SAE 2360 has been synchronized with MIL-PRF-2105E and both are the same.

The viscosity grades (VG) have been defined by SAE J306 (Table 10.3). Performance of gear oils are defined in SAE J2360 (previously MIL-PRF-2105E). API gear oil service designation has identified seven grades from GL-1, GL-2, GL-3, GL-4, GL-5, GL-6, and MT-1. GL-1 is straight mineral oil without additives; GL-2 is for worm gears and contains a fatty material. GL-3 contains mild EP additives. GL-4 is equivalent to old MIL-L-2105 and is usually satisfied with 50% of GL-5 additive level. GL-5 is for severe service in hypoid gears. GL-6 is now obsolete and was earlier proposed for severe service of high offset hypoid gears. MT-1 is a thermally stable gear oil for nonsynchronized manual transmission in heavy-duty service. Details of tests required for each service category are provided in Table 10.4.

API GL-4, GL-5, and MT-1 performance criteria are compared and provided in Table 10.5.

There are other OEM standards of auto gear oils such as Mack GO H, J, J+, and MAN 341, 342, and 3342 standards, but these are essentially based on API criteria with additional tests.

Performance Evaluation of Automotive Gear Oils Automotive gear oil performance is mainly evaluated by the following rig tests.

Gear Oil Corrosion Test: CRC L-33, ASTM D-7038 (Formerly FTMS 791B Method 5326.1) This test evaluates the corrosion inhibition properties of gear oil in an unloaded Dana Model 30 hypoid differential. The cleaned and assembled unit is filled with 1200 ml of oil and 30 ml of water and is then motored at 82.2°C for 4 h. After the test, unit is opened up and rated for corrosion, deposits, and sludge. Lubricant performance to API GL-5 requires no rusting on any working surface and up to 3.2 cm² of rust on the cover plate as a maximum.

TABLE 10.4   Performance requirements for SAE J2360, November 1998 (previously MIL-PRF-2105E lubricants)

SAE VG		75W	80W-90	85W-140
Viscosity at 100°C, (min., mm^2/s)		4.1	13.5	24.0
Maximum		—	18.5	32.5
Viscosity of 150,000 mPa.s, max. temp, °C		−40	−26	−12
Channel point min. °C		−45	−35	−20
Flash point (°C, min.)		150	165	180
Thermal and oxidation stabilities	100°C viscosity increase at 50 h (max. %)	100	100	100
ASTM D-6704 (formerly CRC L-60-1)	Pentane insolubles, max., %	3	3	3
	Toluene insolubles, max., %	2	2	2
	Carbon varnish (min. rating)	7.5	7.5	7.5
	Sludge, max., rating	9.4	9.4	9.4
Moisture corrosion, 7 days	Rust on gear teeth bearing s, max., %	0	0	0
ASTM D-7038 (formerly CRC L-33)	Rust on cover plate, max., %	1	1	1
ASTM D-6121 (formerly CRC L-37)				
High speed–low torque	*Green Gears*	Pass	Pass	NR
High torque–low speed	"Lubrited" Gears	Pass	Pass	NR
ASTM D-7542 (formerly CRC L-42)				
High-speed shock loading axle test	Ring and pinion tooth scoring, max., %	Equal to or better than passing reference oil		NR
Cycle durability, MACK cycling test	No. of cycles	Equal to or better than the average of past five reference runs		
Copper corrosion at 3 h at 121°C, ASTM D-130	Strip rating, max.	3	3	3
Elastomer compatibility, ASTM D-5662		Polyacrylate		Fluoroelastomer
Polyacrylate + fluoroelastomer at 150°C for 240 h	Elongation change, min., %	−60		−75
	Hardness change, points	−25 to +5		−5 to +10
	Volume change, %	−5 to +30		−5 to +15
Foam tendency/stability, ml, max. ASTM D-892	Sequence I/II/III	20/0, 50/0, 20/0	20/0, 50/0, 20/0	20/0, 50/0, 20/0
Storage stability and compatibility, SS&C FTM 791	Method 3340	Pass	Pass	Pass
Field trials[a]		Pass	Pass	Pass

NR, not required, if 80W90 passes in the same base stock. Lower L-37 and L-42 test requirements are required for 75W oils.

[a]Must pass once in a single grade with every additive, 100,000 miles for light-duty and 200,000 miles for heavy-duty axle oils.

TABLE 10.5 API GL-4, GL-5, and MT-1 performance test criteria, comparison

Performance	API GL-4 MIL-L-2105	API GL-5 MIL-L-2105D	API MT-1
Scoring resistance under high-speed shock load conditions	CRC l-19^a test or FTM 6504T; equal to or better than RGO-105	L-42 test; gear/pinion coast-side scoring equal to or better than RGO-110	No requirement
Resistance to gear distress under high-torque, low-speed conditions	CRC L-20 test; no tooth disturbance such as rippling, ridging, pitting, or severe wear	L-37 test; no tooth disturbance such as rippling, ridging, pitting, or severe wear	No requirement
Corrosion resistance in the presence of water	1. CRC L-13 or FTM 5313.1 2. CRC L-21 No evidence in rusting	L-33 test; no evidence of rusting after 7 days of exposure on any working surface; max. 0.5 in rust on cover plate (1% of surface area)	No requirement
Thermal and oxidation stability/component cleanliness	No requirement	L-60-1 test: 100% max. viscosity increase; 3% max. pentane insolubles; 2% max toluene insolubles	L-60-1/ASTM D-5704 test; 100% max. viscosity increase; 3% max. pentane insolubles; 2% max. toluene insolubles; 7.5 min carbon/varnish rating on large gear; 9.4 min sludge rating on all gears
Antiwear	No requirement	No requirement	FZG (A/8.3/90): failing stage 11
High-temperature lubricant stability, cyclic durability test ASTM D-5579	No requirement	No requirement	Equal to or better than reference oil
Oil seal compatibility, 150°C, 240 hr	No requirement	No requirement	ASTM D-5662; polyacrylate elongation change, −60% to 0% Hardness change, −35 to +5 Volume change, −5 to +30 Fluoroelastomer: −75% to 0% elongation change −5 to +10 pts hardness change −5 to +15% volume change

Antifoaming characteristics	CRC L-12 test: readings taken immediately after 5-min aeration; sequence I 23.9°C, 650 ml; sequence II 93.3°C, 650 ml	ASTM D-892: readings taken immediately after 5-min aeration; sequence I, 20 ml max.; sequence II, 50 ml max.; sequence III, 20 ml max.	ASTM D-892: readings taken immediately after 5-min aeration; sequence I, 20 ml max.; sequence II, 50 ml max.; sequence III, 20 ml max.
Copper corrosion, ASTM D-130	3b max. after 1 h at 121.1°C	3 max. after 3 h at 121.1°C	2a max. after 3 h at 121.1°C
Channeling Characteristics	No requirement	FTM 3456.1 modified: SAE 75, −450°C max.; SAE 80W-90 −35°C max., SAE 85W-140 −400°C max.	No requirement
Compatibility with existing gear lubricants	SS&C FED-STD-791	SS&C FED-STD-791	SS&C FED-STD-791
Solubility—measure separated material after centrifuge oil stored for 30 days at room temperature (29.4 + 9.5°C)	FTM 3430: 0.25% wt max. of original nonpetroleum material in sample	FTM 3430: 0.25% wt max. of original nonpetroleum material in sample	FTM 3430,2: compatible with SAE J2360 oils
Compatibility—same as solubility except mixed 50/50 with each of six reference oils	FTM 3440: 0.50% wt max. of original nonpetroleum material in sample	FTM 3440: 0.50% wt max. of original nonpetroleum material in sample	FTM 791 method 3440.1 pass

[a]Equipment no longer available. Not possible to conduct test as per original procedure.

Thermal and Oxidation Stability Test: L-60/L60-1/ASTM D-5704/FTM 2504   This test determines the ability of gear oil to withstand high temperature and resist oxidation. The test oil is placed in special gear case with two spur gears with a copper catalyst strip and run at 1725 rpm for 50 h, at 163°C. Air is bubbled through the sample. At the end of the test, viscosity increase and pentane and toluene insolubles are determined.

High-Torque Axle Test: CRC L-37, ASTM D-6121 (Formerly FTMS 791B Method 6506.1)   This test determines EP and antiwear properties of gear oils in hypoid gears under high-speed–low-torque and low-speed–high-torque conditions. The test is run in a Dana Model 60 hypoid axle (5.86 ratio)-fitted with uncoated drive gear. The pinion is set up to drive two dynamometers from an eight-cylinder, 5.7-l gasoline truck engine. First, a high-speed–low-torque test is run for 100 min, and the gears are visually examined through the inspection plug. A low-speed–high-torque sequence is then run for a further 24 h, and final inspection is carried out. Lubricant performance to API GL-5 is assessed on the basis of tooth surface rippling, ridging, pitting, and wear, together with any deposits or discoloration.

High-Speed Axle Test: CRC L-42, ASTM D-7452 (Formerly FTMS 791B Method 6507.1)   This test evaluates the antiscoring properties of gear oil under high-speed and shock-loading conditions using a Spicer Model 44-1 hypoid axle. The axle is driven by a 5.7-l VB gasoline engine that drives the test axle, a four-speed truck transmission, and two high-inertia dynamometers at a rate to simulate high acceleration. The axle is accelerated through the gears to a speed of 1050 rpm and then decelerated to 530 rpm. The cycle is repeated five times. This high-speed sequence is followed by 10 shock loadings. Lubricant performance to API GL-5 requires scoring to be equal to or better than reference oil RGO-110. (Only the coast side of the gear is rated. The drive-side score is only considered if there is a large amount of scoring.) The drive-side score is only considered if scoring is significant.

Full-scale test of high speed, high-powered helical gear unit has also been reported [35].

Energy-Efficient and Synthetic Automotive Gear Oils

In a gear, only mechanical losses can be reduced by the lubricant. In view of high efficiency of gears, there is a limited scope for energy reduction through gear oil formulation. Reducing the gear oil viscosity by one SAE VG will result in fuel consumption reductions of 0.2–1.5% at high temperatures and 0.4–2.5% at low temperatures. Using friction modifiers in gear oils, fuel consumption reductions of the order of 1.0–5.1% are possible by assuming that 50% friction can be reduced [36]. Use of low-viscosity multigrade gear oils formulated with shear-stable polymer and friction modifier has potential to reduce fuel consumption to a higher level. It has been reported that 3.2% fuel economy was observed in the field trial in 75W-90 GL-5 gear oil as compared to SAE 90 oil of similar performance [37]. Synthetic auto gear oils can also achieve long drain capabilities up to 400,000 km or 3 years' service in commercial vehicles. Several OEMs such as BMW (60,200.0 spec.) and VW specify (75W-90, TL 52157 spec.) the use of synthetic oil for the lifetime of the vehicles. In another study [38], gear oil energy efficiency using mineral oil, PAO, PAG, rapeseed oil, hydrocracked oil using three different VI improvers (PMA, PIB, and PAMA/PAO co-oligomer), and seven EP/ antiwear additive packages have been measured. The following conclusions have been made:

1. Lower-viscosity and higher-viscosity index oils are more energy efficient.

2. High thermal and oxidation stability and sufficient antiwear/EP protection with low friction lead to overall efficiency of automotive gear oils.

REFERENCES

[1] Lauer DA. Industrial gear lubricants. In: Rudnick LR, editor. *Synthetic, Mineral Oils and Bio Based Lubricants, Chemistry and Technology.* Boca Raton: CRC Press; 2006. p 442–443.

[2] Dowson D, Higginson GR. *Elastohydrodynamic Lubrication—The Fundamentals of Roller and Gear lubrication.* London: Pergamon Press; 1966.

[3] Dowson D, Higginson GR. New roller-bearing lubrication formula. Engineering (London) 1961;192:158–159.

[4] Dowson D. Elastohydrodynamics. Proc Inst Mech Eng 1967;182 (Pt 3A):151–167.

[5] Wellauer EJ, Holloway GA. Application of EHD oil film theory to industrial gear drives. Trans ASME J Eng Ind May 1976;98(2).626–634.

[6] Dowson D, Hugginson GR. *Elastohydrodynamic Lubrication.* 2nd ed. Oxford: Pergamon Press; 1977.

[7] Hamrock BJ, Dowson D. Isothermal elastohydrodynamic lubrication of point contact, part III, ASME. J Lubr Technol 1977;99:264–276.

[8] Mang T, Dresel W, Gear lubrication oils. In: Bartels T, editor. *Lubricants and Lubrication.* Weinheim: Wiley-VCH Verlag GmbH; 2001. p 230–270.

[9] Srivastava SP. Industrial gear oil (Chapter 12) and Automotive gear oils (Chapter 13). In. *Modern Lubricant Technology.* Dehradun: Technology Publication; 2007.

[10] Srivastava SP. Additives for gear and transmission oils. In: *Advances in Lubricant Additives and Tribology.* New Delhi: Tech Books International; 2009. p 300–324.

[11] Andress HJ Jr. Derivative of triazole as load carrying additives for gear oil. US patent 4,144,180 A. March 13, 1979.

[12] Olszewski WF. Combination of benzotriazole with other materials as E P agent for lubricants. US patent 3,919,096. November 11, 1975.

[13] Donofrio RJ. Triazole derived acid-ester or ester-amide-amine salts as anti-wear additives. WO patent 1995034614 A1. December 21, 1995

[14] Metric Edition of AGMA/ANSI 2001-D04. *Fundamental Rating Factors and Calculation Methods for Involute Spur and Helical Gear Teeth.* Alexandria: AGMA; 2004.

[15] AGMA 250.04. *Lubrication of Industrial Enclosed Gear Drives.* Alexandria: AGMA; September 1981.

[16] AGMA 925-A03. *Effect of Lubrication on Gear Surface Distress.* Alexandria: AGMA; March 2003.

[17] ANSI/AGMA 110.04. *Nomenclature of Gear Tooth Failure Modes.* Alexandria: AGMA, August 1980.

[18] Saxena D, Mooken RT, Singh MM, Srivastava SP. Energy efficient high performance industrial gear oils. Proceedings of the 7th LAWPSP Symposium; 1990; Mumbai: Indian Institute of Technology.

[19] Corn A, Fatkin J. Graphite in lubrication-fundamental, parameters and selection guide. NLGI Spokesman, L III/4, 137.

[20] Smith RK, Marschek KM. Lubricants containing dispersed molybdenum disulphide improves worm gear efficiency. NLGI Spokesman, X L VIII, 47–54.

[21] Lam WY, Loper J. Titanium containing lubricating oil composition. US patent 2007,0111,908. June 2007.

[22] Pacholke PJ. Solid lubricant additives for gear oils. US patent 4,715,972. December 1987.

[23] Carey JT, Galiano-Roth AS, Wu MM, Haigh HM. Low sulfur and low metal additive formulations for high performance industrial oils. US patent 8,394,746. March 12, 2013.

[24] Singh AK, Chamoli A. Composition of biodegradable gear oil. US patent 8,557,754. October 15, 2013.

[25] Papay AG. E P gear oils today and tomorrow. 29th ASLE Meeting; April 28–May 2, 1974; Cleveland. ASLE Preprint No. 74 AM-2B-3.

[26] Papay AG, Dinsmore DW. Advances in gear additive technology. 30th ASLE Meeting; May 5–8, 1975; Atlanta. ASLE preprint 75 AM-1C-1.

[27] Papay AG. Gear lubrication technology. NLGI Annual Meeting; October 20–23, 1974; Chicago.

[28] Papay AG, Dinsmore DW. Advances in gear additive technology. 30th ASLE Meeting; May 5–8, 1975; Atlanta. ASLE pureprint, No.75 AM-1C-1.

[29] Srivastava SP, Jayprakash KC, Goel PK. Industrial gear oil additive technology. Third LAWPSP Symposium; 1982; Mumbai: Indian Institute of Technology.

[30] Srivastava SP, Jayprakash KC, Goel PK. Development in industrial gear oil technology. 4th International Seminar on Fuels, Lubricants and Additives; March 7–10, 1983; Cairo: Misr Petroleum Co.

[31] Papay A. Industrial gear oils-state of art. Lubr Eng, 1988;44 (3):218–219.

[32] Papay A. Oil soluble friction reducers—theory and applications. ASLE Lubr Eng 1983;39 (7): 419–426.

[33] Adams JH, Godfrey D. Borate gear lubricant-EP film analysis and performance. Lubr Eng January 1981;37(1):16–21.

[34] Bloch HP, Williams JB. Synthetic lubricants measures up to claim. Chem Eng 1995;102: 127–130.

[35] Akazewa M, Tejima T, Narita T. Full scale test of high speed, high powered gear unit-helical gears of 25000 PS at 200 m/s PLV; 1980. ASME Paper 80-C2/DET-4.

[36] Bartz WJ. Gear oil influence on efficiency of gear and fuel economy of cars, Proceedings of the Institute of Mechanical Engineers, Part D. J Autom Eng 2000;214:189–196.

[37] Papay AG. Energy saving gear lubricants. 34th ASLE Meeting; April 8–May 2, 1974; St Louis.

[38] Bartz WJ, Wienecke D. Influence of base oil and formulation of gear oil on friction power losses of gears. In: Srivastava SP, editor. Proceedings of the International Symposium on Fuels and Lubricants, ISFL-2000; March 10–12, 2000; New Delhi. New Delhi: Allied Publisher. p 541–554.

CHAPTER 11

AUTOMATIC TRANSMISSION FLUIDS

AUTOMATIC TRANSMISSION FLUIDS

Most of the modern passenger cars have automatic transmission system to transfer power from the engine to the wheels. In manual transmission, it is carried out through a series of gears, clutch, and axle, which are lubricated by a gear oil. Automatic transmission system has a complex set of equipment consisting of the following five elements:

1. Torque converter
2. Clutch packs
3. Brake band assembly.
4. Epicyclical gear train
5. Hydraulic system to control the above elements

The torque converter is made up of an impeller connected to the engine, a turbine connected to the gearbox, and a stator. Each one of these elements described above has different lubrication requirement. For example, the fluid for torque converter should possess the following important characteristics:

- Good thermal and oxidation stability
- Good anticorrosion properties
- Seal compatibility
- Low viscosity/temperature change characteristics or high viscosity index (VI)
- Nonfoaming

The clutch packs and brake band assembly need the following properties in the fluid:

- Correct frictional properties
- Good thermal and oxidation stability
- Minimal viscosity/temperature change characteristics or high VI
- Antiwear properties

Developments in Lubricant Technology, First Edition. S. P. Srivastava.
© 2014 John Wiley & Sons, Inc. Published 2014 by John Wiley & Sons, Inc.

Epicyclical gear train, consisting of a centrally mounted sun wheel, several planet gears that engage with the sun wheel, and external toothed ring, has the following lubrication requirements:

- Good extreme pressure (EP) and antiwear properties
- Anticorrosion properties
- Adequate viscosity

Hydraulic system requirements for the fluid are as follows:

- Minimal viscosity/temperature change characteristics
- Anticorrosion properties (brass screen)
- Seal compatibility
- Antifoam properties
- Good antiwear properties

An automatic transmission fluid (ATF) thus becomes multifunctional and complex to satisfy the lubrication requirements of various parts of the transmission. A typical ATF formulation requires a large number of compatible chemical additives in carefully selected base oils to meet exacting requirements of OEMs. In recent years, there have been considerable design changes in automatic transmission (continuous variable transmission (CVT)), such as electronic controls, continuously slipping torque converter coupled with compact system and smaller fluid capacity. This has posed greater challenges to the fluid formulators. Automatic transmissions are of the following types:

1. Conventional transmission with multiple gears
2. Metallic belt-type CVT
3. Chain-type CVT
4. Traction drive CVT or toroidal CVT

Conventional automatic transmission based on multiple gears has been in use in the industry for about the last 70 years. The first automatic transmission was fitted in the 1940s, General Motors' (GM) Oldsmobile car. Belt- and chain-type CVT are currently in use in smaller vehicles. The traction drive CVT is a power transmitting mechanism, which uses shear force between the rolling and sliding surfaces of rotating component [1]. Traction is the sum of shear stress in a lubricant film between two rolling elements resulting from the rheological behavior of fluid under high pressure and high shear [2, 3].

The toroidal CVT is the recent development based on a new concept. It has two disks: one connected to the engine (like driving pulley) and the other to the drive shaft (like driven pulley). Two rollers or wheels are placed between these disks and act like belt, transmitting power from one disk to the other. These rollers spin around the horizontal axis and tilt in and out around the vertical axis (hence the name toroidal). This simple tilt of the rollers changes gear ratio smoothly. The basic difference between CVT fluids and conventional ATF is that the conventional oils require low coefficient of friction, while the CVT oils need a minimum or an optimum coefficient of traction.

FUNCTIONS OF AUTOMATIC TRANSMISSION FLUIDS

Automatic transmissions are required to function under a wide range of climatic and load conditions. There is considerable heat generation in transmission which is produced by friction between the moving gears, the friction plates, and the turbulent flow in the torque converter. This heat needs to be dissipated by means of heat exchangers through the fluid. The transmission system is made of several parts of different materials including various elastomers, and the fluid must be compatible with all these materials. The fluid has to transmit power through the fluid coupling or torque converter. The precision control system has to be performed by the hydraulic operation. The friction is to be precisely modified during the engagement and disengagement of clutches and band. ATF is then supposed to lubricant bearings, gears, and all other components of the transmission. Transmission system designs are now diversified, and these are also utilized in the off-highway equipment, agricultural tractors, rail road hydraulics, etc. For each of this class, separate transmission fluid is designed.

SPECIFICATIONS OF AUTOMATIC TRANSMISSION FLUIDS

The specifications of transmission fluids have been conventionally dominated by OEMs like Ford, DaimlerChrysler, and GM. These standards extensively cover the performance characteristics of ATFs. Several major OEMs have developed performance standards for ATF. Generally, the specifications can be divided into factory fill and service fill requirements. This discussion will be limited to service fill requirements as most OEMs require proprietary in-house testing for approval of factory fill fluids. The best known service fill specifications are the GM DEXRON® and the Ford MERCON® specifications. Several OEMs recommend either DEXRON or MERCON ATFs for service fill because these are widely available. The specifications have been evolved over a period of time to cover equipment design and customer demand requirements.

The current GM specification is DEXRON VI. This specification has been evolved over a period of time beginning with DEXRON, DEXRON II, DEXRON IIE, DEXRON III, DEXRON IIIH, and so on. The following is the chronology of transmission fluid development:

1949	Both GM and Ford introduced Type A transmission fluid
1956	GM introduced Type A suffix A fluid
1967	GM introduced DEXRON trademark fluids
1973	GM upgraded to DEXRON II fluid (changes in oxidation, friction cycling, introduction of wear test, and more severe friction retention test)
1987	Ford introduced MERCON trademark and specification
1990	GM improved to introduce DEXRON IIE (improved oxidation and low-temperature fluidity)
1993	GM Improved to DEXRON III (improved flash, fire point, copper corrosion, foam resistance, oxidation resistance, and longer friction retention)
1996	Ford introduced MERCON V specification (improved wear, oxidation, viscometrics, and friction performance requirements)
2005	GM introduced DEXRON VI (improved antifoam, antiwear, low-temperature fluidity, shear stability, oxidation and thermal stability)

DEXRON VI is reported to provide longer life (200,000 miles) and trouble-free service as compared to earlier DEXRON III fluids (100,000 miles) and can be used in all applications where DEXRON IIE or DEXRON III fluids were used.

Daimler-Chrysler introduced the MOPAR® ATF+4 (MS-9602) specification in 1998. This specification is unique in that it specifies the additive package to be used. GM's latest DEXRON VI also specifies approved additives with dosage. If a new additive is required to be approved, additional tests at GM discretion are required to be conducted.

A comparison of Ford MERCON V and GM DEXRON VI is provided in Table 11.1.

JASO, the Japanese Automobile Standards Organization, has developed a specification for common use for Japanese OEMs. The OEMs generally use it as a base level of performance and add their own proprietary tests for factory fill requirements.

TABLE 11.1 Comparison of Ford MERCON V and GM DEXRON VI (GM16444) test requirements

MERCON V	DEXRON VI
Key test requirements	**Key test requirements**
	Elemental analysis, Al, Ba, B, Ca, Cr, Cu, Fe, Pb, Mg, Mn, MO, Ni, P, K, Si, Ag, Na, Sn, Ti, V, Zn, S, Cl, N
Fluid miscibility	Fluid miscibility
Kinematic viscosity, 100, −20, −40°C	Kinematic viscosity 40, 100, 150°C, viscosity grade (VG) 32
	VI
Apparent visc. 100 and 150°C	Blended base oil vis, 4.5 cSt, min.
	Brookfield viscosity at −10, −20, −30, −40°C
	CCS at −30°C
Flash point	Flash and fire point
Copper corrosion	Copper corrosion
Rust test	Rust test
Foam test	Foam test
Modified Noack	Noack evaporation 200°C at 1 h, 10%, max.
FZG wear and vane pump wear	Pump wear ASTM D-2882-00 film thickness
Shear stability	Tapered bearing roller shear
Elastomer compatibility Ford	Elastomer compatibility, GM
Clutch friction evaluation, Ford	Plate friction test, GM
Antishudder evaluation, Ford	Band friction test, GM
Al beaker oxidation test, Ford	Oxidation test, GM-THOT
Cycling test, GM test	Cycling test, GM
Shift feel, Ford	Vehicle performance, GM
Timken MERCON	ECCC vehicle performance, GM
Four-ball wear	Sprag wear test GM
Falex EP test	Low-speed carbon fiber friction test, GM
	Aeration test, GM
	Additional test for new additive chemistry
	Hunting behavior, pitting, carbon fiber durability, and fleet test for 150,000 km, further tests at GM discretion

Detroit Diesel Allison Division of GM, USA, also specifies requirement of ATF for off-highway and heavy-duty commercial vehicles. Allison C series specification was issued in 1953. This was followed by Allison C1 standard in 1959, when the transmission designs were changed. Subsequently, an upgraded version, Allison C2, was issued to meet the enhanced severity of operation. However, Allison C3 came into force in 1977 with better oxidation, frictional, and seal compatibility requirements. This has been one of their popular and long-lasting specifications. In 1989, Allison came out with the present specification, C4 (Table 11.2 and Table 11.3), to accommodate a wider variety of clutch and seal materials.

The major differences between Allison C3 and C4 specifications are as follows:

- C4 calls for a friction test with nonasbestos clutch in addition to graphite of C3 specification, in an SAE no. 2 friction machine.
- C4 oils possess superior thermal oxidation stability as provided in a severe thermal oxidation stability test with higher temperature and airflow.
- C4 specification incorporates new fluoroelastomer compatibility test and has tighter low-temperature viscosity control.
- C4 calls for stricter antifoam, stringent rust and corrosion control and inclusion of chemical analysis as compared to C3 specification.

Recently, Allison also announced the TES-439 specification that has strict requirements for viscosity and oxidation performance. Caterpillar TO-4 specification defines requirements (Table 11.4 and Table 11.5) for transmission and drive train fluids. ZF lists fluids approved for certain applications in their lists of lubricants. Voith also publishes a list of approved fluids by application. For passenger car application, both ZF and Voith require DEXRON or a MERCON license plus additional wear and viscosity shear stability requirements.

TRANSMISSION FLUID FORMULATIONS

Base Oil Selection

In order for the ATF to function properly, the viscosity of the fluid should be controlled within the specified narrow ranges. Use of electric actuators in modern transmissions requires that the ATF viscosity should be low. At low ambient temperatures, the ATF may become viscous and difficult to pump. At higher temperature, it may become too thin to protect the metal surfaces. On the other hand, too high viscosity at operating temperature would lead to frictional loss in pumping the fluid. This situation demands that the ATF have high VI, which can be achieved by using a shear-stable VI improvers (VIIs) and fluidity improvers. VIIs based on polymethacrylates are particularly useful in ATF as these do not severely affect low-temperature fluidity properties as compared to VIIs based on other polymers. Specifications require that the ATF be very shear stable and should not thin out due to mechanical shear in the equipment.

TABLE 11.2 Allison transmission C4 heavy-duty transmission fluid specification

Bench tests	Requirements	Test method
Foaming tendency		GM 6297-M, test M
Foam at 95°C, max.	Nil	
Foam at 135°C, mm max.	10	
Break time at 135°C, max.	23	
Copper corrosion	No blackening or flaking	ASTM D-130, 3 h at 150°C
Corrosion/rust protection	No visible rust on test pins	ASTM D-665, procedure "A" for 24 h
Rust protection	No rust or corrosion on any test surface	ASTM D-1748, 98% humidity, 50 h at 40°C
Elastomer compatibility	Limits are adjusted for each new elastomer batch	GM 6137-M
V1: Volume difference, %	0–20	
Hardness difference, points	−15 to 0	
V2: Volume difference, %	0 to 12	
Hardness difference, points	−7 to +3	
V3: Volume difference, %	0–22	
Hardness difference, points	−14 to 0	
P1: Volume difference, %	0–8	
Hardness difference, points	−10 to 0	
P2: Volume difference, %	0–8	
Hardness difference, points	−11 to +3	
P3: Volume difference, %	0–4	
Hardness difference, points	−8 to + 4	
F1: Volume difference, %	0–3	
Hardness difference, points	−5 to +4	
F2: Volume difference, %	0–4	
Hardness difference, points	−2 to + 5	
N1: Volume difference, %	−2 to +5	
Hardness difference, points	12 to +12	
N2: Volume difference, %	0–6	
Hardness difference, points		
Oxidation stability, C4 oxidation test (THOT)		GM 6297-M (Appendix E)
Viscosity increase, 40°C % max.	100	
Viscosity increase, 100°C%. max.	60	
TAN increase, max.	4.0	
Carbonyl absorbance, max.	0.75	

TABLE 11.2 (Continued)

Bench tests	Requirements	Test method
Wear protection C4 vane pump wear test Total weight loss, mg, max. Clutch frictional characteristics	30	ASTM D-2882 mod. (a) 80 + 3°C (b) 6.9 mPa
C4 graphite clutch test	Midpoint dynamic coefficient and slip	Allison C4 graphite clutch friction test
C4 paper clutch friction test	time must surpass limits set with minimum performance reference oil	Allison C4 paper clutch friction test

TABLE 11.3 **Allison transmission C4 heavy-duty transmission fluid specification**

Test	Requirements	Test method
	Chemical analysis	
	Metal contents	

Barium, boron, calcium,
  magnesium,
  phosphorous, silicon,
  sodium, and zinc
	Report	Emission spectroscopy: ICP

	Nonmetal contents	
Chlorine	Report	ASTM D-808
Nitrogen	Report	ASTM D-3228
Sulfur	Report	ASTM D-4951 or ASTM D-129
Total acid number	Report	ASTM D-664
Total base number	Report	ASTM D-4739 OR D-2896
Infrared spectrum	Report	ASTM E-168
	Physical properties	
Flash point, °C min.	170	ASTM D-92
Fire point, °C min.	185	ASTM D-92
	Viscosity characteristics	
Kinematic viscosity at 40°C	Report[a]	ASTM D-445
Kinematic viscosity at 100°C	Report[a]	ASTM D-445
Apparent viscosity	Report[a]	ASTM D-2605
Brookfield viscosity	Report temperature at 3500 mPa.s	ASTM D-2983
Stable pour point	Report[a]	ASTM D-97

[a]Fluids shall meet SAE J300 VG, and in addition, ATFs must meet GM and Ford requirements.

TABLE 11.4 Caterpillar TO-4 transmission and drive train fluid requirements

Viscometric properties	SAE J300 VG	ASTM D-2983 maximum temperature (°C) for Brookfield viscosity of 150,000 mPa.s	ASTM D-4684 low-temp. pump ability (MRV-TP-1) 30, 000 centipoise max, temp. °C	ASTM D-4683 (or equiv) high-temp high-shear viscosity at 150°C and 10^6 s^{-1} min. mPa.s
SAE J300 requirements plus	10W	−35	−25	2.1
additional low-temp. and high-temp. high-shear requirements as shown opposite	30	−25	−15	2.9
Caterpillar does not	40	−20	−10	3.7
recommend oils that contain viscosity improvers in this application	50	−15	−5	4.5

Wear properties Gears	Average of three separate runs 100 mg max. No single run with more than 150 mg weight	ASTM D-4998 (FZG machine— "A" gears, low speed, 100 rpm, 121°C, load stage 10, 20 h
Gear scuffing	LSP 8 min. (10W and SAE 30) LSP 10 min. (40 and SAE 50)	ASTM D S:82 FZG Visual A Gears 8.3 ms^{-1}, 90°C
Pumps	Total weight loss for all vanes from individual cartridges (not including intravanes), <15-mg ring weight loss, from individual cartridge, <75-mg pump parts, especially rings, should not have evidence of unusual wear or stress in contact areas	Vickers pump test procedure for mobile systems as defined in publication form M-2952-S
Friction properties Link Model 1158 Oil/ Friction Test Machine Dynamic coefficient of friction Static coefficient of friction Energy capability, wear properties (seven friction disk–steel reaction plate combinations evaluated separately: three paper, two sintered bronze. two fluoroelastomer friction disks)	The results of each friction disk–reaction plate combination for the candidate oil must be within the allowable range of variation from the reference test oil	Caterpillar VC 70 standard test method

TABLE 11.5 Caterpillar TO-4 transmission and drive train fluid requirements

Physical properties	Requirements	Test method
Rust control	Less than six rust spots per linear inch of two out of three test specimens	Modified International Harvester BT-9 (175 h under dynamic humidity conditions)
Copper corrosion	1a max.	ASTM D-130 (2 h at 100°C)
Fluid compatibility	No sedimentation or precipitation	Mix 50 ml test oil with 50 ml reference oil; heat to 204°C, cool to ambient; centrifuge for 30 min at 6000G
Homogeneity	No sedimentation or precipitation	Test oil held at −32°C for 24 h, warmed to ambient, centrifuged
Foam, tendency/stability, mls	Sequence I—25/0	ASTM D-892 part 1: No water added
	Sequence II—50/0	Part 2: 0.1% water in oil
	Sequence III—25/0	
Flash point	160°C min.	ASTM D-92
Fire point	175°C min.	ASTM D-92
Elastomer compatibility		
Fluoroelastomer	Av. elongation of elastomer in aged test oil must not be greater than av. elongation with reference oil	ASTM D-471 (240 h, 150 °C)
	Av. elongation with test oil must be less than or equal to av. elongation with reference oil	
Allison C4 elastomer test	See Allison C4 specification	See Allison C4 specification
Oxidation test		
Thermal oxidation stability (THOT)	See Allison C4 specification	GM 6137 October 1990, Appendix E (i.e., DEXRON IIE) (fluoroelastomer input seal, production cooler, 35% silver)
Sludge/varnish on parts	Nil	
Total acid number increase,	4.0 max.	
Carbonyl absorbance diff	0.75 max.	
Further inspection	Fluoroelastomer input seal should not fail. Copper bushings should not undergo mechanical failure due to corrosion attack. Cooler will not be graded	
Viscosity after test		
Kinematic viscosity, mm²/s	Report	ASTM D-445
Viscosity, mpa.s	Report	ASTM D-2983[a]
Viscosity, mpa.s	Report	ASTM D-4684
Viscosity, mpa.s 150°C, $10^6 s^{-1}$	Report	ASTM D-683

[a]At the maximum temperature specified in Section 4 of ASTM for the appropriate VG.

Fluidity modifiers are generally low-viscosity base oils that reduce the low-temperature viscosity of a fully formulated ATF. But these can be used only to a limited extent so that the flash point and fire point of the fluid are within the specification limits.

The low-temperature properties of conventional ATFs based on mineral oils are controlled by its paraffin wax content. As the fluid is cooled, the paraffin wax crystallizes into a three-dimensional network, causing the oil viscosity to rise substantially and eventually solidify. Pour point depressants based on polymers can control or modify wax crystallization process and limit the size of the wax crystal lattice. This improves the low-temperature fluidity of the ATF. Base oils produced by severe hydrotreatment processes have low paraffin wax content and generally have better low-temperature properties than solvent-refined base oils. Hydroisomerization processes for base oil manufacture further improve the base oil quality by producing branched paraffinic molecules that have lower pour point and improved viscosity–temperature relationship as compared to conventional solvent-refined base oils. Synthetic base oils such as polyalphaolefins (PAOs) and polyolesters have still better low-temperature fluidity characteristics and have found favor in formulating modern ATFs. In modern DEXRON VI and MERCON V fluids, base oil selection is important. Fluids could use API group II or group III, PAO or polyolesters, or their combinations to realize low-temperature fluidity, high-temperature viscosity, and oxidation and thermal stability. Fully synthetic transmission oils are now available with better performance. A typical synthetic ATF shall contain 70–80% PAO, 10–20% ester, and about 10–15% additives. The low-temperature and high-temperature properties of this base fluid mixture are such that the need for pour point depressant and viscosity modifiers is minimum or nil. These fluids also have good stability, and the need of antioxidant is also minimized. The presence of ester in PAO provides adequate seal swell characteristics, and a separate additive may not be necessary.

Friction Modification

One of the most important properties of ATF is its frictional characteristics. In most lubrication regimes, the main objective is to reduce friction to a minimum. In ATF, the objective is to control friction within prescribed limits. Automatic transmissions employ wet friction clutches extensively where clutch plates rotate against stationary steel plates. With the increase in pressure, torque increases until the steel plates also start rotating. At lockup, the sliding speed differences between these two plates approach 0, and the dynamic friction is changed to static friction. Static friction is higher than the dynamic friction, and this can lead to uneven shift feel. The mechanical design, materials of construction, and controls for the wet clutch systems vary from OEM to OEM. In order for the clutches to provide smooth engagements, the ATF has to possess frictional properties designed to meet the particular system requirements. The relative levels of static and dynamic frictional properties can be controlled and adjusted by the use of chemical compounds called friction modifiers in appropriate dosage. Friction modifiers are polar molecules and could interfere with other chemical additives such as rust inhibitors and antiwear compounds. A slight change in the dose of friction modifier can influence the antiwear properties substantially.

Care must be taken to select synergistic compounds or noninterfering molecules. The addition of organic alcohols and alkyl aromatics in the fluid increases the coefficient of friction. The high static coefficient of friction allows the clutch to hold without slippage in the reverse. On the other hand, addition of phosphate esters in the fluid reduces the coefficient of friction. Requirements of GM and Ford transmissions are quite different, and different friction modifiers are to be used in the fluids [4].

The belt-type CVT is increasing in number due to their energy efficiency. However, these are mounted only on small automobile of about 1600-cc displacement. The improvement in the transmission torque capacity is a serious problem on the belt-type CVT, where the torque is transmitted by means of a friction force between a belt element and a pulley. Thus, the transmission torque capacity is determined depending on the coefficient of friction between the metals of the belt element and the pulley and the force of the pulley. The coefficient of friction between the metals is influenced by the oil. The lower coefficient of friction between the metals may lead to slippage between the belt and the pulley. On the other hand, an electromagnetic clutch has been used as a starting mechanism on the belt-type CVT. In order to cope with an increasing transmission torque in larger engines, wet clutch or/and torque converter, a lockup clutch has been used. These wet clutch, torque converter, and CVT make use of common lubricating oils, and it is important to design fluid for CVT to meet the requirements of the wet clutch or torque converter as well. In the existing small automobile, the transmission torque is small so that the required level of the coefficient of friction between metals is not high. Thus, if an oil of relatively high coefficient of friction between metals is selected among ATFs, the performance could be satisfactory. However, if the belt-type CVT is mounted in an automobile of a large displacement, a required level of the coefficient of friction between metals increases. The common ATF will not provide satisfactory performance. High coefficient of friction between metals has been obtained by use of the combination of antiwear and overbased sulfonate and phenate metal detergent.

It was found that the friction coefficient between metals decreased at low temperatures. It was also found that although zinc dialkyldithiophosphate (ZDDP) is used for improving the friction coefficient between metals in conventional lubricating oils for belt CVT having a high friction coefficient between metals at 100°C, at lower temperature like −20°C, the friction coefficient could not be increased to a higher level. It is therefore necessary to develop new types of compounds [5] for this application. The combination of a sulfonate, phenate, or salicylate of an alkaline earth metal provided friction coefficient between metals of 0.12 or higher at −20°C as measured by a block-on-disk machine. Sato, Takehisa has also reported similar finding [6, 7] and formulated oils for push belt-type CVT with mineral oil and/or a synthetic oil containing an antiwear, a metal detergent, and an ashless dispersant.

Improved CVT fluids have been reported [8] which are also called elastohydrodynamic (EHD) traction fluids. These fluids are based on synthetic cyclohydrocarbons and are characterized by high traction coefficient. EHD fluids tend to become solid under the influence of pressure (EHD contacts), and the moment pressure is removed, it turns back to fluid. This property makes them a good choice for CVT fluids and protects the components of transmission under EHD contacts. These fluids due to their typical characteristics may find applications in other areas as well.

Rust and Corrosion Inhibition

Rust and corrosion inhibitors are surface-active materials and get attached to the metal surfaces. Alkyl carboxylic acids, dicarboxylic acids, and overbased alkyl–aryl sulfonates may be used as rust inhibitors in ATF. Thiadiazoles and alkaryl triazoles find extensive use in ATF as corrosion inhibitors to protect copper-based alloys used in transmission parts. These are metal deactivators and form chelates with copper and copper alloys. These corrosion inhibitors are also synergistic with antioxidants, since they suppress the catalytic effect of metals. Surface-active compounds must be carefully used to prevent interference with the frictional properties of the ATF.

Antifoam

Polydimethylsilicones and polyacrylates are effective antifoam additives used in ATF. These materials are polymeric and insoluble in ATF and have low surface tensions. To function properly, the antifoam additives are finely dispersed in the ATF. Since there is high shear agitation in automatic transmissions, the dispersion of the antifoam compound is maintained in the ATF. Overtreatment of antifoam additive can lead to excessive air entrainment. Entrained air reduces hydraulic performance of the fluid as the fluid becomes compressible. This can lead to loss of hydraulic pump capacity and line pressure. Also, the entrained air can cause excessive wear due to cavitation and increased oxidation rate. Silicone polymers generally affect air release value adversely, and care must be taken to select the most appropriate antifoam compound.

Oxidation Resistance

Automatic transmission system offers ideal environment for oil degradation through the process of oxidation. High temperatures, presence of various metals, and high air concentration accelerate the oxidation of oil, leading to oil thickening and deposit formation. These deposits can have detrimental effects on the operation of critical parts in the transmissions. Transmission oils are supposed to operate carefree for a long duration and, therefore, require adequate dosage of antioxidants to check the oil degradation through oxidation and thermal degradation. Hindered phenols and alkylated diphenylamines are commonly used antioxidants in ATFs. Effective control of oxidation is critical with extended drain intervals recommended by OEMs. Some of the factory filled ATFs are for life of the vehicle and require high degree of protection against thermal and oxidative degradations. GM uses turbohydronamic oxidation test (THOT) to evaluate oxidation properties of ATF, where the transmission is driven at the third gear at 1755 rpm for 300 h. Ford on the other hand utilizes a more simple aluminum beaker oxidation test. In this test, a 300-ml fluid is heated at $155 \pm 1°C$ for 300 h in a beaker. Copper and aluminum strips are immersed in the fluid. The fluid is also aerated at 5 ml/min. This simple test is more reproducible. Synthetic base oils containing PAO/ester blends with appropriate antioxidants provide good results in oxidation tests [9].

Wear Control

Various organic phosphate esters, alkyl phosphites, amine phosphates, and ZDDPs and their combinations have been used for controlling wear in transmission fluids. The combination of friction modifier and antiwear additives is important and need to be selected carefully to meet individual ATF specification. Sulfurized olefins, sulfurized fatty acids, thiadiazoles, and sulfurized esters are used as EP additives. Vane pump, Timken, SAE machine, FZG, four-ball machine, and Falex EP tests are used for evaluating antiwear and EP properties of ATF. High-temperature high-shear viscosity plays a role in controlling hydrodynamic film strength at high temperature [10]. This property is controlled by base oil viscosity blend and shear-stable VIIs. Again the use of synthetic PAO is helpful in wear protection.

Seal Swell Control

The transmission system uses different kinds of elastomeric materials for gaskets, lip seals, O-ring seals, etc. The OEM specifications of ATF define the limits for various elastomer properties, such as percent volume change, hardness change, tensile strength, and percent elongation. These specifications were developed when normal solvent-refined base oils were used as base stocks for ATF. The trend toward the use of severely hydrotreated base oils (API group II and group III oils) or synthetic/part-synthetic PAOs [11, 12] to improve low-temperature fluidity and the oxidation performance of ATFs has led to the need for increased seal swell since these base oils are poor in seal swell property. Solvent-refined base oils contain aromatic and naphthenic compounds that are effective in swelling elastomers used in transmission system seals. Severely hydrotreated base on the other hand has low concentrations of aromatic and naphthenic compounds as these are removed in the refinery during processing. Hydrotreated base oils in ATF have provided improved oxidation stability and good low-temperature fluidity. However, these are highly paraffinic in nature. This makes them poor solvents for polar additives and poor seal swell agents. In order to meet the OEM elastomer specification requirements, it is necessary to add seal swell agents to the ATF. Aromatic solvents, such as alkylated naphthalene, can be added to improve seal swell characteristics. However, the oxidation performance is then deteriorated as the aromatic solvents are less stable against oxidation. Use of alternate seal swell agents can overcome this deficiency of aromatic solvent oils. Seal swell additives are, therefore, required in ATF formulated from hydrotreated base oils to meet the specifications for elastomer compatibility. Examples of effective seal swell agents are alkylated naphthalenes, sulfones, and alkylated esters, such as phthalates, adipates, and azelates. A proper combination of PAO and polyolester may provide balanced low- and high-temperature performance, thermal/oxidation resistance [13], seal swell characteristics [14], and acceptable flash and fire points.

In formulating ATFs, there are several performance parameters that must be considered simultaneously. The additive formulation must be *balanced* so that the additives used for a particular performance do not interfere with the function of another additive. This sometimes requires a compromise in performance in one area to allow satisfactory performance in another area. The areas where such compromise is required are as follows:

1. Low-temperature viscosity versus high-temperature viscosity
2. Low-temperature fluidity versus volatility and flash point
3. Antifoam versus air entrainment (air release value)
4. Friction versus wear
5. Antiwear versus EP performance
6. Hydrotreated or synthetic base oils (PAO) versus elastomer performance

It is therefore necessary to select first appropriate base oil and then additives that are noninteracting and do not interfere with other properties of ATF. The proper base oil blend, however, is very important since this forms the major constituent of ATF (80–90%). Most of the physical properties of ATF specifications are such that without proper base oil combination, these cannot be met. API group III or PAO–ester blend would be the ideal choice for formulating transmission fluids.

The additive packages used to formulate ATFs contain antiwear additives, EP additives, antifoam, rust inhibitors, corrosion inhibitors, antioxidants, metal deactivator, pour point depressant, friction modifier, VII, antifoam, and seal swell agents. A bis-alkenyl-substituted succinimide has been reported to be of value as friction modifiers, particularly in automatic transmission fluids [15]. Preparation of a branched alkenyl succinimide dispersant has been reported [16]. Detergents and ashless dispersant may also be used to control deposit formation. Antiwear and EP additives are generally based on phosphorus and sulfur chemistry. Zinc dialkyldithiophosphonates, alkyl phosphites, and alkyl phosphates are typical antiwear agents used in ATF. These are also dyed red for the purpose of identification and leak detection. An improved transmission oil composition [17] for use in a push belt CVT is based on hydroprocessed base oil or a synthetic base oil containing viscosity modifier, dispersing agent, friction control agent, and other additives to provide improved friction characteristics, durability, thermal stability, oxidation stability, fuel efficiency, and transmission performance.

REFERENCES

[1] Machida H, Murakami Y. NSK tech. J (Japanese), NO. 669, 9–20. 2000.
[2] Johnson KL, Tevaarwerk JL. Shear behaviour of elastohydrodynamic oil films. Proc R Soc Lond 1977;356:215.
[3] Evans CR, Johnson KL. Regimes of traction in elastohydrodynamic lubrication. Proc Inst Mech Eng 1986;200:C 5, 313.
[4] Frihanf EJ. Automatic transmission fluids—some aspects on friction. SAE Paper 740051; 1974.
[5] Deshimaru J, Takakura Y, inventors. Lubricating oil composition. US patent 2003,0158,053 A1. August 21, 2003.
[6] Kugimiya T, Sato T, inventors. Lubricating oil composition for continuously variable Transmission. US patent 2003,0013,619 AI. January 16, 2003.
[7] Kugimiya T, Sato T, inventors. Lubricating oil composition for automatic transmission. US patent 2002,0187,903 A1. December 12, 2002.
[8] Rose N, Hamid S. The race for better CVT, Lubes N Greases May 2003; 9: 22–27.

[9] Gunsel S, Klaus EE, Duda JL. High temperature deposition characteristics of mineral oils and synthetic lubricant base stocks. Lubr Eng 1987;44:703–708.

[10] Watts RF, Szykowski JP. Formulating automatic transmission fluids with improved low temperature fluidity. SAE Technical Paper, 902144. 1990.

[11] Srinivasan SD, Smith W, Sunne JP, *Synthetic automatic transmission fluids*. Proceedings of the 50th STLE Meeting; 1995; Cincinnati.

[12] Wilermer PA, Haakana CC, Sever AW. A laboratory evaluation of partial synthetic automatic transmission fluids. J Syn Lubr 1985;2:22–38.

[13] Coffin PS, Lindsay CM, Mills AJ, Lindenkamp H, Guhrman J. The application, of synthetic fluids to automotive lubricant development. J Syn Lubr 1990;7:123–143.

[14] Nagdi K. Polyalphaolefins and seal materials. Eur J Fluid Power November 1990:40–48.

[15] Shrestha KS, Shiga M, Fuchi M, Nakagawa T, inventors. Friction modifier and transmission oil, US patent 8,497,232. July 30, 2013.

[16] Loper JT, inventor. Branched succinimide dispersant compounds and methods of making the compounds, US patent 8,575,390. November 5, 2013.

[17] Cha SY, inventor. Transmission oil composition for push belt continuously variable transmission, US patent 8,569,215. October 29, 2013.

PART IV

AUTOMOTIVE LUBRICANTS AND MWFs

CHAPTER *12*

PASSENGER CAR MOTOR OILS

Automotive engine oil constitutes about 55–60% of the total world lubricants and fluids. These lubricants have also attracted maximum attention of engine manufacturers, standardization organizations, and organized users, resulting in defining the automotive oil quality very clearly. However, this extensive activity brought in complexities in the development and evaluation of automotive passenger car oils. Environmental issue also led to extensive improvements in engine design and oil quality. In passenger cars, both gasoline and diesel engines are used. This chapter deals with only gasoline engine oils. Gasoline engines are four-stroke spark-ignition engines. The thermodynamic cycle is completed in four strokes, that is, intake stroke, compression stroke, power stroke, and exhaust stroke. Atomized or partially vaporized fuel and air are drawn into the cylinder and compressed by the forward movement of the piston. The compressed fuel air mixture is ignited by a spark, and the combustion gases develop pressure, forcing the piston to provide driving power. The gases are then exhausted out, and a fresh charge is introduced. The term stroke describes movement of piston within the cylinder. Over the years, this concept of gasoline engine remains unaltered although tremendous improvements have taken place in the modern gasoline engine to provide efficiency, reliability, durability, and comfort.

BASIC CONCEPT

In crankcase engines, lubricating oil is filled in the sump and is splashed or pumped onto the various parts of the engine during its operation. The oil has to carry out several functions in addition to reducing the friction between moving surfaces. It has to provide a seal to the cylinder, cool the engine, transport soot, sludge, and wear particles to the oil filter. Soot, sludge, and other deposits are formed during the combustion process of the fuel [1]. The oil is also subjected to a wide range of temperature, −40°C in arctic region to above 300°C under the piston crown. The blow-by gases, soot, and sludge particles produced by incomplete combustion go into the oil. Most of the engine lubricant problems are, therefore, related to these issues, that is, entry of fuel combustion products in crankcase and decomposition/

Developments in Lubricant Technology, First Edition. S. P. Srivastava.
© 2014 John Wiley & Sons, Inc. Published 2014 by John Wiley & Sons, Inc.

oxidation of lubricant itself at high temperatures in the presence of various metals and materials present in the system.

The main causes of engine malfunction due to fuels and lubricant quality are as follows:

- Deposit and sludge formation at different parts of engine
- Contamination such as dust and dirt, wear particles, coolants, and water
- Fuel and lubricant oxidation products
- Oil thickening due to thermo-oxidation and soot loading
- High oil consumption due to oil evaporation and high wear
- Ring sticking due to deposits in piston grooves

In an engine, the oil is subjected to all the three lubrication regimes, that is, hydrodynamic and elastohydrodynamic (in bearings) to boundary lubrication at TDC and BDC. Various engine components undergo different modes of lubrication (Table 12.1), and engine oil needs an appropriate viscosity at operating temperatures to satisfy hydrodynamic lubrication and a combination of several chemical additives including EP/antiwear to meet boundary lubrication requirements.

It is important to understand how the energy input in the internal combustion (IC) engine is utilized and where is the scope for improvements in some of these areas. Pinkus [2] presented a simplified energy distribution in a typical engine during a combined urban and highway EPA driving cycle (Fig. 12.1).

In a modern engine, some of these numbers have changed due to several new improvements, but this picture provides directional utilization of energy in an IC engine since the basic units still remain the same.

This figure shows that there is tremendous scope of improvement in the design and operation of an engine to save energy and improve fuel efficiency. Engine and transmission friction accounts for 10%, which is the maximum that can be saved through engine oil improvements. In an engine, 66% friction loss takes place in hydrodynamic lubrication regime, and 34% friction loss takes place in boundary lubrication regime. Thus, when viscosity of oil is reduced, advantage is derived in the hydrodynamic region, and with the use of friction modifier—EP/antiwear additive—like molybdenum dithiophosphate or carbamate, benefits are obtained in boundary regime. It has been estimated that by the application of friction modifiers and other tribological factors, fuel consumption reduction of the order of 2.7–5.8% in engine oils and 1.0–5.1% in gears can be realized. The total

TABLE 12.1 Lubrication regimes in IC engine

Components	Lubrication mode	Related oil properties
Bearings, bushings	Hydrodynamic	Oil viscosity
Gears, piston rings, and liners	Hydrodynamic–boundary	Oil viscosity, antiwear, EP additives
Cam, valve train	Mixed to boundary	Antiwear and EP additives

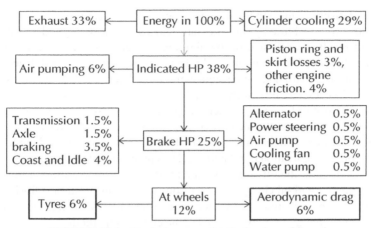

FIGURE 12.1 Typical energy distribution in an IC engine.

reduction could be 3.7–10.9% [3–5]. In another recent study [6], it has been estimated that in passenger cars, direct friction losses (braking friction excluded) are 28% of the fuel energy. In total, only 21.5% of fuel energy is used to move the car. By applying friction-reducing technologies, it is possible to reduce 18% friction in 5–10 years and 61% friction reduction on long-term (15–25-year) basis. These potentials can be realized by using advanced coating and surface texturing technology on engine and transmission components, tire design, low-viscosity oils, shear-stable viscosity modifiers (VMs), nanolubricants [7], and other lubricant additives. Direct fuel injection was initially introduced in diesel engines. This has now been applied to gasoline engines, which has significantly improved fuel economy. This involves 40 times higher pressure as compared to the conventional gasoline engine. Air fuel ratio of 50:1 is now being used. Exhaust gas recirculation up to 40% is also being used along with improved catalysts for emission control. The entire engine management is now computer controlled. The engine design changes and improvement took place after the world oil crisis of 1974 and the U.S.-introduced energy policy and conservation act of 1975, which brought out the concept of Corporate Average Fuel Economy (CAFÉ). According to CAFÉ, the fuel consumption of gasoline vehicles had to be improved from 18 miles/gallon in 1978 to 27 miles/gallon in 1985. Noncompliance would attract penalty. Car manufacturers responded to this by cutting down the weight of vehicle by using smaller cars and using lighter material. Rear wheel drive was converted to front wheel drive, and fuel injection, torque converter, and CVT transmission technologies were introduced. Along with these changes in the engine, oil quality has also improved to cope up with the engine severity.

The life of engine is another important factor to be considered, which is directly linked with the lubricant quality and its condition in engine. Critical engine parts are protected by the lubricant. Insufficient lubrication will lead to cylinder, piston ring, and crankcase bearing wear. High bearing wear will result in decreased oil pressure and less oil delivery to the upper part of the engine. Ring wear or sticking will result in power loss due to blow-by and pressure leakage. Wear of cylinder wall or its glazing will result in excessive oil burning and power loss. All these factors will ultimately reduce the life of engine, and it has to be rebuilt or scraped. Friction and wear properties of the used gasoline engine oils were determined along with some of their chemical characteristics, such as total acid number (TAN), total base number (TBN), and IR spectra for specific chemical groups. The results show that wear rate increases sharply [8] when the increase in TAN above the initial value exceeds about $1 \, \text{mg KOH g}^{-1}$. However, wear is minimized if TBN is greater than about $2 \, \text{mg KOH g}^{-1}$. Therefore, correct application of oil and regular replenishment or change of oil in the engine (depending on the oil condition) are most critical in determining the engine life. For oil quality and oil change, original equipment manufacturer (OEM) recommendation must be adhered to. To keep engine healthy and increase its life, oil condition monitoring and timely change are the cheapest methods.

DEPOSIT IN ENGINE

Hydrocarbons in fuel on combustion in the presence of air should ideally get converted to carbon dioxide and water. If temperatures are favorable, nitrogen in the air will also be converted to NOx, especially in diesel and NG engines. Sulfur present in the molecules will be converted to SOx, which in the presence of water will form sulfuric/sulfurous acids. This combustion, in actual practice, is never ideal, and some amounts of unburnt or partially burnt hydrocarbons are generated. This process leads to the formation of various compounds, soot, lacquer, sludge, varnish, carbon, etc. Generation of soot is more pronounced in diesel engines.

Engine deposits are also formed due to the thermo-oxidative degradation of lubricants and fuels at the engine hot surfaces. This process is accelerated by the presence of free radicals formed as by-products of fuel combustion. These products pass through the piston rings into the crankcase and attack hydrocarbons of the lubricant, which in turn generates peroxides and hydroperoxides. These decompose, thermo-oxidatively to form aldehydes, ketones, and carboxylic acids. Various reactions taking place during the oxidation process have been described earlier. Inorganic acid is also generated by the reaction of nitrogen and sulfur with the oxygen and moisture. The formation of NOx is more pronounced in diesel engines or in long-distance highway driving of gasoline engines. Thus, aldehydes and ketones can undergo condensation in the presence of acids or bases, forming polymeric compounds. These polymers can further oxidize to form sticky materials called lacquer, varnish, sludge, or carbon. Sludge formation takes place when the polymeric material separates at the colder part of the engine like oil screen and rocker arm cover. Sludge is encountered in gasoline engines driven under stop-and-go conditions. Lacquer or varnish or carbon deposit takes place at the hotter parts of the engine such as exhaust valve head, piston crown, and top groove. Soot is mainly derived from the incomplete combustion of diesel fuel. Diesel fuels often contain high-boiling hydrocarbons that do not burn completely in the improper air/fuel mixture environment. These hydrogen-deficient fragments are charged and agglomerate to form bigger particles called soot [1]. Higher soot content can increase the viscosity of the lubricating oil. These have soft or fluffy texture. The dispersant present in the oil tend to disperse the soot in a fine particulate matter and check the tendency of the soot to from aggregate.

Carbon deposits are hard and are formed from the decomposition of the lubricating oil and fuel on hot surfaces. The deposits are usually formed with oily material and found on the piston top land, crowns, piston ring grooves, and valve stems.

Sludge is formed by the oxidation process and by the combination of lubricant oxidation products with blow-by gases, combustion water, and dirt. Low-temperature sludges are found in gasoline engines, while high-temperature sludges are found in diesel engines (above 125°C).

Lacquer and varnishes are formed when oxygenated products in lubricants are exposed to high temperatures. It is found on cylinder and piston walls and in the combustion chamber. Oil thickening can take place when soot and insolubles accumulate in the lubricants beyond a certain limit. Lubricant oxidation also contributes to the oil thickening.

Oil consumption is mainly oil related and depends on the viscosity of oils and its volatility. It is the quantity of oil that escapes through piston rings and valves and burns in the combustion chamber. However, oil consumption in an engine can also increase due to cylinder wear, cylinder bore polish, ring sticking, etc. Appropriate multigrade oils with low volatility reduce oil consumption.

Ring sticking takes place mainly due to the formation of deposits in piston grooves. This results in the loss of oil seal and increases the ingress of blow-by gases in the crankcase.

ENGINE OIL CLASSIFICATION

Engine oil viscosity has been classified by SAE J300 (1999) and has been described in Chapter 2. Six winter grades (0W, 5W, 10W, 15W, 20W, and 25W) have been identified, which are characterized by rheological properties at lower temperature to ensure good lubrication and engine startability at lower temperature. These have also been identified with a minimum kinematic viscosity at 100°C to ensure adequate lubrication, when the engine is running hot. The five monogrades (SAE 20, 30, 40, 50, and 60) have also been identified and characterized by 100°C minimum and maximum viscosities. For example, an SAE 10W oil should meet −25°C cold cranking viscosity and −30°C MRV viscosity and should have a minimum K. viscosity of 4.1 cSt at 100°C. SAE 30 monograde oil should have a kinematic viscosity at 100°C in the range 9.3–12.5 cSt; therefore, a 10W-30 multigrade oil should meet all these criteria, that is, 100°C viscosity in the range 9.3–12.5 cSt, −25°C cold cranking simulator (CCS), and −30°C MRV viscosity requirements.

Monograde oils are formulated by adding a desired performance additive (a mixture of detergent, dispersant, antiwear, antioxidant, etc.) into a base oil of suitable viscosity to get the target viscosity at 100°C of the monograde. Multigrade oil on the other hand is formulated with lower-viscosity base oils, and then the viscosity is lifted by adding a VM polymer (often called viscosity index improver

(VII)) and pour point depressant. Other performance additives are added to meet all the requirements of multigrade oil (requirements of a winter, "W" grade and mono-grade). Figure 12.2 explains how a monograde SAE 30 and multigrade 10W-30/5W-30 are formulated. All the three grades have similar viscosity at 100°C but uses base oils of different viscosities. Performance additive package is also similar in the three grades. The difference lies in the use of VM. The monograde oil does not need VM, but as the multigrade oil becomes wider, more and more polymer is required. The VM provides non-Newtonian character to the oil so that it will have different viscos-ities at different shear rates. Viscosity–temperature characteristics of the oil improve with the use of VM (Fig. 12.3). This figure only shows the trend and is not based on actual viscosity data. Low temperature–viscosities by CCS and MRV also improve,

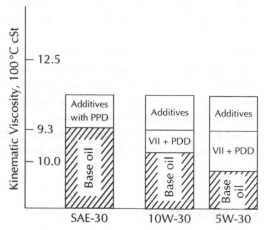

FIGURE 12.2 Composition of SAE 30, 10W-30, and 5W-30 engine oils.

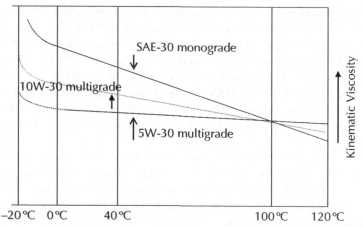

FIGURE 12.3 Viscosity–temperature trend of SAE 30, 10W-30, and 5W-30 engine oils.

and winter grade requirement can be easily met with the use of VM. Generally, monograde SAE 30, 40, or 50 may be adequate for temperate climate, when round-the-year temperatures do not go below 10°C. When the ambient temperatures are consistently high, SAE 50 or SAE 60 may be used.

Multigrade oils, however, offer additional advantage over monograde oils, especially with respect to fuel economy, low-temperature operability, and oil consumption. This is mainly due to their superior rheological properties exhibited by the presence of polymer. This polymer undergoes temporary viscosity loss (TVL) and permanent viscosity loss (PVL) and renders the oil non-Newtonian behavior. Some VMs when subjected to moderate shear rates lose their ability to thicken oil in the immediate vicinity of shear force. As soon as this force is removed, oil reverts back to its original viscosity. This phenomenon called TVL allows the oil to thin out in the high shear area of the engine and thus provide fuel economy [9–11]. This can be further explained by the following viscosity–temperature curves of monograde and multigrade oils. All SAE 30 and 10W-30 or 5W-30 have similar viscosity at 100°C. But the multigrade oils have lower viscosity at temperatures below 100°C. Therefore, friction in the engine at the starting point will be lower, and fuel will be saved. Further, monograde oil will have very high viscosity at lower temperatures, and engine cannot be started easily with this oil, especially when temperatures are below 0°C. On the other hand, multigrade oil has good rheological properties, and engine can be easily started and lubricated. Other important aspect of the multigrade oil is that while it has lower viscosity below 100°C, it has higher viscosity at temperatures above 100°C (Fig. 12.3).

Due to this, oil consumption will be low, because multigrade oil will exhibit higher viscosity in the piston ring area under hydrodynamic condition. However, under the boundary lubrication condition, when the oil is subjected to high shear, viscosity of a multigrade oil drops, so will the friction, and then the multigrade oil will again provide fuel economy benefits. Thus, during the power stroke, when the piston is pushed forward at a great force, the multigrade oil has low viscosity due to shear effect on the oil, but as soon as this force is removed and piston starts downward journey, the oil regains its high viscosity and oil does not leak past the piston rings, thereby providing low oil consumption. A high shear rate viscosity limit at 150°C in the SAE classification protects the engine from wear and therefore also provides a restriction on the use of polymers that are highly shear unstable and also suitable base oils. This simplified explanation helps in understanding the usefulness of multigrade oils over monograde oils. The multigrade oils containing polymeric VIIs (non-Newtonian fluids) reduce engine friction and lower bearing wear [12] as compared to monograde oils. The amount of friction and wear reduction depends on the polymer type and its concentration. This is despite the fact that they have a lower apparent viscosity at high rates of shear. It has been shown that some VIIs form boundary lubricating films [13] of 10–30-nm thickness in rolling concentrated contacts. These films result from the presence of highly concentrated and thus very viscous layers of polymer solution formed on the two rubbing surfaces by polymer adsorption. These boundary films are formed by some types of VII and can persist up to temperatures in excess of 120°C. The implication of this finding is that certain polymeric VIIs can provide wear protection as well.

The viscosity of 10W/40 motor oils formulated with different viscosity index (VI) improvers has been measured at pressures up to 200 MPa (2000 bar) over a wide temperature and shear rate range. The response of viscosity to pressure was found to depend on the chemical nature of the VII at both low and high shear rates. Thus, the viscosities at 150°C of polymer-thickened oils are different at high pressures to that observed in conventional atmospheric pressure viscosities. The bearing weight loss of multigrade oils containing different VIIs in the ALI Bearing Distress Test correlated better with high pressure and high shear viscosities than with high shear viscosities measured at atmospheric pressure [14]. In addition to what has been explained, the polymeric VIIs are also beneficial due to physical adsorption at the metallic surfaces and superior high-pressure viscosities, providing wear protection and giving fuel economy. The key issue is the selection of a proper polymeric VII. Details of additives required in lubricants have been extensively discussed [15] in another publication by the author.

Selection of Base Oils for Engine Oils: API Groups II and III and PAO

Polyalphaolefins (PAOs) have been known to possess superior performance characteristics with respect to VI, pour point, volatility, and oxidation stability as compared to conventional API group I and II mineral base oils. In the modern base oil refinery, VI, pour point, volatility, and oxidation stability can be independently controlled, and API group III oils can be manufactured so that their performance characteristics are close to PAOs. With good availability of such oils at lower price, the performance gap between group III and PAO (group IV) is becoming narrower. API group III or chemically modified base oils are now preferred in many applications where PAOs were used earlier.

Pour Point   Pour points of API group III oils are higher than PAOs of comparable viscosity, but these can be easily corrected in the fully formulated oil by the use of pour point depressant additives. Base oils manufactured with modern isomerization catalysts respond very well to pour point depressant additives and can be lowered to −50°C or below, if necessary. However, technically, it is also possible to manufacture base oils of lower pour point, but it is more economical to correct this property by the use of additives.

Cold Cranking Simulator   Viscosity in engine journal bearings during cold-temperature start-up is an important parameter in determining the lowest temperature at which an engine will start. CCS viscosity is measured by ASTM D-5293 test method, under specified test conditions similar to the low-temperature engine bearings starting. Base oils are characterized by kinematic viscosity and VI. API group III base oils typically have VI comparable to that of 4 cSt PAO. For example, a 4.2 cSt (100°C) group III base oil, with a VI of 129, and PAO 4, with a viscosity of 3.9 cSt and VI of 123, will have similar CCS values. But a similar 4-cSt group II base stock of about 100 VI will have at least twice the CCS value. This performance makes the group III stock quite favorable for formulating fuel-efficient multigrade engine oils in the 0W-20 to 0W-50 or 5W-30/50 grades. Previously, this was possible only with the use of PAOs.

Noack Volatility   Noack volatility of engine oil is measured by ASTM D-5800 and has been found to correlate with oil consumption in passenger car engines. Strict requirements for low volatility are important aspects of engine oil specifications, such as API SM/SN and ILSAC GF4/GF5 in North America. Volatility of API group III base oils is similar to PAOs.

Oxidation Stability   High oxidation and thermal stability are the most important properties of the base oil for formulating long-life or long-drain engine oils. The stability of API group III base oils generally depends on their VI, which is an indication of the presence of highly stable isoparaffinic structures in the base oil [10]. These base oils have only trace amounts of aromatics, sulfur, and nitrogen compounds due to severe hydrofinishing/hydrocracking and hydroisomerization and are therefore quite stable. PAOs are also subjected to hydrogenation to saturate residual olefinic bonds, but any olefinic character left out can lead to stability problems. Chemically modified group III base oils are very similar to PAOs in their stability characteristics and also respond very well to antioxidants.

Thus, there is a wide choice of base oils for formulating engine oils depending on the target specifications. The most modern specifications with long-drain capabilities and fuel efficiency can be met either by API group III base oils or PAOs. Other general passenger car motor oils (PCMOs) can be manufactured with API group II or even with group I base oils.

ENGINE OIL PERFORMANCE

The earliest specification of engine oil was issued by the U.S. Army as MIL-Spec. MIL-L-46152 D/E specifies CRC L-38, Cat 1H2, and sequence IID, IIIE, and VE tests besides other physicochemical tests.

Currently, there are three well-recognized set of specifications:

1. API in the United States
2. ILSAC in association with AAMA and JAMA
3. ACEA in Europe

In addition, there are a large number of OEMs having their overriding specifications, and different countries have separate versions of their own specifications. In Europe, individual OEMs continue to have major influence on engine oil performance requirements for both passenger car and heavy-duty applications. In December 1995, ACEA introduced a new classification system consisting of nine different sequences to define engine oil quality for European automotive oils. The system is based on a physical, chemical, and engine tests similar to those used in the API but using both ASTM and CEC test methods. Performance claims against ACEA requirements have to be supported by data generated under the European Engine Lubricants Quality Management System (EELQMS). This system consists of two codes of practice—one developed by ATC and the other by ATIEL.

In the United States, API administers the licensing and certification of engine oils through a system that meets the lubrication requirements of OEMs. Engine oil performance requirements, test methods, limits for various classifications, and testing procedures are established cooperatively by the OEMs, oil marketers, additive companies, and testing laboratories. API requires that all marketers using the API service symbol obtain a license to use the symbol and sign an affidavit that test data are available to support performance claims.

The following pages outline current service classifications and the performance requirements for gasoline passenger cars. This information is intended as a guide to understand the requirements of modern gasoline engine oils and the testing requirements of such oils. In certain situations, test data generated with one viscosity grade of an engine oil formulation can be extrapolated to another viscosity grade using the same additive technology. This practice is called read-across guidelines. API has established read-across guidelines for passenger car and diesel engine oils. These are quite complex, and readers are advised to refer to the original API guidelines.

ENGINE OIL SPECIFICATION

There are large numbers of standards and test methods by individual OEMs and standard organization. Readers are advised to follow the following:

1. SAE J300 (1999) classification (see Chapter 2)
2. API SN classification and specification
3. ILSAC specification
4. ACEA specification
5. OEM specification

Table 12.2 (API) and Table 12.3 (ILSAC) contain key elements of these documents and must be understood carefully. For actual test limits, original documents must be referred. These tables indicate how the oil severity went on increasing with each category, and as a consequence, test requirements also became comprehensive.

The latest API SN and ILSAC GF5 specifications have been compared in Table 12.4. Data and test limits show that these standards have been fully aligned with each other, and there is virtually no difference between the two. The large numbers of engine and bench tests also indicate the complexity that an oil formulator has to go through to meet the standard. The costs of engine tests have also gone up substantially. In the current scenario, it is extremely difficult for an individual oil company to run a full-scale development program to API SN

TABLE 12.2   Performance engine tests in API category gasoline lubricants

Category obsolete	Engine tests	Category current	Engine and performance tests
SA	None	SJ	L-38, seq. IID, IIIE, VE, limits as per API SG only, Noack volatility 22% max., for 0W20, 5W20, 5W30, 10W30, for all others 20% max., P 0.1% max., foaming, GM filterability, homogeneity, gelation index, and high-temp. deposit TEOST introduced
SB	L-4 or L-38, seq. IV	SL	ASTM BRT, seq. IIIF, VE, IVA, VG, and
SC	L-1; L-38; seq. IIA, IIIA, IV, V		VIII, shear stability seq. VIII-stay-in-grade, volatility loss ASTM D-5800,
SD	L-38; L-1 or Cat 1H; Falcon; seq. IIB, IIIB, IV, VB		max. 15%, P 0.1% max., foaming, GM filterability, homogeneity, gelation index, and high-temp. deposit TEOST limits tightened
SE	L-38; seq. IIC or IID, IIIC or IIID, VC or VD	SM	ASTM BRT, seq. IIIG, IIIGA, VG, IVA, and VIII, shear stability seq. VIII-stay-in-
SF	L-38; seq. IID, IIID, VD		grade, volatility loss ASTM D-5800,
SG	L-38; Cat 1H2; seq. IID, IIIE, VE		max. 15%, P 0.08% max. and min. 0.06%, foaming, GM filterability, homogeneity, gelation index, and high-temp. deposit TEOST limits tightened, EOWTT flow reduction introduced
SH	L-38, seq. IID, IIIE, VE, limits as per API SG only, Noack volatility and phosphorous limits (0.12% max.) were introduced for multigrade oils, homogeneity, foaming, and GM filterability tests also introduced	SN	ASTM BRT, seq. IIIG, IIIGA, VG, IVA, and VIII, shear stability seq. VIII-stay-in-grade, IIIGB, P-volatility, volatility loss, 1 h, 250°C ASTM D-5800, max. 15%, P 0.08% max. and min. 0.06%, foaming, high-temp. foaming, GM filterability, homogeneity, gelation index, and high-temp. deposit TEOST limits tightened, EOWTT, EOFT flow reduction introduced, simulated distillation, 10% max. at 371°C, emulsion retention test, elastomer compatibility

performance level. It is generally advantageous and cost-effective to obtain the recommendation of additive development companies who can offer additive package to several oil companies by running a single test program. Oil companies can then run various viscometrics and optimize base oil combination to get the desired performance.

Sequence tests are engine tests and are required for the development of gasoline engine oils. Some details of these tests have been provided. Further details of these tests and test conditions can be found in the relevant ASTM test.

TABLE 12.3   ILSAC specifications: key bench and engine tests

Category obsolete	Tests	Category	Tests
GF1	L-38, seq. IID, IIIE, VE, API SG limits apply, VI EFEI 2.7% min., HTHS, volatility, GM filterability, P 0.12% max. homogeneity, miscibility and foaming, shear stability	GF4 valid till September 2011	For 0W-XX, 5W-XX, 10W-XX BRT, seq. IIIG, IIIGA, IVA, VG, VIII, P 0.08%, evaporation loss, distillation, GM filterability, TEOST-high-temp. deposit test, gelation index, seq. VIB fuel economy test—three levels, after 16 and 96 h of aging
GF2	For 0W-XX, 5W-X, 10W-XX, engine tests the same as in GF1, Noack evaporation 22% max., P 0.1% max., GM filterability, TEOST-high-temp. deposit test, gelation index, seq. VIA fuel economy test—three levels	GF5 current for 2011 and older vehicles	For 0W-XX, 5W-XX, 10W-XX BRT, seq. IIIG, IVA, VG, VIII, VID, P 0.08%, P-volatility 79% min., sulfur, evaporation loss, simulated distillation, GM filterability, foaming and high-temp. foaming, shear stability, TEOST-high-temp. deposit test MHT-4, 33C, EOWTT, EOFT, gelation index, aged oil CCS viscosity, seq. VIB fuel economy test— three levels after 16 and 96 h of aging. Emulsion retention and elastomer compatibility
GF3	For 0W-X, 5W-X, 10W-X BRT, seq. IIIF, IVA, VG, VIII, API SL limits apply, P 0.1%, evaporation loss, distillation, GM filterability, TEOST-high-temp. deposit test, gelation index, seq. VIB fuel economy test—three levels		

Sequence IID

Sequence IID assesses rusting of parts when engine is operated for 32 h in three stages at low speed and low load (in a 5.7-l General Motors (GM) V8 engine) for short periods during which the oil and coolant temperatures have not reached normal operating conditions. Test is now obsolete.

Sequence IIIF (ASTM D-6984)/IIIG

The test has been developed to evaluate engine oils for protection against oil thickening, piston deposits, sludge formation and cam lobe, and lifter wear during high-speed high-temperature oxidation conditions and has been in use for a long time. The high temperature results in oil viscosity increase due to oxidation. Deposits generated under

TABLE 12.4 Comparison of API SN and ILSAC GF5: engine and bench tests

Test	Performance criteria	API SN	ILSAC GF5
Seq. IIIG	Kinematic viscosity	150	150
	increase at 40°C % max.	4.0	4.0
	Av. wt. piston deposit,	None	None
	merit, min.	60	60
ASTM D-7320	Hot stick rings		
	Av. cam and lifter wear μm		
	max.		
Seq. IVA	Average cam–wear μm max.	90	90
ASTM D-6891			
Seq. VG	Av. engine sludge, merit,	8.0	8.0
	min.		
	Av. rocker cover sludge,	8.3	8.3
	merit, min.		
	Av. engine varnish, merit,	8.9	8.9
	min.		
ASTM D-6593	Av. piston skirt varnish,	7.5	7.5
	merit, min.		
	Oil screen sludge % area	15	15
	max.		
	Hot stick compression ring	None	None
	Oil screen debris, cold stuck	Rate and report	Rate and report
	ring, oil ring clogging		
Seq. VID	SAE XW-20 grade	2.6% min.	2.6% min.
	FEI SUM		
	FEI 2	1.2% at 100-hrs	1.2% at 100-hr
		aging	aging
	SAE XW-30 grade	1.9% min.	1.9% min.
	FEI SUM		
	FEI 2	0.9% min. at 100-h	0.9% min. at
		aging	100-h aging
ASTM D-7589	SAE 10W-30 grade and	1.5% min.	1.5% min.
	above		
	FEI SUM		
	FEI 2	0.6% min. at 100-h	0.6% min. at
		aging	100-h aging
Seq. VIII	Bearing weight loss mg.	26	26
	max.		
ASTM D-6709	Percent loss max. 1 h	15	15
Evaporation loss	250°C, D-5800		
Simulated	Percent loss max. at 371°C,	10	10
distillation	D-6417		
Phosphorous	Percent loss max., D-4951	0.6–0.8	0.8 max.,
P-volatility, seq.	Percent min. retention,	79	79
IIIGB	ASTM D-7320		

(*Continued*)

TABLE 12.4 (Continued)

Test	Performance criteria	API SN	ILSAC GF5
Sulfur	Percent max., D-4951/D-2622, 0W and 5W-XX	0.5	0.5
	For 10W-30	0.6	0.6
TEOST MHT-4	Deposits wt. ASTM D-7097 mg., max.	35 up to 10W-30 and 45 for others	35 up to 10W-30 and 45 for others
TEOST 33C	Deposits wt. ASTM D-6335, mg. max.	–	30
EOWTT and EOFT	Percent flow reduction with 0.6–3% water	50 max.	50 max.
Gelation index	ASTM D-5133	12	12
Foaming	Seq. I, II, III tendency/stability	10/nil, 50/nil, 10/nil	10/nil, 50/nil, 10/nil
High-temp foaming	Tendency/stability	100/nil	100/nil
Aged low-temp. viscosity	Seq. IIIGA test ASTM D-7320	CCS viscosity of the grade	CCS viscosity of the grade
Shear stability	Seq. VIII, 10-h stripped viscosity	Stay-in grade	Stay-in grade
Homogeneity/miscibility		Pass	Pass
BRT	Average gray value	100 min.	100 min.
Emulsion retention		No water separation	No water separation
Elastomer compatibility		Yes	Yes

these conditions have been shown to correlate with field service where piston ring sticking and valve lifter sticking occur. After 64-h test duration in a 3.8-l GM V6 engine, oil viscosity at 40°C and wear metals (Cu, Pb, and Fe) are determined.

Sequence IVA/Nissan KA24E

The 2.4-l Nissan KA24E engine test was initially developed by JASO but later adopted and modified by ASTM. The test runs with specially blended unleaded gasoline for a period of 100 h to evaluate lubricant ability to control camshaft lobe wear. The used oil is tested for viscosity at 100°C, fuel dilution, and wear metals (Cu, Fe, Al). The test simulates stop-and-go driving cycle with significant engine idling. This reduces hydrodynamic lubrication leading to high wear levels.

Sequence VE/VG (ASTM D-6593)

Similar to sequence III, sequence V has been used extensively in the past to assess lubricant formulation for many years. Sequence VE was run on a 2.3-l Ford engine, but the VG is running on a 4.6-l Ford V8 engine. The VG test is run for

216 hours in 54 cycles of 4 h in three stages. The tests have been developed to assess lubricant ability to minimize the formation of sludge, varnish deposits, and overhead valve wear when running under stop/go city driving. Older sequence VE test was also assessing valve train wear, but this is not included in the VG as the 4.6-l engine has roller follower technology, which is not prone to the higher levels of wear as the 2.3-l engine. This aspect has now been taken care and included in sequence IV test.

Sequence VIA/VIB (ASTM D-6837)/VID (ASTM D-7589)

There has been no engine test to evaluate the fuel economy of engine oils prior to the CAFE (Corporate Average Fuel Consumption) legislation in the United States. The first method of measuring fuel consumption was the *5 Car Test*. This was then changed to an engine test and is now being used extensively for a number of years. Sequences VIA and VIB both run on a 4.6-l Ford V8 engine. Sequence VIB (ASTM D-6837) is an engine dynamometer test that measures the lubricant's ability to improve fuel economy of passenger cars and light-duty trucks equipped with a low-friction engine. The method compares the performance of a test lubricant to the performance of a baseline lubricant over five different stages of operation. Sequence VIB/VID also measures fuel consumption in used aged oils.

Sequence VIII/CLR L-38 (ASTM D-6709)

The L-38 test has been in use for several years and is run on a single-cylinder Labeco 700-CC (42.5 cubic inch) engine. It has been developed to measure the oil capability to protect copper–lead-bearing corrosion and deposits of sludge and varnish under high-temperature heavy-duty service. L-38 test was run on leaded gasoline. With engine running with unleaded gasoline, it is now called sequence VIII. Engine operates at 3150 rpm for 40 h.

Ball Rust Test (BRT) (ASTM D-6557)

Sequence IID became obsolete, and there was a need to evaluate engine oils for their rust protection capability. Ball rust test (BRT) uses a laboratory test rig to simulate service conditions for rust formation. In this test, the hydraulic lifter balls are immersed in test engine oil at a controlled temperature. Organic acid and mineral acid are injected into the shaking oil, and the balls are evaluated and rated for surface discoloration and rust formation.

ACEA have formulated very comprehensive standards for gasoline and diesel engine oils. Since 2007, four sets of standards have been prepared and have been again revised in 2008, 2010, and 2012. The service fill designations are provided in Table 12.5 and Table 12.6. There is a separate four sets of standards for engines fitted with aftertreatment devices (Table 12.6). While ACEA follows SAE classification, most of the engine tests stipulated are European engines according to the CEC test procedure. Some of the specific European tests are fuel economy by M111, low-temperature sludge and black sludge formation tendency, oxidation test, and engine test in the presence of biodiesel to address typical European problems.

TABLE 12.5   ACEA service fill oils for gasoline and diesel engines: designations

Year 2007	A1/B1-04	A3/B3-04	A3/B4-04	A5/B5-04
Year 2008	A1/B1-08	A3/B3-08	A3/B4-08	A5/B5-08
Year 2010	A1/B1-10	A3/B3-10	A3/B4-10	A5/B5-10
Year 2012	A1/B1-12	A3/B3-12	A3/B4-12	A5/B5-12

TABLE 12.6   ACEA service fill oils for gasoline and diesel engines: designations

	With aftertreatment devices		
Year 2007	C1-04	C2-04	C3-07	C4-07
Year 2008	C1-08	C2-08	C3-08	C4-08
Year 2010	C1-10	C2-10	C3-10	C4-10
Year 2012	C1-12	C2-12	C3-12	C4-12

OEM SPECIFICATIONS

There are large numbers of OEM specifications, and some of these are as follows. For details, original specification may be referred:

1. GM Dexo1
2. MB 226.5, 226.51, 229.1, 229.3, 229.31, 229.5, 229.51, 229.52 (2009)
3. BMW Long life-01, Long life-01 FE, Long life-04
4. VW 501.01, 502.00, 504.00, 505.00, 505.01, 507.00

JAPANESE ENGINE TESTS

Several Japanese crankcase oil tests based on ASTM procedures have been developed earlier with Japanese engines. These became JASO procedure, and some of these are listed as follows:

JASO M328-95: Nissan KA24E

The test runs on a 1994 2.4-l Nissan engine with unleaded 89–93-octane gasoline, for a period of 100 h. The test period comprises two stages of 50 and 10 min repeated 100 times.

JASO M331-91: Nissan VG 20E, Low Temperature

Running for either 200 or 300 h depending on the quality of the lubricant under test, this assesses the detergency performance of passenger car engines under low- and medium-temperature conditions. The test comprises 200 or 300 cycles, each of which contains three stages with oil and coolant temperatures ranging from 50 and 42 to 117 and 97°C respectively, at speeds from idle to 3500 rpm. Independent of the sump temperature, the rocker cover temperature is controlled between 25 and 45°C.

JASO M333-93: Toyota IG-FE, High Temperature

The Toyota IG-FE procedure evaluates the high-temperature oxidation stability of passenger car engine oils for 48 and 96 h. The performance levels equate to API grades SD–SF. Following the Japanese general practice, the test covers a number of parameters comprising oil viscosity increase, sludge, varnish, ring sticking, oil strainer clogging, oil ring land deposits, oil consumption, and wear. The test uses a 2.0-l Toyota engine running on unleaded 89–93-octane gasoline under steady-state operation. The main test conditions are speed at 4800 rpm and torque at 59 Nm.

JASO M336-90: Nissan SD22

This procedure was developed to evaluate the detergency of automobile diesel engine oils under high-temperature and high-load conditions. Piston deposits, oil ring clogging, combustion chamber deposits, sludge, bearing weight loss, ring weight loss, tappet weight loss, cylinder liner wear, and cam nose wear are reported.

The test can be carried out for either 50 or 100 h depending on the performance level desired. The engine is a 2.2-l naturally aspirated unit from Nissan developing 48 kW and running on 0.5% S diesel. It runs at steady-state conditions at 4000 rpm and with coolant and oil temperatures of 80 and 120°C, respectively.

EUROPEAN TESTS

CEC L-38-A-94: Peugeot TU3M, VTW

This test was developed to evaluate the oil ability to prevent valve train and rocker pad scuffing in gasoline engines. The introduction of overhead camshaft designs to improve valve train rigidity and allow higher accelerations leads to new failures in engines running under low-speed low-temperature and medium-speed high-temperature duty cycles. The cyclic nature of the test evaluates these parameters and runs in two stages of 40 and 60 h using a Peugeot 1.36-l engine.

CEC L-55-T-95: Peugeot TU3M, High Temperature

This test procedure evaluates the capability of gasoline engine oil to resist the formation of deposits, ring sticking, and oil thickening under high-temperature operation. The engine used is a 1.36-l Peugeot in-line four-cylinder engine, which operates at 5500 rpm and maximum power for 11 h 50 min, followed by a 10-min gap. This is repeated eight times to give a total duration of 96 h.

ENGINE OIL FORMULATIONS

Gasoline engine oils contain the following chemical additives to control the problems discussed so far and to meet the performance requirement of the concerned specification.

1. Detergents
2. Dispersants
3. Antiwear
4. Antioxidants
5. Anticorrosion
6. VMs
7. Pour point dispersants
8. Antifoam
9. Antirust
10. Friction modifiers

The dosage of these individual additives would depend on several factors such as quality of base oils, lubricant performance level, and nature of additive whether it is monofunctional or multifunctional. Many additives are multifunctional such as ZDDP (antiwear, antioxidant, and anticorrosion) and certain polymers (VM and pour depressant). Use of synthetic oils such as PAO and ester will also reduce dependence on antioxidant, VM, and pour point depressant. Therefore, each lubricant formulation is balanced with selected base oils and additives keeping the performance characteristics of the oil in consideration. The candidate oil has then to be evaluated in the required engine tests followed by field trial.

The limit on sulfur (0.5 and 0.6%) and phosphorous (0.6 and 0.8%) in the API SN and ILSAC GF5 oils provides challenge to the oil formulators. This means that sulfur in base oil must be controlled, and the choice to do this is either to use API group II/III, GTL base oils, or PAOs for blending such oils. Reduced dosage of sulfur-containing detergents is also required. Limits on phosphorous call for reduced dosage of ZDDP, or alternate antiwear and antioxidant additives have to be used. Thus, to formulate latest gasoline engine oils, new generations of additives are required.

The lubricating composition [16] containing phenolic antioxidants, aminic antioxidants, and an ester compound with a disulfide structure has excellent stability against oxidation, prevents increase in acid value and sludge formation, and has low corrosivity to nonferrous metals.

ENERGY-EFFICIENT FRICTION-MODIFIED OILS

Oil-soluble friction-reducing compounds are polar in nature and have the tendency to get attached to the metal surfaces. These are both organometallic and ashless compounds. Oil insoluble additives like MoS_2 and graphite do possess friction-reducing properties, but their application in automotive engine oil is not favored since they can cause oil filter chocking. There are several aftermarket products using dispersed graphite, MoS$_2$, and nylon, but no oil company uses such products in their fully formulated engine oils. The organic friction-reducing additives generally belong to the following group of compounds: carboxylic acids and their esters; nitrogen compounds like amides, imides, amines, and their derivatives; phosphoric or phosphonic acid derivatives; soluble organometallic; molybdenum dialkyl dithiocarbamate and molybdenum dialkyl dithiophosphates; and organic polymers

Glycerol mono-oleate and oleylamide, derived from unsaturated fatty acids and molybdenum dithiocarbamate/dithiophosphate, have been extensively used as friction modifiers [17, 18] in gasoline engine oils. The oil composition provides satisfactory sequence VIB fuel economy test meeting ILSAC GF4 specification.

Oil-soluble fatty acid esters of a polyol [19] have been found to reduce friction of oils formulated with API group III, GTL, and PAOs. An environment-friendly energy-efficient lubricant [20] can be formulated with sulfated oxymolybdenum dithiocarbamate, an acid amide compound, a fatty acid partial ester compound, an aliphatic amine compound, and a benzotriazole derivative. The product has excellent friction-reducing and anticorrosion (to copper and lead) properties.

CORROSION AND WEAR

Corrosion and wear are due to the oxidation products including sulfur dioxide coming from fuel sulfur. The corrosive effect of sulfur is controlled by the use of overbased sulfonate in oil (having appropriate TBN). This problem is more pronounced in diesel engines due to the generation of both SOx and NOx. However, the introduction of ULSD has reduced SOx-related problems substantially.

VISCOSITY MODIFIER IN MULTIGRADE OILS

VMs play an important role in formulating multigrade oils, and their dosage in the formulated oil is the highest among all other additives used. Their role has been explained earlier. VMs are polymeric compounds that are less soluble or dispersible in base oil blend, but as the temperature rises, their solubility increases and viscosity of oil blend also increases. This phenomenon changes oil viscosity–temperature relationship, and VI of the product increases. The presence of polymer in oil also brings in new aspects since long-chain polymers behave differently under different shear, and the base oil containing polymers become non-Newtonian fluids. Monograde oils without polymer are Newtonian fluids, that is, their viscosity does not change with shear rates. When the shear rate is small, oils may undergo a TVL (temporary viscosity loss) by aligning the polymer chain in the direction of shear. As soon as this shear force is removed, the polymer chains realign themselves to the original shape, and oils return to their original viscosity. These can also undergo PVL (permanent viscosity loss) when higher shear is applied. PVL takes place when the polymer chains are broken permanently by the high shear rates. In this situation, the viscosity loss is permanent, and even on removing shear, oil does not come to its original viscosity. Even the VI will also undergo a change after PVL. Viscosity modification of base oil by polymeric VM is a function of polymer

TABLE 12.7 Fuel economy of 10W-30 oils in sequence VI and VIA tests with different VMs and base oils

Base oil type ———→	API group II	API group I	API group I
VM type ———→	Radial isoprene	Radial isoprene	OCP
Tests	10W-30 oil A	10W-30 oil B	10W-30 oil C
Kinematic viscosity 100°C	10.5	10.2	10.8
SSI of VM	7.5	17.0	39.0
HTHS 150°C	3.1	2.9	3.2
Seq. VI aged viscosity	10.17	9.65	9.40
Seq. VI fuel % economy	1.80	2.22	3.05
Seq. VIA aged viscosity	10.15	9.92	9.73
Seq. VIA fuel % economy	0.79	0.49	0.50

structure, its hydrodynamic volume, or polymer coil size and molecular weight [21]. Higher-molecular-weight products will provide higher degree of thickening but will also make it shear unstable. The increase or decrease in polymer coil or hydrodynamic volume with temperature change is completely reversible [22].

Table 12.7 shows interesting data on sequence VI and VIA fuel economy in 10W-30 engine oil using three different VMs with different shear stability index [23]. In sequence VI, fuel economy results are in the order of VMs' shear instability, that is, the more shear unstable is the polymer, the higher is the fuel economy obtained.

This trend was however not found in sequence VIA engine test since this engine sheared down the polymers to the similar level and the engine runs mainly under hydrodynamic region where viscosity plays the main role. The selection of VM for formulating multigrade gasoline engine oils is, thus, one of the most important exercises and needs to be evaluated for its shear stability, high temperature–high shear viscosity (HTHS), and TVL at low shear (MRV) and at moderate shear (CCS). The right selection of VM would lead to greater fuel economy and lower oil consumption.

SYNTHETIC GASOLINE ENGINE OILS

With conventional mineral base oils (solvent-refined, dewaxed neutral oils), it is difficult to formulate modern energy-efficient, low-volatility, and low-temperature operability PCMO meeting the latest API SN or ILSAC GF5 oils. The process becomes easier with the use of POAs [24–27] and API group II or III category base oils. Usually, a base oil blend of PAO and ester is ideal for such oils. Such blends provide satisfactory elastomer compatibility and high oxidation stability. However, the only limitation is their cost as compared to mineral oils. To reduce cost part, synthetic engine oils based on PAO and mineral oils also find favor. Advantages of synthetic engine oils can be summarized as follows:

1. Provides long life due to higher inherent thermal and oxidation stabilities.
2. Deposits are better controlled with synthetics.

3. PAO/ester blends have lower volatility; thus specification targets can be realized using lower-viscosity base oils.

4. Esters have better friction characteristics; therefore, improved fuel economy can be realized.

5. Low-temperature behavior of synthetics is far superior to mineral oils.

6. Lower dosage of VM and pour depressants is required as compared to mineral oils.

7. Due to such lower dosage, improved shear stability of the blend can be realized.

There are a few top-of-the-line products in the market that are fully synthetic and provide longer life in addition to meeting standard API/ILSAC specifications. Partially synthetic oils are also marketed to a limited extent. API group III and GTL base oils or chemically modified base oil provide tough competition to synthetics as their performance has reached very close to synthetics [28, 29]. However, the market share of fully synthetic oils remains low mainly due to the cost factor; but these have niche market.

Vehicle manufacturers are exploring the use of hydrogen as a fuel for IC engines. Hydrogen is a clean alternative fuel capable of improving energy self-sufficiency. However, the product of combustion of hydrogen is hot water vapors. Hot water vapors can create several performance-related problems, including engine backfire during hot summer days, engine detonation (i.e., misfiring or knocking) due to preignition, reduced sparkplug life due to deposit formation from the lubricant or contaminants, and corrosive or rust attack on piston rings, cylinder heads, and the combustion chamber. Also, combustion of hydrogen, as a gaseous fuel, may lead to higher levels of engine deposits. Following extensive testing, a lubricant formulation suitable for hydrogen-fuelled engines that reduce aforementioned performance problems and provide good engine durability has been reported. Lubricant has been prepared [30] from a synthetic oil blended with 3–6% by weight of a nitrogen-containing dispersant, 1–2.5 weight percent of an overbased magnesium detergent, 1–5 weight percent of an antioxidant, and 0.25–1.5 weight percent of a friction modifier. The blended product has a maximum 0.01 weight percent of calcium, less than 0.01 weight percent of Zn, less than 0.06 weight percent of P, and a sulfated ash level of less than 1.2%.

The modern high-performance, energy-efficient passenger car motor oils are now based on API group II/III, GTL base oils, or synthetic PAO–ester blends with low-ash-containing detergents/dispersants, low phosphorous- and sulfur-containing antioxidants, antiwear, and shear-stable VMs.

REFERENCES

[1] Kalghatgi GT. Combustion chamber deposits in spark-ignition engines: a literature review. SAE Paper 952443; 1995.

[2] Pinkus O, Wilcock DF. The role of tribology in energy conservation. Lubr Eng November 1978;34(11):599–610.

[3] Bartz WJ. Gear oil influence on efficiency of gear and fuel economy of cars, Proc. of Inst. of mechanical Engineers, Part D. J Automob Eng 2000;214 (2):189–196.

[4] Bartz WJ. Some considerations regarding fuel economy improvement by engine and gear oils. Lubr Eng 1983;39 (4):232–240.

[5] Bartz WJ. Fuel economy improvement by engine and gear oils. Petrotech 99, New Delhi; 1999. p 345–360.

[6] Holmberg K, Anderson P, Erdemir A. Global energy consumption due to friction in passenger cars. Tribol Int March 2012;47:221–234.

[7] Martin JM, Ohmae N. *Nano-Lubricants*. West Sussex: John Wiley & Sons; 2008.

[8] Moon WS, Kimura Y. Wear-preventing property of used gasoline engine oils. Wear 1990;139 (2): 361–365.

[9] Langeheim R, Bartz WJ. The significance of effective viscosity in non-stationary loaded journal bearings. ASLE Trans 1981;26 (1):69–79.

[10] Farnsworth G, Bachman H, Overton R. Energy saving with multi grade diesel engine lubricant—an experimental test design in winter urban bus operation. SAE Paper 780371; 1978.

[11] Chamberlin W, Sheaham T. Automotive fuel saving through selected lubricants. SAE Paper 750377; 1975.

[12] Okrent EH. The effect of lubricant viscosity and composition on engine friction and bearing wear. ASLE Trans 1961;4 (1):97–108.

[13] Matthew S, Spikes H, Gunsel S. Boundary film formation by viscosity index improvers. Tribol Trans 1996;39 (3):726–734.

[14] Hutton JF, Jones B, Bates TW. Effects of isotropic pressure on the high temperature high shear viscosity of motor oils. SAE 830030, HS-034 775, International Congress and Exposition, Detroit; February 28–March 4, 1983.

[15] Srivastava SP. *Advances in Lubricant Additives and Tribology*. New Delhi: Tech Books International; 2009.

[16] Katafuchi T, Motoharu I, inventors. Lubricating composition. US patent 81,293,198,129,319. March 6, 2012.

[17] Abraham WD, Kelley JC, Vilardo JS, inventors. Molybdenum, sulphur and boron containing lubricating oil composition. US patent 2003,0166,477 A1. September 4, 2003.

[18] Deckman DE, Poirier M-A, inventors. Lubricating oil composition. US patent 2007,0265,176 AI. November 15, 2007.

[19] Poirier MA, Sutton O, inventors. Fuel economy lubricant compositions. US patent 7,989,408. August 2, 2011.

[20] Kamano H, inventor. Lubricating oil composition. US patent 8,367,591. February 5, 2013.

[21] Covitch MJ. How polymer architecture affect permanent viscosity loss of multi-grade lubricants. SAE Paper 982638; 1998.

[22] Selby TW. The non-Newtonian characteristics of lubricating oils. ASLE Trans 1958;1:68–81.

[23] Dohner BR, Wilk HA, Supp JA. Developing fuel efficient engine oils for passenger car market. In: Srivastava SP, Basu B, editors. Proceedings of the International Symposium on Fuels and Lubricants (ISFL 1997); December 8–10, 1997; New Delhi. New Delhi: Tata McGraw Hill. p 35–40.

[24] Campen M, Kendrik DF, Marikin AD. Growing use of SynLube. Hydrocarbon Processing; February 1982. p 75–82.

[25] Srivastava SP. Synthetic lubricant scenario in 21st century. Proceedings of the 2nd International Conference on Industrial Tribology; December 1–4, 1999; Hyderabad. p 7–17.

[26] Goyal AK, Willyoung RW. Engine oil filter performance with synthetic and mineral oils. SAE Paper 850549; 1985.

[27] Lohuis JR, Harlow AJ. Synthetic lubricants for passenger car diesel engine oils. SAE Paper 850564; 1985.

[28] Kramer DC, Ziemer JN, Cheng MT, Fry CE, Reynolds RN, Lok B, Krug RR, Sztenderowicz ML. Influence of group II and III base oil composition on VI and oxidation stability. AIChE Spring Meeting; March 1999.

[29] Heilman WJ, Chiu IC, Chien JCW. New polyalphaolefin base oil. American Chemical Society Meeting; August 1999; New Orleans.

[30] Bardasz E, Chamberlain WB, inventors. Lubricant for hydrogen-fuelled engines. US patent 8,163,681. April 24, 2012.

CHAPTER 13

ENGINE OILS FOR COMMERCIAL VEHICLES

The diesel engine is highly efficient in converting fuel into energy as compared to gasoline engine. Therefore, it became very popular for land, railroad, and sea transportation. Most commercial vehicles such as trucks, buses, tractors, off-highway equipment, and other stationary engines use diesel fuel. Its application in the passenger car market has not been very favorable due to mainly noisy engine and higher level of exhaust emissions. However, in the past few decades, the diesel engine technology has been evolved to reduce both noise and exhaust emissions. Slow-speed diesel engines have been used in marine engines and general pumping operations. Medium-speed engines found applications in railroad and power generation. High-speed engines exceeding 1000 rpm found use in road transportation and off-highway equipment. Further classification of high-speed engines into light duty and heavy duty is based on power output. Diesel fuel-powered engines are efficient and reliable and have long life.

Diesel engines are based on the principle of compression ignition. Air is first introduced into the combustion chamber with the opening of the intake valves. The air intake is facilitated by the downward movement of the piston that creates a pressure differential through volume expansion. Turbochargers are used to force more air into the combustion chamber to increase air density to burn more fuel for the same displacement. The air, once in the system and the intake valve closed, is then compressed to reach high pressure and high temperature by the upward movement of the piston. Fuel is injected at this point, which autoignites immediately due to the high temperature produced by compressing air. Diesel fuel has this tendency of autoignition at about 210°C, while gasoline has a higher autoignition temperature of about 246°C. Autoignition temperature is that temperature at which the fuel/air mixture ignites automatically without an external spark. That is how diesel engines are known as compression ignition engines. The combustion of the air/fuel mixture creates the expansion that forces the piston to move downward again, producing power output.

The main components of the internal combustion engine are the power cylinders and pistons where the combustion takes place. Other components include the cam and valve train system to control the timing of the combustion, the crankshaft, and connecting rods to receive power and to provide mechanical movement. The gear train operates pumps and accessories to allow the movement of lubricant to different

Developments in Lubricant Technology, First Edition. S. P. Srivastava.
© 2014 John Wiley & Sons, Inc. Published 2014 by John Wiley & Sons, Inc.

parts of engine, providing cooling and power to the engine. All of these components are housed in an engine block with an engine head on the top. The movement of the piston inside the liner from the top to the bottom is called a stroke. Piston rings are used to provide sealing by controlling the oil film on the liner wall. Multiple rings are used for this purpose. The lowest ring is called the oil ring to limit the quantity of oil reaching to the upper end of the liner to prevent excessive oil accumulation leading to deposits. The top ring is used to prevent the leakage of compressed air or combustion gases through the rings so that the pressure in the cylinder is maintained. There are usually additional rings between the oil ring and the top ring to facilitate the process. The liner is cooled by coolant flowing between the liner and the block, since most heat is generated in this area. Further cooling is provided by crankcase oils. Most of the heat is rejected in the exhaust gases. The firing of different cylinders is synchronized by the design of the cam and the movement of the valve train for smooth operation. The power generated by the combustion is sent to the transmission for useful work.

The basic concept and working of an engine is also described in the previous chapter dealing with PCMO and generally remains unchanged in diesel engines. However, the major difference lies in the use of fuel; diesel engines run hotter, deposit and soot formation due to thermo-oxidative process is more pronounced, and the formation of SOx and NOx is higher in diesel engines. Recent diesel engine designs are largely driven by emission regulations. Major refinement in bowl design, air handling, and fuel injection has led to progressive reductions in NOx and particulate matters (PM). Meanwhile, higher power density, improved fuel economy, and extended oil drain intervals are becoming an integral part of modern diesel engines. Currently, exhaust gas recirculation (EGR) and aftertreatment devices are necessary to achieve higher level of emission control. All these parameters provide challenges to diesel lubricant formulations. These different conditions also make diesel engine oils different from gasoline engine oils, and their testing procedure/engine tests are, therefore, different. The physicochemical parameters such as viscometrics [1], low-temperature flow properties, high-temperature high-shear (HTHS) viscosity, and multigrade philosophy remain guided by SAE J-300 and similar to gasoline engine oils.

PERFORMANCE CHARACTERISTICS: DIESEL ENGINE OILS

The primary function of a crankcase lubricant is to reduce friction between moving metal surfaces. In addition, it has to cool the engine parts and handle contaminants such as acids, sludge, soot, deposits, and oxidation products. The oil should also possess good rheological properties so that it is pumped to various parts under different and varying ambient and operating temperature conditions. The cold flow properties and HTHS viscosities are important to provide lubrication as defined in SAE J-300. When the loads are high, lubricant viscosity alone cannot sustain an oil film. Under boundary lubrication conditions, antiwear and extreme pressure (EP) additives are required to form a solid film of low shear. To obtain continuous good performance, the engine parts and all metal surfaces must be kept clean. This function is carried out by detergents in oil formulation. Further, the debris and contaminants like soot generated during the combustion process must be dispersed and removed by dispersant. The detergency provided by additives like metal sulfonates, phenates, or salicylates cleans deposits from the surfaces and keep them in suspension. Calcium sulfonates and magnesium phenates are the most common. Most formulations consist of either calcium sulfonate alone or a combination of the two.

In the early 1980s, there was an increased use of turbocharging to obtain higher-power outputs from existing engines and to produce newer engines with higher efficiencies. The higher pressures around the top piston ring led to a polishing effect in engines operating under severe duty cycles. The high top ring pressures and soot-loaded oil produced a mirrorlike appearance on the cylinder walls of affected engines. The impact of this mirrorlike finish was that the oil control ring became ineffective and resulted in very high oil consumptions. This problem could be solved by controlling soot agglomeration in oil and dispersing it finely in oil by using specific combination of dispersants.

As already discussed, diesel engines operate at relatively higher temperatures and use higher-boiling fuels containing higher amount of sulfur. Sulfur level has, however, come down with the use of ultralow-sulfur diesel (ULSD) fuel (15 ppm sulfur max.). There is high soot loading and higher chance of forming deposits [2] in the diesel engines as compared to gasoline engines. The diesel engine oils, therefore, require higher level of detergency, dispersancy, and antioxidant characteristics. With the decrease in sulfur content in diesel fuels, the problems related to SOx and acid formation have been reduced. For example, the TBN of diesel engine oil operating on ULSD can be safely reduced to a lower level. Diesel engines are now equipped with exhaust gas aftertreatment systems to allow them to comply with emission legislation. Some of these systems are sensitive to the combustion products of the fuel and lubricants such as phosphorus coming from the lubricant and sulfur and sulfated ash coming from the combustion of both fuel and lubricant. In order to ensure the durability of these aftertreatment systems, special lubricants are developed that have low levels of sulfated ash, sulfur, and phosphorus (low SAPS). Lubricating oil formulations with no or minimum phosphorus have been developed, and wear measurements in a heavy-duty diesel (HDD) engine showed low cylinder liner wear. EGR has been found to be very effective in reducing emissions of oxides of nitrogen, for light-duty diesel engines. However, EGR results in a sharp increase in particulate emissions in HDD engines. The effects of soot-contaminated engine oil on the wear of engine components were examined in a three-body wear machine. Results show that diesel soot interacts with oil additives and reduces [3] antiwear properties, possibly by abrasive wear mechanism. Analysis also showed that the phosphorous level plays a dominant role on the antiwear performance of the oil. The effect of dispersant was not very significant, but higher dispersant levels reduce wear. The effect of sulfonate was not noticed within the concentrations studied. Ball-on-flat-disk-type wear tests also revealed the increased wear due to the presence of soot. SEM studies of wear scar diameters suggest that soot is abrasive in nature. However, according to Round, soot reduces antiwear additives' effectiveness by preferential adsorption of active antiwear components and does not allow them to reach surfaces, rather than removing the surface by abrasion [4]. It was also found that ZDDP is one of the most effective antiwear additives. Superior soot dispersing characteristics can be obtained in diesel engine oils by using a combination [5, 6] of a high-molecular-weight dispersant and a linked aromatic oligomer. High level of basic amine from dispersants leads to the deterioration of seals in the engine. Therefore, it is desirable to formulate lubricants with good soot dispersancy properties using lower amounts of high-molecular-weight nitrogen-containing dispersant. These compounds [7] have been used in engine oils along with overbased calcium sulfonate, phenates, and salicylates. Such additive combinations possess good corrosion resistance against SOx produced in diesel engines. Emission regulations have become stricter all over the world. For reducing NOx and PM in the exhaust gas from diesel engines, combustion improvement technologies, such as high-pressure injection, EGR systems, and exhaust gas treatment technologies, like oxidation catalysts, diesel particulate filters (DPFs), and NOx storage reduction catalysts (NSRC), have been developed. DPFs are used for capturing and removing PM in exhaust gas with a filter and have various structures. These particulate filters have to be regenerated to remove PM by oxidization and combustion. Recently, continuous regenerative DPFs have been developed. Engine oils based on a combination of several magnesium detergents lead to the reduction of clogging in DPF.

CONTAMINANT CONTROL

The most detrimental contaminant in today's diesel engines is oil soot. Soot is generated during the combustion process and brought to the bulk of the oil primarily by the scraping of the soot-containing oil from the ring/liner area. A secondary source of oil soot is blow-by. Because of the way soot is brought into the oil, the soot rate in the oil is not directly proportional to exhaust soot or particulates. Soot in the oil is correlated with the end of fuel injection event. When the injection timing is retarded, or if the combustion event is prolonged, combustion will continue while the piston is already moving downward, exposing more oil film on the cylinder wall to the combustion products. Soot trapped in the oil film ends up in the sump as oil soot. While soot is generally described as a carbonaceous core covered with adsorbed fuel/oil fractions, soot is actually defined by the method used to measure it. The oil industry currently uses thermal gravimetric analysis as a standard procedure to measure the amount of soot in the oil (ASTM D-5967, 1997).

Modern engines meeting the latest emission regulations can reach over 5% soot level before an oil change. This number has further risen with the advent of long-drain oils and EGR for the control of emissions. The accumulation of soot in oil leads to oil thickening through agglomeration of soot particles and increased wear of parts under boundary lubrication. Oil thickening can lead to loss of oil pressure, which leads to loss of oil flow to critical components. Soot in oil also affects cold flow performance of the oil. The increase in viscosity at low temperatures leads to startability problems under cold conditions. On the other hand, HTHS viscosity also increases with soot, providing a thicker oil film. Typical soot particles range from below 0.1 μm to over 1 μm.

TABLE 13.1   Tolerance of soot by oils meeting API category

Oil specification	CC/CD	CE-CF-4	CG-4	CH-4	CI-4 and CJ-4
Soot tolerance (wt.%)	1	1.5	3.8	4.8	≥6

These particles cannot be removed easily by lube filters. The solution is to use an appropriate dispersant (like succinimides) or multifunctional viscosity index improvers or both (like functionalized OCPs or PMAs) to keep soot particles finely dispersed and prevent them from agglomeration.

At the same time, antiwear additives need to be rebalanced to maintain wear performance. As discussed earlier, soot particles are capable of depleting antiwear additives by absorption on their surface. This aspect needs careful consideration, and a balanced selection of additives for diesel engines has to be done. When soot is properly dispersed, not only the oil viscosity will stay within limits, wear is also controlled. Typical soot levels tolerated by various API oils are given in Table 13.1. If soot is not properly dispersed and it is allowed to agglomerate, the larger soot particle may become abrasive and cause higher wear.

LUBRICANT PERFORMANCE: FRICTION AND WEAR REDUCTION

All internal combustion engines involve the three lubrication regimes, that is, hydrodynamic lubrication, elastohydrodynamic/mixed film lubrication, and boundary lubrication.

For hydrodynamic lubrication, the rheological properties of oil are important and are defined by kinematic viscosity at 40 and 100°C (SAE J-300). In addition, cold flow performance (ASTM D-5293 and ASTM D-4684) and HTHS viscosities (ASTM D-4683) are also defined at low and high temperatures. For mixed film and boundary lubrication, antiwear and EP additives and their response in base oil determine the performance of the lubricant. The most commonly used antiwear additive is ZDTP. They are generally of three types: primary alkyl, secondary alkyl, and aryl. With the restrictions imposed on phosphorus and sulfur in current engine oils, substitute of ZDDP is being worked out. In diesel engines, soot creates additional factor for higher wear. Valve train wear in diesel engines has been found to be related to the soot aggregation at the inlet of the cam-tappet contact [8]. Aggregation is related to the interparticle interactions and, more particularly, to the dielectric constants of the soots and of the lubricant. Cummins M-11 EGR engine test has been used for studying soot generation and wear characteristics. This M-11 engine can generate soot levels as high as about 9% [9]. The higher the EGR rate, the more soot in the oil and the most detrimental effect is found in valve train wear. Some engine tests, including field tests, have been carried out to investigate the contribution of soot to valve train wear. Soot seems to act as an abrasive on the antiwear solid film formed by the oil [10] on the metal surface, and this film contains Ca, O, P, and S.

DIESEL ENGINE OIL SPECIFICATION

Diesel engine oil specifications have been defined by both OEMs and organizations such as API, ACEA, U.S. Military, and others. Various national standards have also been developed, but these are generally variants of some of the following standards:

API CA to CJ	Specification
ACEA	Standard
U.S. Military	Standard
Japanese	Standard
Global Engine Oil	Specifications
OEM	Standards

Key performance characteristics of some of the important specifications are provided in Table 13.2, Table 13.3, Table 13.4, and Table 13.5.

In Europe, engine manufacturers continue to dominate and specify their own oil specifications. The standards issued by the Comite des Constructeurs d' Automobiles du Marche Commun (CCMC) have been in place for several years and were quite

TABLE 13.2   Performance engine tests in API-category diesel engine lubricants

Category obsolete	Engine and bench tests	Category current	Engine and bench tests
CA	L-4 or L-38, L-1 (0.35% min. S fuel)	CH-4	Cat 1P, Cat 1K, T8-E, T-9, M-11, RFWT, Seq. IIIE/F, EOAT, bench corrosion test
CB	L-4 or L-38, L-1 (0.95% min. S fuel)	CI-4	Cat 1R, Cat 1K, Cat 1-P, T-8E, T-10/12, M-11, Seq. IIIF/G, RFWT, HEV1, HTHS, MRV, Noack evaporation, bench corrosion, shear stability, foaming, elastomer compatibility
CC	L-38, LTD, Cat 1-H2, Seq. IIC or IID	CI-4 plus (2006)	Mack T08E replaced by T-11 Other tests same as CI-4
CD	L-38, Cat 1-G2	CJ-4	Mack T-12 (ASTM D-7422)
CD II	L-38, Cat 1-G2, 6 V-53T		
CE	L-38, Cat 1-G2, T-6, T-7, NTC-400		ISM (ASTM D-7549) Mack T-11 (ASTM D-7156)
CF	L-38, Cat 1M-PC		ISB (ASTM D-7484)
CF2	L-38, Cat 1M-PC, 6V-92TA		Cat 1-N (ASTM D-6750)
CF4	L-38, Cat 1-K, T-6, or T9, T-7, or TBA (D-5967), CBT (D-5968)		RFWT (ASTM D-5966) Seq. IIIF/G (ASTM D-6984) EOAT (ASTM D-6894)
CG4	L-38, Cat 1N, T-8, Seq. IIIE, 6.2L, bench corrosion test		HTCBT (ASTM D-6594) Shear stability (ASTM D-7190) MRV, foaming, elastomer compatibility 1.00% max. sulfated ash (ASTM D-874) 0.12% max. phosphorus (ASTM D-4951) 0.40% max. sulfur (ASTM D-4951)

TABLE 13.3   ACEA service fill for heavy-duty diesel engine oil: designations

Year 2007	E2-96 issue 5	E4-07	E6-04 issue 2	E7-04 issue 2
Year 2008	E4-08	E6-08	E7-08	E9-08
Year 2010	E4-08 issue 2	E6-08 issue 2	E7-08 issue 2	E9-08 issue 2
Year 2012	E4-12	E6-12	E7-12	E9-12

TABLE 13.4   ACEA service fill for heavy-duty diesel engines: key requirements

Test	Test method	E4-12	E6-12	E7-12	E9-12
Visc. grade	—	SAE J-300 requirements			
Shear stability	ASTM D-6278	Stay-in-grade			
HTHS visc. (cSt)	CEC-L-036-90	≥3.5			—
Evaporation loss	Noack, 250°C 1 h (%)	≤13			—
Sulfated ash (%wt.)	ASTM D-874	≤2.0	≤1.0	≤2.0	≤1.0
Phosphorous (%)	ASTM D-5185	—	≤0.08	—	≤0.12
Sulfur (%)	ASTM D-5185	—	≤0.03	—	≤0.4
Foam test	ASTM D-892	10/nil, 50/nil, 10/nil			10/0, 20/0, 10/0
High temp. foaming	ASTM D-6082	Seq. IV(150°C)-200/50			—
Elastomer compatibility	CEC-L-039-96	Refer to standard for details			
Oxidation, induction time (min)	CEC-L-085-99, PDSC	Report			≥65
Corrosion	ASTM D-6594	Refer to standard for details			
TBN (mgKOH/g)	ASTM D-2896	≥12	≥7	≥9	≥7
Low-temp. pumpability	CEC-L-105, MRV	SAE J-300 requirements			
Bore polish	CEC-L-101-08	Refer to standard for details			
Cam wear (µm)	CEC-L-099-08 OM-646 LA	≤140	≤140	≤155	≤155
Soot in oil	Mack T-8E, ASTM D-5967	Refer to standard for details			
Soot in oil	ASTM D-7156, Mack T 11	—	—	—	See standard
Soot-induced wear	Cummins ISM, merit	—	—	≥1000	
Wear, liner, rings, bearing	Mack T-12, merit	—	≥1000	≥1000	≥1000

TABLE 13.5   Global diesel engine oil specification DHD-1: key test requirements

Bench tests	Engine tests
Corrosion test	Cat 1R
Elastomer compatibility	Cummins M11-HST
Foam test and high-temp. foam test	Mack T-9/T-12 and Mack T8-E
PDSC oxidation induction time	6.5L RFWT
HTHS visc.	HEUI
Noack volatility	MB OM 441 LA
Sulfated ash (%)	Mitsubishi 4D 34 T4, 160h

popular. These are now obsolete. Long-drain diesel engine oils such as meeting CCMC D5 specification became common in Europe and some other counties [11]. Mercedes Benz has the most comprehensive testing and approval requirements for diesel engine oils. Volkswagen, Volvo, MAN, and many others have their own test requirements. In 1996, CCMC has been replaced by the Association des Constructeurs Europeans de l' Automobile (ACEA), and a series of specifications have been issued by ACEA:

1. ACEA E1–E3 in 1996
2. ACEA E4 in 1998
3. ACEA B1–B5 in 2002, 2004, 2007, 2008, 2010, and 2012 for light-duty diesel engines
4. ACEA E2–E5 in 2002 for HDD engines
5. ACEA E6 and E7 in 2004 for HDD engines
6. ACEA E2 06, E4-07, E6-04 issue 2, and E7-04 issue 2 in 2007
7. ACEA E4, E6, E7, and E9 in 2008, 2010, and 2012
8. ACEA C1–C4 for engines with aftertreatment devices in 2007, 2010, and 2012

These specifications employ both U.S. and European engine tests for evaluating various categories of oils as provided in Table 13.3 and Table 13.4. Key performance characteristics of global diesel engine oil specification DHD-1 are shown in Table 13.5.

OEM SPECIFICATIONS

There are large numbers of OEM specifications for diesel engine oils, and some of these are as follows. For details, original specification may be referred:

1. JASO—2008 diesel engine oils, DH-1-05, DH-2-08, and DL-1-08
2. GM—Dexos 2
3. MB—DEO V-2012.2, MB 226.9, 228.0/.1, 228.2/.3, 228.31, 228.5, and 228.51
4. MAN—270/271, M-3275, 3277, 3477, and 3575
5. VOLVO—VDS, VDS-2/3/4
6. CUMMINS—20078 and 20081
7. Caterpillar—ECF-1a, ECF-2, and ECF-3
8. MACK—EO-NPP-03 and EO-O-VDS-4

U.S. military specifications have been the guiding factor in developing diesel engine oil specification, and API standards have been derived from these. Following are U.S. military specifications that have been followed in several countries:

MIL-L-2104A/B/C/D/E/F

MIL-PRF-2104G and MIL-PRF-2104H (2004)

MIL-L-46152D and MIL-L-46152E

TABLE 13.6 U.S. military diesel engine oil specifications: key bench and engine tests

Specification	Engine and bench tests	Specification	Engine and bench tests
MIL-L-2104E	L-38, Cat 1G2, Seq. IID, IIIE, VE, Cat TO-2, Detroit Diesel 6V-53T, Allison C3 (seal), time/torque	MIL-PRF-2104H	CCS, MRV, HTHS, evaporation loss, S. ash, foaming, aeration, EOHT, shear stability, Cat 1K, Cat 1P, Cummins 11 EGR, Mack T8E, used oil pumpability, D-4686, 2-cycle engine tests, Allison friction and wear, Cat TO-4, TO-4 M
MIL-L-2104F	L-38, Cat 1K, Seq. IIIE, Mack T-7, 6V-92TA, Allison C4 friction, seal, Cat TO-3 and TO-4		

Key performance tests specified in MIL-L-2104E/F/H specifications are provided in Table 13.6. For complete specifications and test limits of each engine tests, original standard may be referred.

EUROPEAN ENGINE TESTS

These CEC test procedures have been developed in order to meet the approval requirements of ACEA. However, some of the tests have been developed as low-cost screen tests, and some remained due to significant useful data bank generated. Engine tests around the world are being evolved continuously due to various improvements taking place in the engine design to meet emission norms. Several of the following tests are now not available due to hardware changes or because the test has become obsolete.

CEC L-O2-A-78: Petter W1

This is one of the very earliest test procedures to measure high-temperature gasoline engine performance to evaluate copper–lead bearing corrosion due to oil oxidation.

A single-cylinder spark-ignition Petter engine of 470 cc is run at steady state for 36 h. This test is no longer used.

CEC L-12-A-77-MWM-B, KD 12E, DIN 51361

This is also one of the early single-cylinder diesel engine tests running on 1.0% S fuel to compare oils with respect to their influence on piston cleanliness. A single-cylinder 850-cc naturally aspirated, direct injection (DI) diesel engine is run for 50 h under steady-state conditions. This test has been used as a screener for the long-duration Caterpillar IG2, with which a good correlation was established. However, with the introduction of newer tests and the use of lower sulfur fuels, this test is not used now.

CEC L-24-A-78: Petter AV-B

This is a supercharged single-cylinder diesel engine running with higher stress levels for 50 or 100h, using 1.0% S fuel, and has been used as a screen test like MWM-B.

CEC L-42-A-92: Mercedes OM 364 LA

This 300-h test has been evolved from the earlier OM352A and OM364A and measures bore polishing in highly stressed supercharged DI diesel engines using 0.3% S diesel fuel. The test is now obsolete.

CEC L-46-T-93: Volkswagen Intercooled, Turbocharged Diesel (Not Available Now)

The introduction of a number of indirect injection (IDI) diesel passenger cars in the early 1980s resulted in a number of failures due to ring sticking. This is a detergency test evaluating piston ring sticking, piston cleanliness, and oil degradation in a 50-h test under full load and maximum speed.

CEC L-51-A-98: Mercedes OM602A (Not Available Now)

As IDI diesel engines became more refined and were also used more for larger cars used under highway conditions, oil thickening began to be seen in some applications. This then led to increased wear and deposits. The OM602A was developed to assess these criteria. The severe conditions were obtained by running low-speed/low-temperature and high-speed/high-temperature/full-load operation for 200h in repetitive 1-h cycles comprising 23 individual steps.

CEC L-52-T-97: Mercedes OM441-LA

The OM441-LA test, based on an 11-1 V6 turbocharged, intercooled DI diesel engine measures piston cleanliness, cylinder wear, oil consumption, and oil thickening under severe operating conditions. The engine runs for 400h on low-sulfur diesel fuel (0.05%) under alternating stages of steady-state and cyclic operation.

Mercedes OM-611 DE22LA

This test was developed to evaluate wear, sludge, and deposits. This engine is currently not available.

CEC-L-099-08: Mercedes OM 646 LA, 2.2-l VTG Turbocharged DI Engine

This test has been developed to evaluate bore polish, wear, and engine sludge.

OM-501-LA: Euro V V6 11.9-Liter Turbocharged

This test evaluates piston cleanliness and oil consumption.

CEC L-53-T-95: Mercedes M111E Sludge

In the mid-1980s, there were reports of high levels of black sludge in engines subjected to significant periods of idle followed by sustained high-speed operation. Originally, Mercedes Ml02E engine test was developed to measure oil's capability to overcome black sludge problem, and this was superseded by the M111E-based engine test, which measures both sludge and camshaft wear. The Mercedes 2.0-l engine is run for 286h during which it covers four stages: cold operation, engine quality assessment, full-load operation, and finally an alternating section.

CEC L-54-T-96: Mercedes M11E Fuel Economy

The M111E fuel economy test was developed to measure the reduction in fuel consumption with modern multigrade engine oils. It is known that fuel consumption can change considerably after an engine is stopped, and so the test uses a *flying flush* technique during which period the engine oil is changed while the engine is still running.

The engine is operated firstly on the reference oil, followed by the candidate oil and finally a repeat of the reference oil, in order to identify any engine drift. A comparison is made of the fuel consumption between the candidate oil and the reference oil. This test procedure provides a measure of the fuel consumption benefits of fresh oil only.

CEC L-56-T-95/98: Peugeot XUD11 A/B-TE

The test was developed originally on the A-TE version of this engine, a 2.1-l turbocharged IDI unit from Peugeot, but was converted to the BTE version. Running for 75h, the test comprises 150 cycles of 2-min idle and 27min at rated power and speed, with oil and coolant temperatures of 110 and 100°C respectively. Oil viscosity increase, sludge, and wear are measured at the end of the test.

CEC-L-093: Peugeot DV4 TD

The test measures piston cleanliness and soot dispersion by measuring viscosity increase at 6% soot in oil.

CEC L-78-T-97: Volkswagen TIC DI Diesel

The VW DI diesel test was developed to assess piston deposits, ring sticking, and oil degradation in high-speed DI diesel engines for passenger cars. A 1.9-l turbocharged and intercooled DI diesel engine is run on 0.25% S fuel, for 60h comprising stages of 30min of idle and 150min of maximum torque at rated speed.

NORTH AMERICAN ENGINE TESTS

The U.S. engine tests were originally developed to meet the demands of the equipment manufacturers and then the requirement of U.S. Military. As the lubricant standards have been evolved, these are now widely used for qualifying lubricants to a particular specification and also as a tool for new lubricant development programs.

Originally, Caterpillar 1H2 and 1G2 tests were quite popular for lower category of oils that are now obsolete. Most of the North American tests revolve around Caterpillar, Mack, and Cummins engines.

Caterpillar 1K (ASTM D-6750)

The Caterpillar 1K test procedure utilizes a supercharged single-cylinder 1 Y540 engine using one-piece aluminum piston and runs for 252 h under steady-state conditions, with min. 0.35% sulfur fuel. It measures engine deposits, oil consumption, liner polish, oil deterioration, and wear.

Caterpillar IN (ASTM D-6750)

The Caterpillar IN test and the Cat 1K use identical procedure, but Cat 1N uses 0.05% S fuel to evaluate the performance of oils.

Caterpillar IP (ASTM D-6681)

The HDD engine (IY3700) runs on 0.03–0.05% S fuel for 360 h and measures oil consumption and piston deposits under high-speed high-temperature conditions.

Caterpillar 1M-PC (ASTM D-6618)

This test using a single-cylinder supercharged diesel engine (1Y73) was developed to correlate with HDD engines operated at medium severity. It uses 0.4% S diesel and runs for 120 h at a lower temperature than other Caterpillar tests. The test evaluates ring sticking, ring and cylinder wear, and piston deposits.

Caterpillar 1R

The test is carried out in a single-cylinder DI (with four valves) 1Y-3700 engine for 504 h with 0.3–0.5% S fuel. The test evaluates piston deposits, ring sticking, liner wear, and oil consumption under high-speed and supercharged test conditions.

Caterpillar C-13

The test is run on a 2004 Caterpillar six-cylinder, 13-l engine for 500 h to evaluate piston deposits and oil consumption. This test is based upon a modified C-13 on-highway, six-cylinder, 445-horsepower engine with ACERT technology and closed crankcase ventilation.

Mack T-7

This test was developed to evaluate viscosity increase in lubricants in medium-speed, turbocharged diesel engine using Mack EM6-285 six-cylinder engine.

Mack T-8/T-8E (ASTM D-5967)

Mack developed the T-8 procedures in conjunction with ASTM in order to investigate the effect of high soot levels on viscosity increase in turbocharged, intercooled, DI, six-cylinder diesel engines. The procedure uses a Mack E7 engine of 12-l capacity running on 0.03–0.05% S fuel for either 250 (T8) or 300 (T8E) hours. Lubricant performance is evaluated by measuring the increase in viscosity due to soot loading to 3.8% in Mack T8 or 4.8% in the case of the T-8E.

Mack T-9

The T-9 test uses a six-cylinder Mack 12-l V-Mac 11 engine with the objective of evaluating bearing, piston ring, and liner wear in turbocharged, intercooled DI engines.

The test runs for 500 h in two stages of 1800 rpm (for 75 h) and 1250 rpm (for 425 h).

Mack T-10 (Obsolete) (ASTM D-6987)

The test uses a Mack E-tech 460 diesel engine with EGR and runs for 300 h in two phases. The first 75-h stage is the soot generation phase, and in the next 225 h, peal torque is generated. Test evaluates oil performance with respect to piston, bearing, and liner wear.

Mack T-11(ASTM D-7156)

The test uses a six-cylinder Mack E-tech V-Mac III diesel engine with EGR and runs for 252 h with variable timing to hit three different soot windows at 96, 192, and 252 h. This test evaluates the soot level loading performance of engine oils in EGR environment by measuring viscosity increase in oil after the tests. Percent soot in oil leading to 12-cSt viscosity increase at 100°C is reported.

Mack T-12

The test uses a six-cylinder Mack E-tech V-Mac III diesel engine with heavy EGR and runs for 300 h in two stages of 100 and 200 h to evaluate bearing corrosion, ring/liner wear, and oil consumption.

Cummins M11 HST (Obsolete)

This test is intended to assess the performance of heavy-duty engine oil to control valve train wear and deposit formation under operating conditions of soot production in a turbocharged, intercooled DI diesel engine. The test uses an 11-l Cummins M11-330E engine and consists of four stages, each of 50 hours' duration, with engine speeds of either 1800 or 1600 rpm.

Cummins M11 EGR (Obsolete)

Cummins ISB

This test utilizes the Cummins 5.9L ISB medium-duty diesel engine equipped with EGR and DPFs. The 350-h, two-stage test is designed to evaluate valve train wear and aftertreatment device compatibility with oil.

Cummins ISM 425

This is a six-cylinder, turbocharged diesel engine with EGR and is used to evaluate soot-related valve train wear, top ring wear, engine sludge, and filter plugging in EGR and high-soot environment. The test is run for 200 h in four stages.

GM Roller Follower Wear Test

The use of IDI diesel engines in small delivery vans in the United States resulted in a number of cases of high valve train wear. The test is run on GM 6.5-l diesel engine, and the pin wear used to hold the roller follower is measured. The test runs on 0.05% S diesel for 50 h at 1000 rpm and 30–34 kW power with oil and coolant temperatures of 120°C.

DD 6V-92TA (ASTM D-5862)

There are several 2-T diesel engines in use in the United States, and this test, a successor to the 6V-53T, assesses the oils' ability to inhibit ring and liner wear. The test runs on a 9-l Detroit Diesel V6 turbocharged engine for 100 h in six cycles using 0.4% S diesel fuel.

Ball Rust Test (ASTM D-6557)

This is a bench test to measure corrosion-preventive properties of oil in the engine parts.

DIESEL ENGINE OIL FORMULATION

The diesel engine oils therefore contain the following chemical additives to meet the performance test requirements:

1. Detergents
2. Dispersants
3. Antiwear/EP
4. Antioxidants
5. Anticorrosion

6. Viscosity modifiers for multigrade oils
7. Pour point dispersants
8. Antifoam
9. Antirust
10. Friction modifiers

The dosage of these individual additives would depend on several factors such as quality of base oils, lubricant performance level, and nature of additive whether it is monofunctional or multifunctional. Many additives are multifunctional such as ZDDP (antiwear, antioxidant, and anticorrosion) and certain polymers (viscosity modifier and pour depressant). Use of synthetic oils such as PAO and ester will also reduce dependence on antioxidant, viscosity modifier, and pour point depressant. Therefore, each lubricant formulation is balanced with selected base oils and additives keeping the performance characteristics of the oil in consideration. Since diesel engine oil formulation contains large number of chemical additives, additive compatibility with each other and base oil blend must be checked carefully. Incompatibility will lead to additive ineffectiveness. The candidate oil has then to be evaluated in the required engine tests followed by field trial. These additives listed appear similar to the additives used in passenger car motor oils, but differ considerably. For example, in diesel engines, detergency and dispersancy levels have to be much higher to address deposits, bore polish, and soot thickening. Usually, more than one detergent based on sulfonate/phenate/phenolate/salicylate has to be used to meet engine test requirements. Soot has never been a problem in gasoline engine oils. Restriction on sulfated ash, phosphorus, and sulfur in diesel engine oils puts another restriction on the selection of performance additives. Magnesium-based detergents are now preferred over calcium based to control sulfated ash. Lower level of phosphorus-containing additive ZDDP or other nonphosphorus antiwear compounds is being used. Sludge in a turbocharged diesel engine can be reduced by the incorporation of an aminic antioxidant [12] in combination with or without a phenolic antioxidant. HDD engine oil with low SAPS (TBN 7–15) has been formulated [13] with 0.5 mass% of an ashless sulfur-free phenolic antioxidant, aminic antioxidant, and an overbased magnesium detergent.

Friction modifiers, if used in diesel engines, have to be different from the one used in gasoline engine oils due to higher temperatures encountered in diesel engines. Usually, a robust friction modifier such as molybdenum dithiocarbamate is used.

DIESEL ENGINE OIL DEVELOPMENT TRENDS

Emission controls continue to drive the development of engine technology and engine oils. The continual reduction in NOx and PM in diesel engine saw all-round improvements in fuels, lubricants, and engines. Diesel fuel sulfur level came down from 500 to 15 ppm in ULSD. Introduction of heavy EGR and exhaust aftertreatment devices including DPFs led to the development of higher-stability (thermal and oxidation) lubricants with lower sulfur, phosphorus, and sulfated ash. These changes have the implication of increased cost of vehicles, fuels, and lubricants. OEMs in North America have used heavy EGR with clean gas induction (about 40%), closed crankcase, and DPF with active regeneration to remove soot and other PM. In closed crankcase ventilation, the harmful gases generated in the crankcase are taken into the engine intake system, where these are burnt along with the fuel. EPA requires that DPFs operate for at least 150,000 miles before these need cleaning. Engine emissions also must comply for 435,000 miles. Diesel engine oil additive based on molybdenum compound [14] improves the combustion property of a PM trapped with DPF. The efficiency of removal of PM with this additive is improved, and the life of DPF is increased.

The modern diesel engine generates more soot and higher peak temperature due to higher levels of EGR. These conditions would require that the engine oil should have higher thermal and oxidation stability. And to protect aftertreatment devices, SAPS must be within the API CJ-4 limits. Despite these limitations, the oil must control piston deposits, oxidative thickening, oil consumption, high-temperature stability, soot dispersion properties, antifoaming, and shear stability. Reaction of a carboxylic acid-containing polymer with certain aromatic amines and polyols results in ester-containing dispersant viscosity modifiers with improved soot-handling performance [15] in HDD engines.

API CJ-4 is the latest specification for HDD engine oils and is an upgraded version of API CI-4 plus. To provide protection of the aftertreatment devices, SAPS limits have been set in API CJ-4 oil. These chemical limits include the following:

1.00% maximum sulfated ash (per ASTM D-874)

0.12% maximum phosphorus (per ASTM D-4951)

0.40% maximum sulfur (per ASTM D-4951 or ASTM D-2622)

ACEA 2012 specifications (C1–C4) for diesel engines having aftertreatment devices have the toughest limits for SAPS. These are as follows:

≤0.05% sulfated ash for C1–C4

0.05% max. phosphorus for C1

≤0.2% sulfur for C1 and C4

The sulfated ash and sulfur limits in ACEA oils are just half of the values provided in CJ-4 oils, and phosphorous limits are less than half of CJ-4 limits. To formulate ACEA C lubricants, a new additive chemistry and API group III or IV base oils are required.

In addition to these chemical limits, a volatility limit of 13% maximum as determined by the Noack Volatility Test Method ASTM D-5800 has also been set, which further limits the choice of base oils.

Low SAPS diesel engine oil can be formulated [16] with an alkaline earth metal-containing detergent, a nitrogen-containing ashless dispersant of an average molecular weight of 4500 or more, a phenolic or amine oxidation inhibitor, and a basic nitrogen-containing compound oxymolybdenum complex.

SOURCE OF SULFATED ASH

Sulfated ash in the oil is derived from the additives such as metallic detergents based on sulfonate/phenate/salicylate (Ca and Mg compounds) and antiwear additive based on ZDDP or other chemical compounds. Ash is formed when the oil undergoes combustion and can get deposited onto the piston ring grooves and piston crown land, causing wear. The ash particles can be carried downstream with the exhaust and will collect on the DPF leading to filter blockage. Ash particles cannot be removed by filter regeneration since these are not combustible. A lubricating oil composition for diesel engines, which is low in ash content and is metal-free, has very high detergency and antiwear properties. It is suitable for diesel engines equipped with devices for the aftertreatment of exhaust gas. The additive system is based on alkenyl succinimide and boron compounds [17].

SOURCE OF SULFUR

Sulfur poisons the catalyst in the DPF and makes it ineffective. This can lead to the conversion of SOx to sulfates, which increases particulate emissions. Sulfur is present in base oils as well as in lubricant additives. Solvent-refined oils have higher levels of sulfur, but hydrotreated API group II or III base oils have much less sulfur content. Polyalphaolefins and other synthetic oils do not have sulfur. Antiwear additive ZDDP, friction modifiers, and detergents have sulfur in their molecule. API CJ-4 oil specifies sulfur level of 0.4% max, which can be achieved by controlling both base oil sulfur and additive sulfur. Obviously, ZDDP and phosphorus/sulfur base friction modifiers such as MoDTP need to be reduced to control both S and P. If possible, overbased sulfurized phenate and sulfonate also need to be reduced. Fortunately, due to sulfur reduction in fuels, higher oil TBN are not required and detergent level can be safely reduced. These changes need careful formulation balancing.

SOURCE OF PHOSPHORUS

The main source of phosphorus in diesel engine oil is antiwear additives based on ZDDP. Some amount of phosphorus may also come from corrosion inhibitor, friction modifier, and antioxidants. Typical API CI plus oils contain 0.15% phosphorus, but CJ-4 oils have to have less than 0.12% P. Phosphorus in known to deactivate noble metal catalyst and can thus damage the catalyst in due course of time resulting in increased NOx, CO, and HC levels.

LONG-DRAIN ENGINE OILS

Soot accumulation in diesel engine oil is the main factor that governs engine oil drain interval. Efficient soot elimination from crankcase oil can be a practical way to achieve longer drain interval. Combination of high-performance oil filter and low-soot-dispersancy oil can result in an effective measure to trap soot efficiently. Based on this principle, long-drain oil with low soot dispersancy has been reported [18]. Detergency of this oil was evaluated by JASO diesel engine oil detergency test followed by a satisfactory 60,000 km field test in an engine provided with a high-performance filter. The other approach to formulate long-drain oil is to have oil with high soot dispersancy, detergency characteristics with synthetic base oil of high stability, and high-performance additives.

In Europe, long-drain diesel engine oil has been specified by several OEMs such as MB (229.5), Volvo (VDS-3), MAN (3477), Scania DF (100,000-km drain), and others. Such long-drain oils are commercially available in Europe and elsewhere.

ENGINE TECHNOLOGY AND LUBRICANTS

To formulate API CJ-4 and ACEA 2012 HDD engine oils with aftertreatment devices, a complete reformulation strategy needs to be adopted with newer additives having less of sulfur, phosphorus, and metals in them. API group I base oils will be inadequate for these products. Synthetic base oils such as PAO and esters or their combination would be ideal. All modern commercial diesel engine oils have now been converted to multigrade, the most popular being 15W-40 or 10W-40. Lower grades would be required in colder climatic zones. If possible, ashless detergents/dispersants/antioxidant and antiwear additive technology would have to be utilized. Magnesium-based detergents such as magnesium salicylate, magnesium phenates, magnesium sulfonate, and their combinations are useful in improving the life of DPF. It is claimed [19, 20] that high-TBN magnesium detergents with other conventional additives in engine oil reduce the burden on DPF by 50% after 450 h of operation, as measured by the VW DPF test. Magnesium detergents have lower ash due to their lower atomic number. Ashless detergents have been used in combination [21] with ashless antioxidants and other engine oil additives to reduce overall ash level in the finished oil. Metal-free diesel engine oil has been reported, which has low ash and yet possesses high detergency and good antiwear properties and, therefore, is most suitable for diesel engines equipped with the aftertreatment of exhaust gas.

The future tighter limits on emissions and use of ULSD will result in further engine design modifications and additional aftertreatment devices such as the following:

1. Lean NOx catalyst (LNC)
2. Lean NOx traps (LNT)
3. NSRC
4. DeNOx catalysts
5. NOx absorbers
6. Diesel oxidation catalyst (DOC)
7. Selective catalytic reduction (SCR)

These technologies and their improved versions will bring in further restrictions on lubricating oil limits and a new service classification; possibly, API CK-4 will come up in the near future. Thus, the challenges to develop improved lubricating oils for diesel engines will continue.

SYNTHETIC ENGINE OILS

There have been efforts to develop long-drain, superhigh-performance, energy-efficient diesel engine oils [22]. The basic approach is to have low-viscosity multigrade oil for energy efficiency and low oil consumption. Higher energy efficiency can also be achieved by the use of a robust friction modifier that can withstand higher temperatures in diesel engine. For long-drain capability, high oxidation stability and thermal stability are the basic requirements for the base oil blend. This could be achieved by either using all synthetic base oils based on PAO or esters or their blend [23, 24]. Part-synthetic oils (<15%) and mineral oil have also been developed, but this approach only helps in meeting low-temperature viscometrics rather than any advantage in performance characteristics. Commercially, several companies around the world are marketing fully synthetic 5W-30, 5W-40, and 15W-40 diesel engine oils suitable for modern diesel engine fitted with DPF and emission control equipment. Some of the products are of API CJ-4 performance level. However, the cost of these products is higher than the mineral oil-based products, but the synthetic oils are claimed to provide high drain intervals (three times or more) to offset the initial high cost. Synthetic diesel engine oils are currently favored only in a niche segment.

REFERENCES

[1] Majumdar SK, Rao AM, Srivastava SP, Bhatnagar AK. Determination of rheological properties of automotive engine oils. Ninth National Symposium on Analytical Techniques for Fossil Fuels and Lubricants; December 22–24, 1992; New Delhi.

[2] Hamblin PC, Rohrbach P. Automotive engines-deposit control using ashless antioxidants. In: Srivastava SP, editor. Proceedings of International Symposium on Fuels and Lubricants (ISFL-2000); March 10–12, 2000; New Delhi. p 405–411.

[3] Gautam M, Chitoor K, Durbha M, Summers JC. Effect of diesel soot contaminated oil on engine wear—investigation of novel oil formulations. Tribol Int 1999;32 (12):687–699.

[4] Rounds FG. Society of Automotive Engineers International Engineering Congress and Exposition; February 23, 1981; Detroit. SAE Paper 810499.

[5] Gutierrez A, Bloch RA, Diggs NZ, Girshick FW, Martella DJ, Stevens MG, Emert J. Lubricating oil composition. US patent 2002,0115,575 A1. August 22, 2002.

[6] Gutierrez A, Bloch RA, Diggs NZ, Girshick FW, Martella DJ, Stevens MG, Emert J. Lubricating oil composition. US patent 2005,0130,852 A1. 2005.

[7] Katafuchi T, Nakano T. Succinimide compound, process for producing the same, lubricating oil additive, and lubricating oil composition for internal combustion engine. US patent 2002,0103,088 A1. August 1, 2002.

[8] Colacicco P, Mazuyer D. The role of soot aggregation on the lubrication of diesel engines. Tribol Trans 1995;38 (4):959–965.

[9] Yamaguchi ES, Untermann M, Roby SH, Ryason PR, Yeh SW. Soot wear in diesel engines. J Eng Tribol 2006;220 (5):463–469.

[10] Nagai I, Endo H, Nakamura H, Yano H. Soot and valve train wear is passenger car diesel engines. SAE 831757 Reprint.

[11] Majumdar SK, Taneja GC, Bharadwaj A, Christopher J, Mathia R, Chhatwal VK, Tyagi BR. Extended drain interval of DHD-1. Fifth International Symposium on Fuels and Lubricants (ISFL-2006); March 7–10, 2006; New Delhi. Auto lube 1. p 96.

[12] Cook SJ, Adamczewska JZ. Aminic antioxidants to minimize turbo sludge. US patent 8,476,209. July 2, 2013.

[13] Diggs NZ, Gutierrez JA, Alessi ML. Lubricating oil compositions. US patent 8,513,169. August 20, 2013.

[14] Katafuchi T. Additive for diesel particulate filter. US patent 7,989,406. August 2, 2011.

[15] Gieselman MD, Pudelski JK, Eveland RA, Preston AJ. Ester dispersant composition for soot handling in EGR engines. US patent 8,581,006. November 12, 2013.

[16] Nakazato M, Takeuchi Y, Muramatsu T. Low sulfated ash, low sulfur, low phosphorus, low zinc lubricating oil composition. US patent 8,361,940. January 29, 2013.

[17] Katafuchi T. Lubricating oil composition for diesel engine. US patent 8,575,080. November 5, 2013.

[18] Naitoh Y, Hosonuma K, Tamura K, Miyahara M. Investigation into extended diesel engine oil drain interval (II). SAE Paper 912340; 1991.

[19] Kurihara I, Kagaya M. Lubricating oil composition for diesel engine. US patent 2007,0179,070 A1. August 2, 2007.

[20] Busse P, Leonhardt H, Sant P, Willars M. Lubricating oil composition. US patent 2007,0129,266 AI. June 7, 2007.

[21] Koishikawa N. Lubricating oil additive and lubricating oil composition containing the same. US patent 2007,0155,636 AI. 2007.

[22] Ripple DE, Fuhrmann JF. Performance comparison of synthetic and mineral oil crankcase lubricant base stocks. J Syn Lubr 1989;6 (3):209–232.

[23] Kennedy S, Ragomo MW, Lohuis JR, Richman W. A synthetic diesel engine oil with extended laboratory test and field service performance. SAE Paper 952553; 1995.

[24] Lohuis JR, Harlow AJ. *Synthetic lubricants for passenger car diesel engines.* SAE Paper 850564; 1985.

CHAPTER 14

TWO-STROKE AND SMALL ENGINE LUBRICANTS

A two-stroke engine completes the thermodynamic cycle in two strokes or in one crankshaft revolution only, that is, upward and downward movements of the piston. In these engines, the end of combustion stroke and the beginning of compression stroke take place simultaneously. The intake and exhaust functions are carried out at the same time. In four-stroke engines, this cycle is completed in four strokes. There are fewer moving parts in two-stroke engines and are thus cost-effective and provide high specific power to weight ratio as compared to four-stroke engines. These engines have been very popular in small engine devices such as handheld chain saw, mopeds, scooters, lawn mowers, and outboard motors. Majority of these engines are air cooled. Marine outboard engines are, however, water cooled. Due to their simple design, these have even been used in some cars, tractors, and large industrial and marine engines. Several car makers such as SAAB, Suzuki, DKW, and others utilized 2T engines in the past. The lower cost of rebuilding and maintaining the engine made them extremely popular. In Asian countries including India, the population of two-stroke scooters and motorcycles became very large due to these reasons.

Two-stroke engines are lubricated by a total loss system. Usually, oil is mixed in the ratio of 1:20–1:50 with the gasoline, and large part of oil is burnt in the combustion process, and about 20–25% of oil is exhausted. Use of 1:100 low-dosage two-stroke engine oils has also been described [1]. Some of the leaner oil ratio 2T oils are synthetic in nature, based on esters or diesters to meet the high-rpm and high-load operation conditions. The emission level in these engines is, therefore, high in smoke and unburnt hydrocarbons. The high emission of 2T engine has led to the strict emission norms for the two-stroke engines. Due to this control, there is now a shift to use four-stroke engines for scooter and motorcycles to meet the emission norms. In a conventional two-stroke engine, the fresh fuel/air mixture scavenges the cylinder after the combustion. Due to simultaneous charging and emptying, about 25–30% of the air/fuel mixture is also exhausted without burning, causing high smoke and emission. There have been efforts to introduce catalytic converter in two-stroke engines to control emissions [2, 3]. The efforts to improve two-stroke engines by introducing direct or indirect fuel injection system have significantly reduced the emission and improved fuel economy. Direct injection (DI) has considerable advantages in two-stroke engines, eliminating some of the waste and emissions caused by

Developments in Lubricant Technology, First Edition. S. P. Srivastava.
© 2014 John Wiley & Sons, Inc. Published 2014 by John Wiley & Sons, Inc.

carbureted two strokes where a part of the fuel/air mixture entering the cylinder is exhausted out from the port, unburnt. Since the fuel does not pass through the crankcase, a separate source of lubrication is needed in DI engines.

SPECIFICATIONS AND CLASSIFICATIONS

Air-cooled two-stroke engine oil performance is classified according to the API, Japanese Automobile Standards Organization (JASO), International Organization for Standardization (ISO), and Thai Industrial Standards Institute (TISI). For outboard engines, oil performance is classified according to the National Marine Manufacturers Association (NMMA, USA).

Air-Cooled Engines

API-TA, API-TB, and API-TC are the oldest performance categories for two-stroke air-cooled engines, although API-TC has been the most popular category. Other categories are now obsolete.

API-TA *(Obsolete)*   This was the proposed classification for two-stroke engine oils required for small engines, typically less than 50cc. Engine tests for this classification could not be developed when the Coordinating European Council (CEC) withdrew support for this category.

API-TB *(Obsolete)*   This classification for two-stroke engine oils was proposed for the engines of scooters and other highly loaded small engines, typically between 50 and 200cc.

API-TC *(Current)*   This classification is designed for various 2T engines, typically between 150 and 500cc, such as those on motorcycles, snowmobiles, and chain saws with high fuel:oil ratios—but not for outboards. API-TC addresses ring sticking, preignition, and cylinder scuffing problems. At present, the API-TC detergency test is run using previously used cylinders and slightly oversized pistons because Yamaha no longer provides the test parts. Also, new low-ash reference oil is being used instead of the original ashless oil because the ashless oil had experienced hot ring sticking problems. Engine test requirements are shown in Table 14.1.

API-TD *(Obsolete)*   Designed for water-cooled outboard engines, this classification used the identical engine test to that in the NMMA TC-W category. API-TD has been superseded and is no longer accepted by the NMMA, who now recommend oils meeting the requirements of TC-W3 for water-cooled outboard engines.

In 1994, the JASO defined a specification that describes three quality levels (FA, FB, and FC) for low-ash air-cooled two-stroke engines. The candidate oil is rated against reference oil in four performance criteria: detergency, lubricity, exhaust port blocking, and smoke level. JASO's main thrust is the low-smoke requirement, which is especially stringent for the FC performance category. This requirement arises from the need to address smoke problems in Asian cities including India. The ISO Global

TABLE 14.1 Two-stroke classification: API-TC

	Engine	Parameter	Limits
API-TC (CEC TSC –3)	Yamaha CE 50S Yamaha CE 50S Yamaha 350 M2	Tightening, mean torque drop Preignition, occurrences Piston varnish Ring sticking Piston deposits Piston scuffing	< Ref. oil 1 max. in 50h test Better than or equal to ref. oil

TABLE 14.2 Two-stroke classification: ISO/JASO

ISO	Test method	EGB	EGC	EGD
JASO		FB	FC	FD
Properties	Procedure	Limits		
Kinematic viscosity at 100°C	D 445	6.5 cSt min.		
Flash point (°C)	D 83	70 min.		
Sulfated ash (wt.%)	D 874	0.25	0.25	0.18
Engine tests		Standard index min.		
Lubricity	JASO M-340	—	—	95
Initial torque		—	—	98
Detergency, Honda DIOAF-27	JASO M-341 60-min test Fundamental parts	85 min.	95 min.	—
Piston skirt	JASO M-341 180-min test Fundamental parts	—	—	—
				125
Piston skirt		—	—	95
Exhaust smoke Suzuki SX 800R	JASO M-342	45	85	85
Exhaust blocking Suzuki SX 800R	JASO M-343	—	90	90

Notes: All limits are relative to the reference oil, JATRE-1.

Test engines

Honda DIO AF27: Lubricity, torque index, detergency, piston skirt varnish

Suzuki SX800R: Exhaust smoke, exhaust blocking

Specification for two-stroke oils has also been developed. This specification standardized the requirements described in the JASO and established an additional, higher international performance level. The difference between the JASO and ISO specifications is the inclusion of a performance category higher than JASO FC and an additional requirement to evaluate piston varnish. These oils will be required to provide higher detergency and ring stick protection than the JASO oils. ISO and Japanese classification is provided in Table 14.2. JASO FA is now obsolete. JASO FD equivalent to ISO EGD has now been introduced.

Some details of the engine tests are provided in the following.

JASO M 340-92: Honda Lubricity   The Honda Lubricity test uses a 49-cc Honda scooter engine running at wide-open throttle steady state for five cycles during which cooling air is switched on and off. As the air supply is terminated, the temperature of the engine increases and the engine torque decreases. The reduction in engine torque is then used as a measure of lubricity. The fuel for the 2T mix is unleaded 89–93-octane gasoline.

JASO M 341-92: Honda Ring Stick   This is a detergency test run on a 2T engine, using a 49-cc Honda motor scooter engine running at 6000 rpm and maximum power. The fuel for the 2T mix is unleaded 89–93-octane gasoline. After completion, ring sticking, piston, and cylinder head deposits are rated.

JASO M 342-92: Suzuki Smoke   Running on a generator set connected to a load rather than a dynamometer, this test measures the white smoke produced by 2T engines in a simple procedure. A comparison of the smoke produced by the candidate oil is made with that of the reference oil in back-to-back tests on the same day using a 2T fuel based on unleaded 89–93-octane gasoline.

JASO M 343-92: Suzuki Exhaust Blocking   Similar to the Suzuki Smoke Test, this procedure uses a Suzuki generator set for the test. The main aspect of this test is the requirement to run two engines under the same procedure at the same time but on different oils, in order to allow a comparison in the time taken for a predetermined increase in exhaust back pressure to be reached under the same ambient conditions.

CEC L-79-T-97: Honda AF27-2T   Currently, this is the only two-stroke motorcycle engine test developed by CEC for detergency. CEC investigated the performance capabilities of oils formulated to meet the requirement of JASO 2T tests, and it was considered that there was a need for a more severe detergency test to meet the requirements of the European manufacturers and duty cycles. As a result, this test was developed, using the existing JASO procedure and a 49-cc Honda AF27 scooter engine. The test runs for 3 h to assess the ring sticking and engine deposit-forming tendency.

The TISI category applies to oils marketed and used in Thailand and other Asian countries. Oil performance with respect to detergency, lubricity, and smoke for this category is evaluated in 125-cc Kawasaki engine. Oils that meet the JASO FC smoke index requirement are judged to meet smoke requirement of this test. TISI classification is provided in Table 14.3.

In addition to these performance classifications, the miscibility and fluidity characteristics of oils must be satisfactory to serve in either a premix (lubricant into gasoline) or lubricant injection system. These characteristics are described in SAE Recommended Practice, *Two-Stroke-Cycle Engine Oil Miscibility/Fluidity Classification—J 1536*, which covers four grades of oil:

F/M 1 for tropical climates

F/M 2 and 3 for temperate climates

F/M 4 for Arctic climates

TABLE 14.3 Two-stroke classification: TISI 1040

Bench tests	Parameter	Limits
Tests	Viscosity (100°C, cSt)	5.6–16.3
	Viscosity index	95 min.
	Flash point (°C)	70 min.
	Pour point (°C)	−5 max.
	Sulfated ash (wt.%)	0.5 max.
	Metallic element content (wt.%)	Report
Kawasaki KH 125M	Piston seizure and ring scuffing at fuel:oil ratio of up to 200:1	No seizure
	Detergency (general cleanliness) fuel:oil ratio 40:1	
	Ring sticking	8 merit min.
	Piston cleanliness	48 merit min.
	Exhaust port blocking	None
Suzuki SX 800R (JASO M 342-92)	Exhaust smoke, fuel:oil ratio 10:1	85 min.

Note: Since mid-1991, all two-stroke oils used in Thailand are required to meet TISI-1040 requirements.

TABLE 14.4 SAE miscibility for two-stroke engine oils

SAE grade	Allowable Brookfield viscosity[a] (cP, max.)	References oil[b] for miscibility
1	3,500 at 0°C	VI-GG
2	3,500 at −10°C	VI-FF
3	7,500 at −25°C	VI-D
4	17,000 at −40°C	VI-II

[a]Both miscibility and Brookfield tests must be run.
[b]Results for candidate oil must not exceed those for reference oil by more than 10% to pass miscibility test.

Fluidity is assessed by its low-temperature Brookfield viscosity, which must be below certain absolute values for each grade. Miscibility is determined by observing how readily it mixes with gasoline. Test requirements are given in Table 14.4.

LUBRICANT FORMULATION

The two-stroke engine oil requirements are quite different from the normal four-stroke engine oils. The following properties are important in 2T oils:

1. Provide lubricity
2. Control scuffing or scoring
3. Should have adequate detergency to keep the engine part clean
4. Reduced deposit formation

5. Control smoke from the exhaust
6. Keep spark plug clean
7. Should be miscible with gasoline at operating temperature

Lubricity is mainly achieved by the use of high-viscosity oil like bright stock up to 10%. The use of heavy base oil like bright stock leads to the formation of high smoke and exhaust port blocking. Detergency properties are provided by calcium petroleum sulfonate or by an ashless alkenyl succinimide. Solvents like mineral turpentine oil or kerosene fraction are used to improve fluidity and miscibility in gasoline. Other performance additives like antiwear/antioxidants may also be added. Total loss system will not require antioxidants since the oils are not recirculated. In new ISO EGB/API-TC and JASO FB oils, severity of lubricity and detergency requirements were increased. The limits for smoke and exhaust system blocking were also introduced. These requirements could be satisfied by new improved oil containing higher detergents/dispersant levels and substituting bright stock with synthetic material polyisobutylene. Modern oils use a combination of low- and high-molecular-weight polyisobutylene as a part or full replacement of bright stock to control both lubricity and smoke. Higher level of detergency and dispersancy is achieved by using improved additives in combination with polyisobutylenes of appropriate molecular weights.

The severity levels further increased in ISO EGC and JASO FC. There was a greater thrust in reducing smoke and reducing exhaust port blocking. This could be made possible by completely eliminating bright stock from 2T engine oil formulation and increasing PIB dosage to a higher level of 20–25%. Improved detergents and dispersants along with some antiwear additives had to be used. Kitano [4] has carried out field trials with ISO EGD and API-TC oils in Indian 2T engines for 10,000 km and concluded that ISO EGD oils provide better protection with respect to exhaust port blocking and wear. Exhaustive emission studies have been carried out in both small two-stroke and four-stroke engines [5–9].

Many snowmobile engines are designed with power valves to enhance performance. This requires an oil to be exceptionally clean burning to prevent the *gumming up* and sticking of the power valves in the open or closed position. Several OEMs recommend synthetic oils to reduce valve sticking. Synthetics have advantages over petroleum-based 2T oils in the areas of high-rpm protection, lower-subzero-temperature operation, cleaner burning with less smoke, and lower carbon deposits on pistons, rings, and power valves.

With tighter emission norms in larger 2T engines such as in snowmobiles and outboard engines, direct-fuel-injected (DFI) engines were developed. These DFI engines reduce emissions by directly injecting fuel into the combustion chamber, after the closure of the exhaust port. This arrangement eliminates the short circuiting of the fuel/oil mixture in the earlier designs. Thus, in DFI engines, a fuel/oil mixture of as high as 30% (which was scavenged earlier) is saved, and emissions are also reduced, improving both fuel efficiency and emission control, which was the main deficiency of two-stroke engine. This change, however, required improved lubricants with respect to scuffing, ring sticking, and carbon

TABLE 14.5 Two-stroke classification: NMMA TC-W3

Test	Parameter	Limits
Kinematic viscosity at 40°C		
Cloud and pour point		
Flash point		
Nitrogen content and TBN		
Brookfield viscosity at −25°C		≤7500 cP
Rust percent		Equal or better than ref. oil
Engine tests		
OMC 40HP, 98-h test	Piston varnish and top ring sticking	Equal or better than ref. oil
OMC 70HP, 98-h test	Piston deposit and second ring sticking	Equal or better than ref. oil
Mercury 15 HP, 100 h	Two consecutive passes	Equal or better than ref. oil
Yamaha CE 50S, lubricity and tightening	Torque loss	Equal or better than ref. oil
Yamaha CE 50S, preignition	Preignition , occurrences	Equal or better than ref. oil
AF-27 lubricity test	Torque loss	Equal or better than ref. oil
NMMA rust test	—	Equal or better than ref. oil
Miscibility at −25°C	CP inversions	Not more than 10% inversions than ref.
NMMA filterability	Decrease in flow rate (%)	20% max.

buildup. Carbon buildup in ring groove area required special attention in these engines, which was solved by OEMs by adopting special coating and metallurgy for rings and cylinder walls. Improved lubricants have also been developed to provide higher level of engine cleanliness (reducing ring groove fill and carbon deposit) by using low ash and low-nitrogen-containing detergents/dispersants in the oil formulations [10].

Water-cooled engines: Performance requirements for outboard engine oils have been defined by NMMA. These include:

- NMMA (BIA) TC-W—now obsolete
- NMMA TC-W2—now obsolete
- NMMA TC-W3—now obsolete
- NMMA Recertified TC-W3

In 1994, the industry decided that all existing TC-W3-certified oils must be recertified by passing two consecutive Mercury Marine 15-hp engine tests. In addition, some base oil/bright stock read-across provisions were allowed. Tests required for obtaining a recertified TC-W3 qualification for new oil are outlined in Table 14.5. NMMA grants reblending/rebranding licenses for only recertified TC-W3 category. All other previous categories are now obsolete.

FOUR-STROKE FC-W OILS FOR OUTBOARD ENGINES

EPA in the year 2006 decided that outboard engine manufacturers must reduce hydrocarbon emissions from these engines by 75% as compared to 1996 level. This was only possible by producing such engines in four-stroke technology. OEMs and NMMA developed engine test for this new nomenclature oils of FC-W category. The most critical parameters for this category of oils are antiwear and antirust protection under sea environment.

NMMA also developed FC-W specification for four-stroke outboard engines. Oils seeking approvals must meet minimum API-SG performance in addition to corrosion inhibition and antiwear requirement of an outboard engine (115-hp engine). A pass is determined by the inspection of cam lobes, caps, journal, bearings, piston rings, con rod bearings, cylinder bore, main bearing, crank journal, and fuel pump lobe and by comparing with reference oil.

BIODEGRADABLE OUTBOARD TWO-STROKE ENGINE OIL

The International Council of Marine Industry Association (ICOMIA) has finalized a standard for biodegradable lubricants, ICOMIA-27-1997, for water-cooled outboard motors. The product must show very low toxicity for algae, fish, and *Daphnia* as per ISO and OECD standards in addition to being biodegradable. Such products can be formulated only with synthetic esters and compatible additives to provide performance. The product also must meet TC-W3 specification.

FOUR-STROKE MOTORCYCLE OILS

Traditionally, four-stroke motorcycles utilized oils designed for spark-ignited four-stroke passenger car engines. However, due to the inclusion of fuel economy requirements in the latest API categories, many oils have been formulated to contain friction modifiers. In a motorcycle, the wet clutch is also lubricated by the same oil. The presence of friction modifier in oil causes clutch slip and power loss. The Japanese (Yamaha, Suzuki, and Kawasaki) have been the major manufacturers of four-stroke motorcycles around the world, and they recognized the need for a separate oil standard for these engines. Therefore, JASO established new specifications for four-stroke motorcycle oils. In addition to API categories, the oils are classified according to their frictional characteristics as determined in an SAE No. 2 test procedure defined by JASO.

In 1999, JASO MA (high-friction) and JASO MB (low-friction) standards were drawn. In 2005, these standards were revised and two more MA categories were included. These were MA1 and MA2. For these four categories, minimum and maximum dynamic and static coefficients of friction have been specified. Stop time index (STI) was also specified. Another important addition was the limit on phosphorus to make the oils catalytic converter compatible.

These specifications are given in Table 14.6, Table 14.7, and Table 14.8. In view of this development, ISO plan to formulate worldwide ISO standard for 4T oils has been deferred.

There were limits for DFI, SFI, and STI earlier in 2006. These have now been revised and tightened in 2011 and new limits are provided in Table 14.8.

TABLE 14.6   Classification standards for JASO T 903, 2011, four-stroke motorcycle oils

Standard	Performance level
API	SG, SH, SJ, SL, SM, and SN
ILSAC	GF1, GF2, GF3, and GF4
ACEA	A1/B1, A3/B3, A5/B5, C2, C3, and C4

TABLE 14.7   Physical/chemical properties for JASO T 903, 2006 and 2011, four-stroke motorcycle oils

Property	Value	Test method
Sulfated ash (% mass, max.)	1.2	JIS K 2272-85
Phosphorus (% mass, min.)	≥0.8 and ≤0.12	JPI-5S-38
Evaporation loss (% mass, max.)	20	JPI-5S-41-93
Foaming tendency/stability (ml, max.)		JPI K 2518
Sequence I	10/0	
Sequence II	50/0	
Sequence III	10/0	
Shear stability, FOT kinematic	xW-30>9.0	
viscosity (mm^2/s)	xW-40>12.0	JPI-5S-29-88
	xW-50>15.0, other stay-in grades	30 cycles in diesel injector
HTHS viscosity (mPa-s, min.)	2.9	JPI-5S-36-91

TABLE 14.8   Four-stroke classification JASO T 903, 2011

JASO T 904	Dynamic friction characteristic index (DFI)	Static friction characteristic index (SFI)	Stop time index (STI)
JASO MA	1.30≤DFI<2.5	1.25≤SFI<2.5	1.45≤STI<2.5
JASO MA1	1.30≤DFI<1.8	1.25≤SFI<1.70	1.45≤STI<1.85
JASO MA2	1.85≤DFI<2.5	1.70<SFI<2.50	1.85≤STI<2.50
JASO MB	0.50≤DFI<1.3	0.50≤SFI<1.25	0.50≤STI<1.45

REFERENCES

[1] Malhotra RK, Raje NR, Chaube DM, Srivastava SP. Development of low dosage 2 strike engine oils and catalytic converter. Proceedings of the 1st International Symposium on Fuels and Lubricants (ISFL 1997); December 8–10, 1997; New Delhi. New Delhi: Tata McGraw Hill. p 327–332.

[2] Sithanathan M, Saranathan SD, Sharma GK, Raje NR, Srivastava SP. Study of exhaust smoke and emission of Indian two stroke engine lubricants. Proceedings of the 2nd International Symposium on Fuels and Lubricants (ISFL 2000); March 10–12, 2000; New Delhi. p 713–718.

[3] Sithanathan M, Singh S, Saranathan SD, Sharma GK, Raje NR, Srivastava SP. Studies of performance characteristics of two strike engine lubricants. 4th Petrotech, Ps2/L/5; January 9–12, 2001; New Delhi.

[4] Kitano SK. ISO EGD and API TC oil evaluation in field test. In: Srivastava SP, editor. *Proceedings of the 1st International Symposium on Fuels and Lubricants (ISFL 1997)*; December 8–10, 1997; New Delhi. New Delhi: Tata McGraw Hill. p 319–326.

[5] Sun X, Brereton GJ, Morrison K, Patterson DJ. Emissions analysis of small utility engines. Milwaukee: SAE Off-Highway and Power Plant Congress. SAE Paper 952080; 1995.

[6] Sun X, Assanis D, Brereton GJ. Assessment of alternative strategies for reducing hydrocarbon and carbon monoxide emissions from small two-stroke engines. Detroit: SAE Congress. SAE Paper 960743; February, 1996.

[7] Brereton GJ, De Araujo A, Bertrand E. Effects of ambient conditions on the emissions of a small carbureted four-stroke engine. SAE Paper 961739; 1996.

[8] Fischer H, Brereton GJ. Fuel injection strategies for minimizing cold-start hydrocarbon emissions. SAE Paper 970040; 1997.

[9] Brereton GJ, Bertrand E, Macklem L. Effects of changing ambient humidity and temperature on the emissions of carbureted two- and four-stroke hand-held engines. SAE Paper 97P-382; 1997.

[10] Cleveland WKS, Petric JL, Svarcas LR, inventors. Lubricant composition suitable for direct fuel injected crankcase-scavenged two-stroke engines. US patent 2005,0130,856 AI. June 16, 2005.

CHAPTER 15

RAILROAD, MARINE AND NATURAL GAS ENGINE OILS

These three types of lubricants are being discussed together, since they are all internal combustion engine oils but differ quite a bit from passenger car motor oils and conventional diesel engine oils. Their specifications are highly OEM driven and have not been covered by API or ASTM or ISO. The formulation of railroad (RR), marine, and NGEO takes different approaches due to some very specific problems encountered in each one of them. These engines are large in size and highly cost conscious in their operation, requiring oils and fuels at the most competitive prices. Therefore, synthetic oils have not entered this field mainly due to cost considerations.

RAILROAD OILS

Railway traction has changed from robust steam locomotive to electric locomotive and to diesel–electric locomotive. Steam locomotives have been phased out due to environmental concerns and their low efficiency. The use of electricity for traction has the advantage that the electric motor generates maximum torque at the start-up, and this property is advantageous to a stationary train to move on. The wheel or axles can be driven individually by separate motors to each axle. Alternately, one large motor can drive several wheels through a system of connecting rods. Diesel–electric locomotives are advantageous when continuous electric overhead supply is not available or it is only partially available. Diesel–electric locomotive is thus a power-generating and electric transmitting equipment using diesel engine power and an alternator with controls to regulate speed and torque. Various other modes of power transmission from the diesel engine such as mechanical, hydraulic, and pneumatic have not been very successful in a high-HP locomotive. Lower-HP engine can, however, use mechanical or hydraulic transmission.

The steam locomotives require very high-viscosity cylinder oils to lubricate piston and cylinder. The electric locomotive does not need an engine lubricant; only greases are required to lubricate traction motor and gear pinion. However, diesel–electric locomotive will need additional diesel engine lubricant called RR oils. The difference between an electric locomotive and a diesel locomotive is that the latter does not depend on the external source of power; it carries its own power plant.

The use of diesel became widespread after World War II in all transportation sectors. Medium-speed diesel–electric locomotives also became popular. These

Developments in Lubricant Technology, First Edition. S. P. Srivastava.
© 2014 John Wiley & Sons, Inc. Published 2014 by John Wiley & Sons, Inc.

engines were constantly improved for increased performance, greater efficiency, cost effectiveness, and emission reduction. The higher performance of medium-speed diesel engines was achieved through higher compression ratios, improved turbocharging, greater fuel injection rates, electronic fuel injection, and higher firing pressures with only about 10% increase in piston displacement volume. During this change, the oil sump volume remained unaltered. Thus, the average amount of work performed by a locomotive for the average amount of engine oil used that is the oil stress (measured by net tons-km/liter of the engine oil) had grown twofold. This meant very higher severity on the engine oil. Both two-stroke and four-stroke diesel engines up to 4400 BHP range were available for railway applications. General Motors (GM), however, introduced a new 6000-BHP, 16-cylinder four-stroke engine, which is low oil consuming as well. Lowered oil consumption means that oil temperatures, soot, and oil degradation products increase faster in the used oil, leading to engine deposit and filter choking problems. This will result in frequent change of engine oil. Such development will eventually replace older two-stroke high-oil-consuming engines worldwide. All these improvements led to the development of higher-performance RR oil.

Traditionally, the additive system for medium-speed diesel engine oil has been different from other diesel engine oils with respect to the antiwear additive. The most common and popular antiwear additive zinc dialkyldithiophosphate (ZDDP) was not used in RR oils. This is because a major OEM, GM Electromotive Division (EMD), had in the beginning used silver metallurgy in the engine's upper connecting rod bearings, which get corroded by the active sulfur in ZDDP. The non-zinc-containing antiwear additives are costly and add to the overall cost of RR oils. However, silver bearings have now been phased out in favor of a copper–lead matrix bearing many years back. Currently, most of the new medium-speed diesel engines can operate on oils containing ZDDP. For example, in India, all the ALCO locomotives have been operating successfully with zinc-containing oil. However, there are still some old GM locomotives in operation that require zinc-free oils. This situation should change soon in the future and cost-effective zinc-containing RR oils will be common. Modern medium-speed diesel engine oils are typically higher-ash, dispersant-detergent-containing products of SAE 40 or SAE 20W-40 viscosity grade. Some 15W-40 oils have also been used by certain locomotives.

The earlier medium-speed diesel engine oils used medium-viscosity-index base oils (naphthenic base oils). These have now been replaced with solvent-refined high-viscosity-index oils or hydrocracked, low-sulfur, highly saturated paraffinic base stocks. These base oils are thermally stable and offer improved oxidation stability in the fully formulated oils. Several RR have now changed to multigrade 20W-40 oils that contain polymeric viscosity index improvers and provide low oil consumption and fuel economy.

LMOA CLASSIFICATION OF RAILROAD OILS

LMOA *Generation* Oil Definitions

The Locomotive Maintenance Officers Association (LMOA), an organization consisting of railroaders and rail venders based in North America, has been concerned that engine oil performance meet the demands of the railway users. In the early 1970s, the LMOA began a classification system for railway lubricating oils based on performance requirements and introduced the LMOA oil *Generation* designations.

Since the 1940s, five generations of railway diesel engine oils have been defined. LMOA Generation 1 oils, first marketed in 1940, were low-alkalinity oils, less than 7 total base number (TBN by ASTM D-2896), and were typically straight mineral oils or oils containing some detergent and antioxidants. Turbochargers were first used in medium-speed railway diesel engines in 1956. As engine performance increased, these oils lost alkalinity faster, resulting in lead corrosion and bearing failures. These problems led to the development of Generation 2 oils that were defined in 1964 requiring a minimum TBN of 7. Such oils contained an antiwear agent, higher level of antioxidant, and ashless dispersants to take care of extra sludge in oil and to improve oil filter life.

Generation 3 engine oils, introduced in 1968, contained higher dispersant and detergent levels over Generation 2 oils. This was necessary to reduce piston ring wear due to faster depletion of alkalinity in the oil. Generation 3 engine oils required an alkalinity of 10 TBN since the engines were now turbocharged and producing higher levels of soot and deposits.

The severity on oil continued to increase with more powerful engines. In 1976, Generation 4 oils of 13 TBN were introduced with improved alkalinity retention, dispersancy, and detergency. The normal useful life of Generations 3 and 4 oils coincided approximately with the 92-day locomotive inspection interval mandated by the U.S. Federal Railroad Administration (FRA). The higher severity levels in oils required a combination of highly alkaline detergent/antioxidant compounds based on phenate chemistry to control viscosity increase, engine deposits, and alkalinity losses.

With the introduction of new low-oil-consuming locomotives, need was felt for further improved lubricant having longer life. The LMOA Generation 5 engine oil category was introduced in 1989 to cope up with this situation. The Generation 5 oil definition demands oil drain intervals of 6 months (two FRA inspection intervals) in new low-oil-consuming locomotives operating for 16,000 km/month using diesel fuels with a sulfur content of 0.3–0.5 wt.%. The used oil must retain a minimum of 1.0 TBN (ASTM D-664-84) and not exceed a 30% increase in kinematic viscosity at 100°C.

The General Electric (GE) Company, a major manufacturer of low-oil-consuming, four-stroke railway diesel engines, has not recognized LMOA Generation 5 definition. GE refers to such oils as "GE Generation 4 *Long Life* Engine Oils" and recommends an oil change period of 3 months in severe duty service.

The introduction of Generation 5 oils required a different approach to the lubricant formulation. For longer life, oil required higher oxidation protection, higher detergency and dispersancy, and improved antiwear protection. Chlorine-containing antiwear additives were no longer desirable from environmental considerations. Alkalinity reserve in the oil had to be higher. The introduction of ULSD and emission norms created a new situation and requirement for low SAPS, 9 TBN multigrade oil emerged; forming the basis for a new category of modern RR oils.

PERFORMANCE EVALUATION OF RAILROAD OILS

There are no generally accepted tests for assessing RR oil quality. Each engine builder has its own preferences based on field experience, and only proven quality in an extended field test can lead to product approval. Most commercial oils are compounded with additives to meet the API CD/CF4 performance with adequate reserve alkalinity to combat the acidic effect of sulfur in the fuel. Traditionally, zinc has been limited in RR oil formulation due to engine components damage (silver bearings). Earlier, only monograde SAE 40 oils were used, but multigrade oils (15W-40 and 20W-40) are now preferable due to its improved fuel economy and lower oil consumption. Performance evaluation of RR oils has been described in greater detail [1].

LMOA Generation 5 oil should meet the following requirements:

- Allow drains of 180 days minimum for low-oil-consuming engines averaging 16,000 km/month and consuming fuel (0.3–0.5% sulfur) at a rate of 20,000 gallons/month
- Pass OEM oxidation, corrosion, and friction tests
- Meet OEM engine test requirements
- Pass OEM field test requirements

Details of LMOA generation and OEM's engine tests required are provided in Table 15.1 and Table 15.2.

GM and GE need the following (Table 15.2) bench tests before accepting the oil for field tests.

Lubricant requirements of different RR builders are shown in Table 15.3.

The introduction of low- and ultralow-sulfur diesel (ULSD) fuel as well as Environmental Protection Agency (EPA) emission regulations [2] has once again changed the RR oil scenario. In March 2008, the EPA, USA, finalized the emission norms for diesel locomotives of all types—line-haul, switch, and passenger rail. These rules will lead to the reduction of particulate matter (PM) emissions from these engines by 90% and NOx emissions by 80% when fully implemented. This final rule sets new emission standards for existing locomotives when they are remanufactured.

TABLE 15.1   RR diesel engine oil performance categories

Designation	Dispersant level	Detergent level (TBN D-2896)
Generation 1	Nil	4–5
Generation 2	Moderate	7
Generation 3	Moderate	10
Generation 4	High	13
Generation 5	High	13 and over

TABLE 15.2   OEM engine tests

Bench	Engine	Field testing[a]
EMD silver corrosion	25-h EMD 2-567C	3–10 late model
GE oxidation test	750-h GE 7FDL	locomotives, 1 year
GE oxidation test	480-h Caterpillar 1G2	and 100,000 miles
Pass bronze friction test		

[a] Approval for field testing is granted by OEM after the candidate oil has completed bench and engine tests.

TABLE 15.3 RR engine builder lubricant requirements

Engine builder	SAE grade	TBN (ASTM D-2896)	Sulfated ash (% mass) max	Zinc content	API classification	Road test requirement
GM EMD	40 or 20W-40	10–20	—	10ppm	—	3–10 locomotives, 1 year
General Electric, USA	40 or 20W-40	13–20	—	—	—	3 locomotives, 100,000 miles
	30	—	1.5	0.005%	SE/CC, SE/CD	Required
	40	—	1.5	0.005%	SE/CD	Required
MTU	15W-40	—	1.8	0.005%	SE/CD	Required
GE Locomotives Canada (ALCO 251 engines)	40	7–13	—	—	CD	—
Sulzer	40	—	—	—	CD	Required
SEMT Pielstick	40	10 min	—	—	CD	Required

TABLE 15.4 Some key EPA emission regulations for locomotives

Duty cycle	HC, g/hp-h	NOx, g/hp-h	PM, g/hp-h	CO, g/hp-h	Smoke, %	Minimum useful life, hours/years/miles
Line-haul (2005–2011)	0.30	5.5	0.10	1.5	20/40/50	$(7.5 \times hp)/10/-$
Switch (2005–2010)	0.60	8.1	0.13	2.4	20/40/50	$(7.5 \times hp)/10/-$
Line-haul (2012–2014)	0.30	5.5	0.10	1.5	20/40/50	$(7.5 \times hp)/10/-$
Switch (2011–2014)	0.60	5.0	0.10	2.4	20/40/50	$(7.5 \times hp)/10/-$
Line-haul (2015+)	0.14	1.3	0.03	1.5	—	$(7.5 \times hp)/10/-$
Switch (2015+)	0.14	1.3	0.03	2.4	—	

Detailed explanation of terminology and methods used may be referred to the original EPA standard [2].

Tier 3 emission standards for newly built locomotives, provisions for clean switch locomotives, and idle reduction requirements for new and remanufactured locomotives have also been decided. Tier 4 standards for newly built engines based on the high-efficiency catalytic aftertreatment technology will be applicable from 2015. These initiatives along with the introduction of ULSD RR oils will move toward low-SAPS engine oils. High alkalinity (TBN) was required to neutralize sulfuric acid formed during the combustion of high-sulfur fuel. With low-sulfur fuels, higher TBN above 10 mg KOH/g are not necessary. RR oils are, therefore, reformulated with 9 TBN and low SAPS content. GM EMD and GE have already approved such oils for use in their new engines. The balanced detergent, dispersant, and antiwear system permits such 20W-40 multigrade, 9 TBN oil to provide superior soot, sludge, viscosity, energy efficiency, and wear control resulting in improved engine cleanliness and lower wear and particulate emissions. It has been found that formulations containing salicylate detergent in addition to the conventional components showed decreased levels of lead corrosion and better TBN retention. Mixture of two carboxylate detergents with different TBN and a dispersant based on polyalkenyl succinimide has been reported to provide improved performance [3]. Reduced SAPS in the engine oils lowers particulates to meet EPA tier 3 and 4 standards. Key elements of EPA emission standards are provided in Table 15.4.

CHALLENGES IN RAILROAD OIL FORMULATIONS

The RR diesel engine operation is quite different from the other engines used in trucks or in stationary engines, and that is how the lubricant formulation technology is also different. There are also no industry-acceptable tests for assessing RR oil quality. Each engine builder has preferences based on their field experiences. Further, one of the OEM GM EMD had silver bearing in their locomotives; therefore, the excellent sulfur-containing antiwear additive ZDDP (universally used in gasoline and diesel engine oils) was eliminated completely from RR oil formulation, and non-sulfur-containing antiwear additives were used (chlorinated paraffin and esters and phosphorous-based compounds). This approach made the product costly. Further, unlike other diesel engines, locomotives idle for a long duration. Line-haul cycle operates at 38% idling and switch duty cycle operates at about 60% idling. Such idling wastes a large amount of fuel and avoidable emissions are generated. This fuel wastage can, however, be avoided by using a small auxiliary D.G. set during idling phase. Emission controls for locomotives have been finalized by EPA. Currently, tier 3 norms are applicable and tier 4 norms shall apply after 2015. Silent features of EPA norms are provided in Table 15.4. These norms are progressively being made tighter. Tighter control on NOx means retarding of injection timing, which will delay the fuel combustion. Higher piston top ring reduces transient emission, but this is also responsible for bringing in soot-loaded oil into the crankcase. These changes require that the oil must be capable of handling higher level of soot, sludge, and deposits [4, 5]. Sludge and deposits in diesel engines are formed by nitro-oxidation mechanism. NOx is produced in the combustion chamber and increases as the temperature increases. Blow-by gases can also enter into crankcase due to worn-out rings and piston liners and can produce sludge due to nitration. The tighter limits on PM put restriction on sulfur in fuels, and sulfur in ULSD is already 15 ppm max. Lower sulfur levels in fuels direct that there is no need for high alkalinity (TBN) in oils yet high level of detergency and dispersancy is required to combat deposits, sludge, and soot. High level of sludge is responsible for filter chocking and higher wear from oil thickening. Thus, as the locomotive engine design changes to accommodate emission control mechanism and fuel quality measures, the oil formulations have also to be adjusted and balanced to counter severity imposed on it. Recently, newly formulated 9 TBN 20W-40 grade RR oils have appeared in the market and approved by OEM for use in new high-performance locomotives using low-sulfur fuels. With very low sulfur levels in API group II/III base oils and ULSD, the alkalinity levels may further come down in future oils. These 9 TBN and low SAPS multi-grade RR oils form the basis of new LMOA generation VI oil.

From the discussion so far, the modern diesel RR oil should possess the following characteristics:

1. Oils would have to perform under most severe conditions of temperature and yet provide long life. This would be possible by using higher amount of high-temperature antioxidants based on new chemistry.

2. The threat from the oxides of nitrogen and sulfur has to be countered by the use of overbased detergents such as phenates, sulfonates, salicylates, succinates, and phenolates or their mixtures.

3. The detergency level of oil has to be of high order to protect engine parts from deposits and keep them clean during the operation of the engine.

4. Similarly, the dispersancy characteristic of the oil has to be such that oil thickening does not take place due to the presence of fine soot generated during combustion. This is controlled by ashless dispersants based on PIB succinimide of appropriate structure. The ratio of detergent/dispersant is also important to control both detergency and dispersancy [6]. One of the RR oils is based [7] on sulfonate detergent of 20–100 TBN and a phenate detergent of 200–350 TBN, a hindered phenolic antioxidant, and C_7–C_9 esters of hindered phenolic. The lubricant composition is free from ZDDP and has a sulfated ash content of less than 1.0 wt.%. In this formulation, phenate is greater than 15% of the detergent molecule.

5. For longer engine life, wear protection is important. In conventional diesel engine, this property is imparted by ZDDP. However, most RR oils are formulated without ZDDP. A different approach is required to formulate RR oils with good antiwear properties. Earlier oils contained chlorinated paraffins as antiwear additives. These have now been replaced [8–10] by complex esters. One of the silver wear protection additives has been reported [11] to be a mixture of hydrocarbylamine salt of a dialkyl, a phosphoric acid, and a hydrocarbylamine salt of an alkyl acid phosphate.

6. Energy efficiency and oil consumption in a locomotive are important properties to cut down costs. These have been controlled by the use of multigrade oils [12] such as 20W-40 and 15W-40 oils. British rail was the first to use multigrade oils after a detailed trial [13]. This has been followed by most railways around

the world. However, in order to achieve fuel efficiency with multigrade oils, it is important to select a balanced shear stable polymer [14, 15]. Two to seven percent fuel economy in GM EMD locomotive and 2–13% fuel economy in GE engine [16] have been reported with the use of 20W-40 oil. In ALCO locomotives, 5% fuel economy and 14% less oil consumption have been [17–19] reported in a trial after 150,000 km on 20W-40 oil. The entire ALCO locomotive fleet in India is now on multigrade oils.

7. Further, fuel economy can be realized with the use of a robust friction modifier and using low-viscosity oils.

MARINE LUBRICANTS

A marine ship uses a variety of engine oils, industrial oils, and greases. While most of the industrial oils and greases are similar to the ones used on land-based equipment, engine oils form different category due to the specific engine design and high-sulfur residual fuels used in these engines. However, many land-based power generation units also use residual fuels and similar engine design (as that of marine). Consequently, these land-based power generation units also use marine engine oils for their lubrication. Broadly, marine ships would need the following types of lubricants:

- Crosshead engine cylinder oil
- Crosshead engine system oil
- Trunk piston engine oil (TPEO)
- High-speed diesel engine oil
- Turbine oils
- Gear oils
- Hydraulic oils
- Air compressor oils
- Open gear/wire rope lubricant
- Rust prevention oils
- Coolants
- Synthetic lubricants (compressor/gear oils)
- Greases
- Stern tube lubricant

Marine engine oils have been classified depending upon their speed into the following three categories:

1. Slow speed: 50–250 rpm, 260–980-mm bore size, 500–7500 bhp/cylinder, and two strokes
2. Medium speed: 400–2000 rpm, 180–580-mm bore size, 150–1800 bhp/cylinder, and four strokes
3. High speed: greater than 1000 rpm and four strokes

Slow-speed two-stroke crosshead engines are the main marine propulsion engines. MAN B&W and Sulzer dominate the market for these engines. Together, these constitute about 90% of the world market. In these engines, the lubrication of cylinder and crankcase is carried out separately by two oils. The cylinder oil generally requires SAE 50 and sometimes SAE 60 viscosity grade with high TBN of 70 mg KOH/g. System oils are, generally, SAE 30 viscosity grade with a low TBN of 5–10. The high-TBN requirement in cylinder oil is mainly due to the high sulfur in fuel oil and to neutralize the acid formed during the combustion process. The TBN also denotes high level of oil detergency, which keeps the piston rings, lands, and grooves free from deposit. Excess alkalinity (means presence of calcium carbonate in detergent molecule) can also lead to ash deposit formation on piston crown and exhaust valves, especially when using low-sulfur fuels. Recently, thermal load on combustion chamber has increased due to increase in engine firing pressures from 70 to 150 bar. This calls for cylinder oil of good thermal stability up 270°C. The high firing pressure and high cylinder liner temperature increase the risk of blow-by, oxidation, deposits, and adhesive wear. On the other hand, higher temperatures reduce the risk of sulfuric acid condensation leading to reduced acidic corrosion as compared to older engines. In older engines, corrosive wear is one of the main problems. However, in the field, both old and new engines are operating with the same oils, and it is expected that the oil will provide protection to both types of engines with respect to cleanliness, rings, and cylinder liner wear over a period of 20,000 h of operation. Two-stroke cylinder oil is evaluated for cylinder liner wear in a three-cylinder crosshead Bolnes 3DNL test engine at 130-bar combustion pressure. The conditions in the engine require a very balanced detergent/dispersant combination with high alkalinity reserve and better neutralization rates. Overbased sulfurized phenates/salicylates provide antiwear properties and also antioxidant properties. These can be utilized in combination with overbased sulfonates [20, 21] in marine oils with benefits. 10–55 TBN oil containing a detergent, a dispersant, and an antiwear additive is used in the crosshead engines [22].

CRANKCASE LUBRICATION

Crosshead engine crankcase (with oil-cooled pistons) is lubricated with an 8 TBN detergent oil. The water-cooled piston engines are, however, lubricated with a rust- and oxidation-inhibited (R&O) oil. R&O oils were mainly preferred due to their capability to separate out from water easily. However, presently, detergent-type oils can be produced with good water-separating characteristics using good-quality base oils, and therefore, an oil of this type can replace both the oils mentioned earlier. The incorporation of PTO/PTI gears on crosshead engine oils calls for EP properties in these oils. Such oils require an FZG gear rig pass up to 11 stages. Usually, SAE 30 oils are satisfactory for these applications.

MEDIUM-SPEED TRUCK PISTON ENGINE OILS

These engines have a common sump for the crankcase and cylinder and therefore single oil is required to lubricate the system. Trunk piston engines have no separation between the cylinder and crankcase, and the system oil is exposed to blow-by gases and fuel leakage. Such oils therefore require higher level of dispersancy to cope up with these contaminants. TPEO have splash lubrication system, and therefore, excess oil goes into the cylinder lubrication; because of this, TBN of the oil can be lower than the crosshead engines. In crosshead engine, cylinder oil is regulated according to the operating conditions, requiring higher TBN. TPEO, therefore, contains detergent (overbased), dispersant, antioxidant, antiwear (generally ZDDP), anticorrosion, and antifoam additives to meet all the lubrication requirements. These oils are generally SAE 30 or SAE 40 viscosity grade with TBN in the range of 12–55 depending upon the sulfur in the fuel and engine oil consumption.

In the medium-speed trunk piston engines also, there has been a series of design changes to improve power output. This has resulted in new problems such as bore polish, liner lacquering, undercrown deposits, faster TBN depletion, hot corrosion of piston crown, oil scraper ring clogging, increased oil consumption, increased piston deposits, and fouling of purifier heaters. Both OEMs and oil producers have tackled these problems successfully. Bore polish is caused by the deposit buildup on piston crown land. When these deposits start touching the liner, bore polish starts and oil consumption increases. This has been resolved by fitting an anti-bore polish ring, which also controls oil consumption. Specific lubricants have also been developed, which do not allow deposit buildup and control bore polishing. Hot corrosion on the piston crown has been controlled by introducing piston cooling gallery. Fuel pump pressures in modern medium-speed engines have been increased up to 1600 bar. This has resulted in the higher leakage of fuel into the lubricant. Generally, about 2% of heavy fuel has been found in the used oil. In some cases, fuel as high as 10% has been found.

The fuel oil produced in the refineries contains both aromatics and asphaltenes, which are insoluble in paraffins but are soluble in aromatics. When this fuel oil is leaked into the paraffinic lubricating oil, asphaltenes tend to separate out from the fuel and float on the oil surface. Asphaltenes are black in color and sticky in nature and thus get deposited on various engine parts, creating multiple problems. This problem can be resolved by improving oil that can solubilize asphaltenes or by improving oil purification parameters to remove asphaltenes continuously. Several studies have been reported [23–28] on the development of fuel-oil-compatible marine oils by investigating the interaction of detergent–dispersant–antiwear additive combinations. A right composition of overbased sulfonate–phenate and PIB succinimide has to be selected to obtain asphaltene dispersancy properties.

Wartsila NSD, Pielstick, MAN B&W, and Mack are the major manufactures of these engines.

Marine engine oil requirements for both crosshead and trunk piston engines have been summarized in Table 15.5.

TBN DEPLETION IN MEDIUM-SPEED TPEO

During the operation of the engine, alkalinity (TBN) of the oil is depleted due to acid neutralization, and it is maintained at half of its original value by the oil top-up. Acid neutralization prevents corrosive wear of rings and liner. Most OEMs recommend a TBN value of 20 mg KOH/g oil during service. TBN depletion of oil in service

TABLE 15.5 Marine engine oil requirements

Low-speed crosshead		Medium-speed trunk piston crankcase
Cylinder	System	
Sulfuric acid neutralization	Lubrication of bearings, crankshaft chains, and gear	Sludge, deposit, lacquer piston deposit control, and ring stick prevention
Detergency and thermal stability	Detergency and thermal stability for piston undercrown cooling	Stability in the presence of fuel contamination. Control of asphaltenic deposits
Good film strength and scuffing protection	Good water separation properties	Thermal stability and oxidation control
Good spreading characteristics	Control of drip oil and combustion trash	Bearing corrosion protection
Piston, ring, and port cleanliness	Release of insolubles and water to purifiers	Good water tolerance or good demulsibility
Antiwear properties	Good film strength	EP properties
System oil compatibility	Rust and oxidation prevention	Rust control, alkalinity retention, acid neutralization
SAE 50 grade	SAE 30 grade	SAE 30/40 grade

depends on oil top-up rate or oil consumption and sulfur content of the fuel. With the introduction of anti-bore polish ring, oil consumption and consequently oil top-up rate have come down. This situation causes the TBN to drop faster. At this stage, it was necessary to introduce oils of 50 TBN to maintain in-service TBN of 20 minimum.

A detergent based on overbased alkaline earth metal C10–C40 hydrocarbyl-substituted hydroxybenzoates has been found [29] to control asphaltene precipitation in marine engines. Overbased alkaline earth metal C_{22} hydrocarbyl salicylate [30] has also been used for controlling asphaltene precipitation or formation of black paint.

HIGH-SPEED DIESEL ENGINES FOR AUXILIARY EQUIPMENT

These are mainly used on board vessel for emergency equipment like generators, fire pumps, air compressors, and life boats. Generally, oils of API CF-4/CG-4 performance level with 10W-40 viscometrics are preferable for such applications. We shall not discuss these further, as they are covered under the topic on diesel engine oil.

MARINE OIL APPROVALS

The performance of lubricants for marine engines is not defined by a standard classification system. Standard engine test methods or testing protocols are not available. Generally, API CF+ lubricants are necessary to provide good performance.

Oil performance levels and the approval process are driven by major engine builders, most of whom publish list of lubricants approved for use in their engines. They recommend their customers to use lubricants appearing on the approval list. In general, OEMs require a full-scale ship trial lasting 5000 running hours (about 1 year) to approve a lubricant. An engine builder approves only formulation-specific lubricants.

Each OEM has a performance trial and approval procedure, which has to be followed by the oil industry to obtain full approval of their lubricants for the engines. Other industrial oils like turbine, hydraulic, gear, compressor, and refrigeration oils are, however, accepted based on the standard specifications of the oils or based on the equipment builders' specification/approval. These are also not being discussed here. Reader can refer to relevant chapter describing industrial lubricants.

ENVIRONMENTAL CONCERNS

The U.S. EPA has adopted stringent exhaust emission standards for large marine diesel engines to control emissions from all ships that affect the U.S. air quality. International standards for marine engines and their fuels are provided in Annex VI to the International Convention on the Prevention of Pollution from Ships (Treaty called MARPOL). U.S. coasts have been designated as an Emission Control Area (ECA) through an amendment to MARPOL Annex VI. It is estimated that by 2030, this strategy is expected to reduce annual emissions of NOx in the United States by about 1.2 million tons and PM emissions by about 143,000 tons. The international air pollution requirements of Annex VI of MARPOL establish limits on nitrogen oxides (NOx) emissions and require the use of fuel with lower sulfur content. The requirements apply to vessels operating in U.S. waters as well as ships operating within 200 nautical miles of the coast of North America, also known as the North American ECA. NOx and sulfur limits in MARPOL Annex VI are provided in Table 15.6. Existing ECAs include the following:

1. Baltic Sea (SOx, adopted in 1997/entered into force in 2005) and North Sea (SOx, 2005/2006)
2. North American ECA, including most of the U.S. and Canadian coast (NOx and SOx, 2010/2012)
3. U.S. Caribbean ECA, including Puerto Rico and the U.S. Virgin Islands (NOx and SOx, 2011/2014)

TABLE 15.6   MARPOL Annex VI: NOx emission limits, g/kWh

Tier	Year	$n < 130$ rpm	$130 \leq n < 2000$ rpm	$n \geq 2000$ rpm
Tier I	2000	17.0	45. n −0.2	9.8
Tier II	2011	14.4	44. n −0.23	7.7
Tier III	2016[a]	3.4	9. n −0.2	1.96

[a] In NOx ECAs (tier II standards apply outside ECAs).

TABLE 15.7 MARPOL Annex VI: fuel sulfur limits, % m/m

Year	SOx ECA	Global
2000	1.5	4.5
2010	1.0	—
2012	—	3.5
2015	0.1	—
2020[a]	—	0.5

[a] Alternate date is 2025, to be reviewed in 2018.

Annex VI has now been revised and comes into force on July 1, 2010. By October 2008, Annex VI was ratified by 53 countries.

NOx emission standards are provided in Table 15.6. Tier I and tier II standards are global, and tier III standards apply only in NOx ECAs.

Tier II standards are expected to be met by combustion process optimization such as fuel injection timing, pressure, fuel nozzle flow area, exhaust valve timing, and cylinder compression volume. Tier III standards require specific NOx emission control measures such as various forms of water induction into the combustion process (with fuel, scavenging air, or in-cylinder), exhaust gas recirculation, or selective catalytic reduction.

MARPOL Annex VI regulation provides limits (Table 15.7) on the sulfur content of fuel oil to control SOx and indirectly PM. Special fuel quality provision exists for SOx control areas (SOx ECA). Limits for 2020 SOx ECA will be reviewed in 2018.

Heavy fuel oil (HFO) is allowed provided it meets the applicable sulfur limit (i.e., there is no mandate to use distillate fuels).

Thus, future fuel and lubricants for marine applications would be greatly influenced by emission standards and restrictions imposed by MARPOL. Land-based power-generating units also employ same engine and use same lubricants. There may be only slight variation with respect to their installation since there is no space constraint with land-based DG sets. There may be liberal fuel settling and purifier system. There could also be separate oil centrifuge for the used oil in circulation.

MARKETING OF MARINE LUBRICANTS

Unlike land-based equipment, ocean-going vessels are mobile, and the initial lubricants filled in the equipment have to be made available at different ports around the world to service the vessels. Marine lubricant suppliers have to keep the inventory of the product range in a wide range of ports around the world. Each supplier publishes an international port directory where the products are available and provide information regarding the delivery methods available (drum, truck, barge), notice period required to execute the order, and details of local/country specific restrictions, if any. The ship operators can, therefore, plan their product requirement suitable to reduce their operating costs.

TABLE 15.8 Heat of combustion for some liquid and gaseous
fuels (higher value)

Fuel	HHV MJ/kg	HHV BTU/lb	LHV MJ/kg
Hydrogen	141.8	61,000	119.96
Methane	55.5	23,900	50.0
Ethane	51.9	22,400	47.8
Propane	50.35	21,700	46.35
Butane	49.5	20,900	45.75
Gasoline	47.30	20,400	44.40
Kerosene	46.20	19,862	43.00
Diesel	44.8	19,300	43.4

Source: NIST Chemistry web book.

NATURAL GAS ENGINE OILS

Among the liquid and gaseous hydrocarbons, methane has the highest heating value (Table 15.8) and is only next to the cleanest fuel hydrogen. It also has the highest hydrogen-to-carbon ratio (CH_4, i.e., 4:1). With the larger availability of natural gas (NG) around the world, its applications are increasing ranging from internal combustion engine to power generation, fertilizer production, gas transmission, air-conditioning, and host of other applications. Its increased use is also in the direction toward the use of ultimate cleanest fuel hydrogen. We have been consistently moving toward the hydrogen-rich fuels (wood to coal to petroleum liquid fuels to ethanol/methanol to gas). NG is a clean burning fuel, and the usual problems associated with diesel such as soot, sludge, deposit formation, and emissions are better controlled in natural gas engines (NGEs). Methane is the main constituent of NG and is abundantly available at gas fields/oil fields, coal beds (coal bed methane), and deep sea bed as gas hydrates. The amount of methane in NG varies from 80 to 99% and is associated with ethane, propane, butane, nitrogen, and carbon dioxide. In certain cases, hydrogen sulfide (H_2S) is also present, which is corrosive in nature due to the high sulfur content.

NG can be subdivided into the following four groups:

1. **Dry gas**: Dry gas contains up to 99% methane with very small amounts of carbon dioxide (CO_2) and nitrogen (N_2). It does not contain heavier hydrocarbons such as butane and pentane.

2. **Wet gas**: Wet gas contains up to 90% methane, heavier hydrocarbon ethane (C_2H_6), (4–5%) propane (C_3H_8), (2–3%), butane (C_4H_{10}), and (1–2%) some liquid hydrocarbons.

3. **Sweet gas**: Sweet gas contains less than 10 ppm of H_2S or any other sulfur compounds. It may contain up to 25% ethane.

4. **Sour gas**: Gases containing sulfur compounds especially hydrogen sulfide (6–7%) are termed as sour gas. Sour gas contains less methane and much higher proportions of heavier hydrocarbons.

Generally, sweet and dry NG is preferred for use in NGEs. Biogas derived from anaerobic or biochemical digestion of organic industrial and domestic waste, sewage water, agriculture, and food wastes also contains methane as the main constituent with some amount of carbon dioxide and H_2S. Landfill gases sometimes may contain a variety of solvents and even chlorofluorocarbons from refrigeration waste and on combustion can generate harmful/corrosive hydrochlorides/fluorides.

NG by virtue of containing a higher amount of hydrogen as compared to gasoline and diesel fuels has higher calorific value and on combustion will produce more heat. These engines will therefore run hotter. Liquid fuels like gasoline evaporate faster along with the intake air and thus cools the engine. This advantage of gasoline is lost with the use of NG. Further, NGEs also run on lean air/fuel ratios to control NOx and thus also have less air in the combustion zone to cool or dilute combustion gases. NGEs employ three-way catalysts to further control NOx. These factors make NGEs different than gasoline or diesel engines. Gas engines thus provide lower NOx emissions due to three-way catalysts employed and other design parameters, lower CO_2 emissions, and very low PM as compared to diesel engines. However, these engines have volumetric disadvantages providing lower power output than an equivalent-size diesel engine. NGEs also need elaborate fuel supply pipeline and pressurized storage tanks.

CLASSIFICATION OF GAS ENGINES

U.S. design engines are predominantly of two-stroke design, and European engines are predominantly of four-stroke design. NGEs are grouped into three categories according to their ignition systems and induction methods:

1. Two-stroke (spark-ignited (SI)) engine
2. Four-stroke (SI) engine
3. Four-stroke (dual fuel) engine

Two-Stroke Spark-Ignited Engine

In the two-stroke gasoline engine, the air/gas mixture is carbureted into the cylinder and ignited by a spark at the end of compression stroke. Generally, SI gas engines have lower volumetric efficiencies than similar gasoline engines, due to the fact that gas and air mixture is compressed rather than only air in the cylinder. This is counterbalanced by the operation of SI gas engines at higher compression ratios, that is, 12 to 13/1.

Four-Stroke Spark-Ignited Engine

In the four-stroke engines, spark ignition operates on the same principle as that of a normal gasoline engine. The air is trapped in the cylinder during compression. Fuel is injected at high pressures into the cylinders at the end of the compression stroke. The compression ratios are higher or similar to diesel engines.

Four-Stroke Dual-Fuel Engine

The term dual-fuel engine is used to describe compression ignition engine that can operate either on a combination of gas with diesel fuel for ignition or as a full diesel engine only. In a dual-fuel gas engine, the air/gas mixture is admitted into the cylinder during the induction stroke and ignited at the end of compression stroke by the injection of pilot diesel fuel, usually about 5–10% of normal diesel fuel volume.

Dual engines operate at lower compression ratios than diesel engine due to the fact that air/gas mixture cannot be compressed to the 16/1 compression ratio as in the case of diesel engine. The basic differences between gas engines and gasoline/diesel engines are as follows:

1. Gas engines operate at lower compression ratios than diesel engines but operate at higher compression ratios than petrol engines.

2. Volumetric efficiency is generally lower than petrol engines.

3. Combustion temperatures in gas engines generally tend to be higher than diesel engines.

4. Gas engines normally have more advanced ignition timing than gasoline engines.

5. More sophisticated valve and seat materials and wider valve seats are used in gas engines due to high temperatures and soot-free nature of gas combustion.

Stationary gas engines are generally midsize, medium-speed engines used for pipeline gas compression, electrical power generation, and cogeneration. These engines are further described as low-pressure and high-pressure engines:

1. Low-pressure-gas, spark-ignition engines are similar to low-pressure dual-fuel engines. They use a sparkplug instead of pilot fuel to ignite the gas/air admixture as in SI engines.

2. High-pressure-gas, pilot-fuel ignition engines operate on the gas diesel principle where a charge of pilot fuel (usually distillate fuel and typically 5% of the total fuel charge) is injected through a fuel valve just prior to TDC, initiating the combustion process. The balance charge (usually NG) is then injected into the cylinder at high pressure (about 250 bar). The gas ignites as it enters the cylinder, allowing for a clean combustion process. Such engines may also operate on distillate fuel if the gas supply is interrupted.

3. Low-pressure-gas, pilot-fuel ignition engines operate on a similar principle to high-pressure-gas engines except that the gas is mixed with the air charge at a lower pressure either in the inlet air manifold or in a prechamber before entering the engine cylinder. A distillate pilot fuel is then injected into the cylinder to initiate combustion of the gas/air admixture. These engines are generally known as dual-fuel engines and handle inert gases more efficiently than high-pressure engines. However, these also have a reduced scavenge period, which reduces thermal efficiency.

Stationary gas engines are either stoichiometric or lean-burn engines. Stoichiometric engines are often fitted with three-way catalysts to control NOx, hydrocarbon, and CO emissions. Lean-burn engines usually use a precombustion chamber in which a rich mixture is ignited and charged into the main chamber to ignite the lean fuel/air mixture. These engines are designed to meet NOx emission requirements without an exhaust catalyst. The reduced NOx is due to a lower combustion temperature caused by the excess air. An oxidation catalyst may be used on lean-burn engines if lower hydrocarbon and CO emissions are required.

The quality of gases burned in stationary gas engines can vary significantly. Typical gases include NG (predominantly methane), sour gas (high sulfur), sewage gas (containing hydrogen sulfide), and landfill gas (containing corrosive organic halides as well). Each gas has different characteristics and requires different properties from the lubricating oil; therefore, gas engine oils must be chosen carefully for specific applications. Gas engine oils are generally classified according to their sulfated ash levels and typical base number. Two-stroke engines require relatively low-ash or ashless products, while four-stroke engines using severe fuels require higher-ash products. Table 15.9 provides this classification. Table 15.10 lists some of the major gas engine manufacturers.

Exhaust valve recession and valve guttering are two important problems encountered in NGEs. Valve recession is the pounding of valve back into the cylinder head by its repeated closing and opening action. Due to this action, both valve and valve seat are damaged. This problem is caused either due to insufficient oil feed down the valve seat or due to insufficient ash level in the oil. Many operational and mechanical parameters may also be responsible for valve and valve seat recession.

TABLE 15.9   Classification of gas engine oils

Classification	Sulfated ash content, wt.%	TBN typical	Application
Ashless	<0.1	1–3	Two strokes
Low ash	0.1–0.6	3–6	Four strokes
Medium ash	0.6–1.0	6–12	Four strokes and sour gas
High ash	>1.0	>12	Severe fuels

TABLE 15.10   Major gas engine builders

North America		Europe		Japan	
Caterpillar	Dresser Industries	Caterpillar	MAN B&W	Caterpillar Mitsubishi	Niigata
Cooper Industries	CLARK	Crosley	Mirrlees Blackstone	Cooper	Nissan
AJAX	DRESSER-RAND	Deutz–MWM	SEMT Pielstick	Daihatsu	Waukesha–NKK
Cooper–Bessemer	Waukesha	Dorman	Ruston	Kinmon	Yanmar
SUPERIOR	Worthington	Jenbacher	Sulzer	MAN	
Fairbanks Morse	Cummins	GMT	Wartsila	MHI	

Nitration is another problem in four-stroke stoichiometric engines operated with low oil sump temperatures. Two-stroke engines generally do not have nitration problem since oil is continuously exhausted out along with the nitration products and crankcase oil is topped up with fresh oil. Under nitration conditions, oil viscosity, acidity, insoluble, varnish (in hot engine areas), and sludge (in cooler engine areas) increase. NOx reactions with oil are quite complex but essentially generates free radicals that propagate further chain reactions. These can be represented by the following simple reactions; in realty, these are quite complex.

$$RH + NO_2 \rightarrow R* + HONO$$

$$HONO \rightarrow HO* + NO*$$

Chains may be terminated by the reaction of free radicals with NOx and nitroalkanes or alkyl nitrates may be formed:

$$R* + NO_2 \rightarrow RNO_2 \text{ or } RONO$$

Formation of NOx is related to air/fuel ratio and peak combustion temperature. NGEs offer good environment for the formation of NOx. The gas formed may either leave the combustion chamber with the exhaust (as in two-stroke engines or turbocharged engines) or may react with oil available on the cylinder walls or may reach crankcase sump along with the blow-by gases. There are no chemical additives to check NOx reaction with oil or hydrocarbon molecules, but certain carefully selected antioxidants can destroy free radicals generated and reduce the harmful effect of nitration.

Another concern in gas engine is the formation of lacquer and deposits, which is the result of oil and fuel oxidation and polymerization at higher temperatures. Lacquer is known to form both on piston and cylinder liner. The piston lacquer is the result of high temperature. Cylinder liner lacquer is formed over the range of top piston ring travel that is on the cylinder wall below the liner dead space and above the top compression ring at the bottom of the piston travel. Lacquer formation can be controlled by using thermally stable base oils such as API group II or III base oils. Some salicylate-based detergents and molybdenum compound-based formulations have been reported to control lacquer formation. A composition containing metal salicylate (100–190 TBN), diarylamine, hindered phenol, polyalkelene succinimide, boron-containing polyalkenyl succinimide, ZDDP, and metal phenate (100–300 TBN) has been reported to possess NOx-resistant properties and provide long engine life [31].

Ashless lubricants are generally used in two-cycle engines, while low-ash lubricants are used in four-cycle engines [32]. Medium- and high-ash lubricants are more commonly used with severe fuels such as landfill, sour, and sewage gases containing hydrogen sulfide. The ash content must be optimum to prevent valve recession. The ash in oil prevents valve damage and in fact provides lubrication between hot valve and seat. However, very high ash can cause ring sticking, plug fouling, catalyst masking, or port blocking (in two-stroke engines). Currently, there are no standard classification procedures or specification for gas engine oils; therefore, approval for NGEO is granted only upon completion of an extensive field trial.

In addition to ash content, other important parameters for gas engine oils are phosphorus content, resistance to oxidation and nitration (due to high operating temperatures), and corrosion inhibition (with landfill gases). The limiting phosphorus content of lubricants is becoming important because of its poisoning effect on catalyst. To avoid catalyst poisoning and extend catalyst life, catalyst manufacturers have specified restrictions on the phosphorus content of lubricants. The limits vary depending on the catalyst manufacturer and type. Nonselective reduction catalysts are used on conventional (stoichiometric) engines, while selective reduction catalysts are used on lean-burn engines. Emission regulations are setting strict limits on NOx and HC emissions from stationary gas engines, which can be met by installing catalytic converters on the engine exhaust systems. Table 15.11 and Table 15.12 provide some of the OEM's requirement of NGEO.

TABLE 15.11 Some key element of OEM NGEO specifications

Properties	Cummins CES 20074	Caterpillar	Detroit Diesel	John Deere
API level	—	—	—	CF
Viscosity grade	15W-40	—	15W-40	15W-40
Sulfated ash %	0.4–0.6	0.4–0.6	<0.5	<1.0
Phosphorous, ppm	650–850	—	<800	600
Zinc, ppm	700–900	—	—	700
Calcium, ppm	900–1300	—	—	2500
TBN D-2896	4.5 min.	5.0	—	10.4
TAN D-664	0.5–1.5	—	—	—
Cu corrosion	3 max.	—	—	—
Engine test	CNG-8.3 G	—	Seq. III E, L-38	—
Field test	Yes	—	—	—

TABLE 15.12 Requirements of ash and performance level: other OEMs

OEM	Ash, wt.%	API level
Dresser-Rand		
Jenbacher		
Stoich	<0.5	CC
Leanox	0.6–1.0	CD
MAN D.E		
Deutz–MWM		
New	<0.5	CC
Old	<0.75	
Ruston	<0.5	—
Wartsila	0.3–0.6	—
Superior–natural aspirated	<1.0	—
Turbocharged	0.5–1.0	
Waukesha		
Most	0.35–1.0	CD
Special	0.95–1.7	CF

From the foregoing, NGEO have to essentially carry out the following functions:

Protect exhaust valves

Control oxidation and nitration

Reduce friction and wear

Prevent corrosion and rusting

Keep engine clean

Control lacquer formation

These functions can be obtained by using oils [33–35] with specific ash content as specified by OEM, with adequate detergent and dispersant, antioxidant, and antiwear additives. Thermally stable base oils are preferable. The oil should be free from chlorine-containing additives and low in phosphorous content.

REFERENCES

[1] Thompson JL, Anderson RL, Hutchison DA. Perspective on bench and laboratory engine testing of railway diesel lubricant. Lubr Eng 1988;44 (9):768–774.

[2] www.epa.gov/otaq/locomotive.htm

[3] Li Y-R. Diesel engine oils. US patent 2013,0157,911. 2013.

[4] Tobias MF. Current trend and impact of emission regulation on the U S rail road industry. International Symposium on Fuels and Lubricant (ISFL-2004); October 27–29, 2004; New Delhi. SAE Paper 2004.28.004.

[5] Logan M, Middleton NA, Palazzotto JD. Rail road diesel engine cleanliness: impact of the engine oil additive formulation; SAE Paper 961094; May 1996; Michigan.

[6] Ramakumar SSV, Rao AM, Srivastava SP. Studies on additive–additive interaction: formulation of crankcase oils toward rationalization. Wear 1992;156:101–120.

[7] Devlin CC, Hutchison DA. Extended drain diesel lubricant formulations. US patent 8,377,856. February 19, 2013.

[8] Tobias MF, Tobias M, Nelson K, Yamaguchi E, Small V. Crankcase lubricating oil composition for protection of silver bearings in locomotive diesel engines. US patent 2007,0021,312 AI. January 25, 2007.

[9] Hutchison DA, Moore LD. Chlorine-free silver protective lubricant composition. US patent 4,948,523. 1990.

[10] Hutchison DA, Staufer RD. Medium speed diesel engine lubricating oils. US patent 5,174,915. 1992.

[11] Tobias MF, Nelson KD, Yamaguchi ES, Small VR. Crankcase lubricating oil composition for protection of silver bearings in locomotive diesel engines. US patent 8,084,404. December 27, 2011.

[12] Logan MR, Shamah E. Multigrade rail road lubricants fuel saving and field performance in 2-stroke and 4-stroke engines. Lubr Eng 1989;45 (10):625–631.

[13] Morley GR, Eland JA, Dunn K. British rail switch to multigrades, Ind Lubr Tribol July/August 1984: 124–130.

[14] Majumdar SK, Arikkat P, Rao AM, Srivastava SP, Bhatnagar AK. Effect of multigrade engine oils on engine wear in railroad applications. In: Singhal S, editor. Proceedings of the Xth National Conference on Industrial Tribology, Volume II; 1993; New Delhi: Tata McGraw Hill. p 173–178.

[15] Majumdar SK, Arikkat P, Rao AM, Srivastava SP. A statistical method for establishing wear control benefits with a properly balanced multi grade diesel engine oil. Proceedings of the XIth National Conference on Industrial Tribology; January 22–25, 1995. New Delhi: Tata McGraw Hill. p 243–248.

[16] Thomas F, Ahluwalia JS, Shamah E. Medium speed diesel engine part II- lubricants their characteristics and evaluation, Trans Am Soc Mech Eng 1984;106:879–892.

[17] Majumdar SK, Tiwari OP, Chaubey DM, Rao AM, Srivastava SP, Koganti RB, Bhatnagar AK. Development of high performance railroad diesel engine oil for improved oil consumption control. Lubr Sci 1997;9 (4):435–447.

[18] Majumdar SK, Koganti RB, Rao AM, Srivastava SP, Bhatnagar AK, Bhattacharya P, Kabir Ahamed S, Mishra R, Singh KK. Evaluation of high performance multi-grade rail road oil. In: Srivastava SP, editor. Proceedings of the International Symposium on Fuel and Lubricants (ISFL-2000); 2000; New Delhi: Allied Publisher. p 601–608.

[19] Majum dar SK, Bharadwaj A, Mathai R. Medium speed diesel engine oil: a approach to selection of VI improvers. Proceedings of the 4th International Symposium on Fuel and Lubricants. SAE Paper 2004.8.003; 2004; New Delhi.

[20] Ing A. Verhelst, Modern marine lubricants in aging engines. Proceeding of the 1st International Symposium on Fuels and Lubricants (ISFL-97); December 8–10, 1997; New Delhi. p 219–224.

[21] Nagamatsu H. Diesel engine lubricating oil composition for large-bore two-stroke cross-head diesel engines. US patent 2007,0149,420 AI. June 28, 2007.

[22] Chambard L, Kosidowski L. Method of lubricating a crosshead engine. US patent 8,377,857. February 19, 2013.

[23] Ramakumar SSV, Aggarwal N, Rao AM, Saranathan SD, Srivastava SP. Superior high performance crankcase oil for medium speed diesel engines. Proceedings of the 1st International Symposium on Fuels and Lubricants (ISFL-97); December 8–10, 1997; New Delhi. New Delhi: Tata McGraw Hill. p 41–46.

[24] Ramakumar SSV, Sarangpani N, Paul S, Aggarwal N, Rao AM, Srivastava SP, Bhatnagar AK. Wettability characteristics of medium speed diesel engine oils. Petrotech 2001, L072; January 9–10, 2001; New Delhi.

[25] Ramakumar SSV, Aggarwal N, Bathala VK, Chhatwal VK, Tyagi BR, Singh F. Development of range of India's first ever indigenous marine lubricant technology. 6th Petrotech-2005; January 15–19 2005; New Delhi.

[26] Ramakumar SSV, Aggarwal N, Rao AM, Sarpal AS, Srivastava SP, Bhatnagar AK. Studies on additive–additive interaction: effect of dispersant and antioxidant additives on the synergistic combination of over based sulphonates and ZDDP. Lubr Sci 1994;7:1.

[27] Ramakumar SSV, Aggarwal N, Rao AS, Tyagi BR. Performance of marine cylinder lubricants towards rationalization. Petrotech 2003, LO 41; January 9–10, 2003; New Delhi.

[28] Bastenhof D. Large diesel engines-fuels and emission an outlook on tomorrow. 1st International Symposium on Fuels and Lubricants (ISFL-97); December 8–10, 1997; New Delhi. New Delhi: Tata McGraw Hill. p 255–260.

[29] Bertram RD, Dowding PJ, Watts PD. Detergent comprising C10 to C40 hydrocarbyl substituted hydroxybenzoates for reducing asphaltene precipitation. US patent 8,404,627. March 26, 2013.

[30] Bertram RD, Dowding PJ, Watts PD. Method of reducing asphaltene precipitation in an engine utilizing a C.22 hydrocarbyl substituted salicylates. US patent 8,399,392. March 19, 2013.

[31] Sougawa Y, Shimada M, Shiomi M. Lubricating oil composition containing overbased metal salicylate, amine antioxidant, phenol antioxidant, polyalkenylsuccinimide and zinc dialkyldithiophosphate. US patent 6,147,035. November 2000.

[32] Smreka NK. Development of the new generation low ash oil additive package for N G engines. ASTM Paper 91-ICE-A.

[33] Pyle WR, Smreka NK. SAE Paper 932780; 1993.

[34] Logan MR, Carabell KD, Smreka NK. Why gas engines need gas engine oils. Proceedings of the 2nd International Seminar on Fuels and Lubricants (ISFL); March 10–12, 2000; New Delhi. New Delhi: Allied Publisher. p 413–418.

[35] Batala VK, Majumdar SK, Chaube DM, Rao AM, Srivastava SP, Bhatnagar AK. High performance potential of a natural gas engine oil. 3rd International Petroleum Conference and Exhibition, Vigyan Bhawan, New Delhi, Petrotech 1999; New Delhi; 1999. p 369–372.

CHAPTER 16

METALWORKING FLUIDS

Metalworking fluids (MWFs) constitute very large number of products for a variety of operations involving metal removal such as cutting, grinding, milling, threading, broaching, turning, sawing, drilling, deep hole drilling, boring, tapping, etc. and metal forming such as rolling, drawing, forging, stamping, and forming of different metals. For each of these operations with a particular metal, different fluid is required for a satisfactory product and good tool life. Every manufacturing industry starting from a small workshop with a lathe machine to a large automobile unit carrying out sophisticated operations needs these fluids. There is also equally large number of small and big organized manufacturers and suppliers of MWFs that it is not known how much quantity of these fluids are manufactured [1]. Through these operations, a crude metal piece is shaped into desired geometric specifications. Metalworking operations can be broadly divided into two categories: metal removal processes and metal-forming processes. The MWF plays the important role of removing heat and providing lubrication between the tool and workpiece. These fluids are thus referred to coolant as well. In industry, there are also several metal removal operations that are carried out under dry condition without using a coolant. However, the use of coolant is more prevalent to get a good-quality workpiece and extended tool life. MWFs are neat oils or water-based fluids/emulsions used during the machining and shaping of metals to provide both lubrication and cooling. In addition, these fluids must protect metals from corrosion and should be able to remove metal chips formed during the operation effectively. These machining operations are such that the exposure of these fluids to the human being cannot be completely avoided. MWFs are mostly applied by jet, spray, or hand dispenser, and generate mist during the operation, which can be inhaled by the operators. The fluids also come in direct contact with skin, specially hands and forearms. Consequently, there are several health risks involved in the applications of MWFs, which in turn restrict the use of hazardous chemicals in MWFs; even the selection of base oils is important from the point of health risks.

The main health risks from MWFs are irritation of the skin or dermatitis, irritation of the upper respiratory tract, breathing problems, allergies, and occupational asthma. Similar risks are also involved with any fluids used for washing machines and equipment. Metalworking industry is therefore unique in the sense that the methodology required to develop MWF is quite different than the methodology followed for the development of industrial oils or automotive oils. In addition, there are no well-defined specifications with test methods to evaluate MWFs. These products remain performance

Developments in Lubricant Technology, First Edition. S. P. Srivastava.
© 2014 John Wiley & Sons, Inc. Published 2014 by John Wiley & Sons, Inc.

guided rather than specification driven, and each manufacturer or user applies its own test methodology to evaluate products.

MECHANISM OF CHIP FORMATION IN MACHINING DUCTILE MATERIALS

A typical simplified picture of cutting action is depicted in Figure 16.1. During continuous machining, the uncut layer of the work material just ahead of the cutting tool (edge) is subjected to compression as indicated in the figure. The force exerted by the tool on the chip arises out of the normal force, N, and frictional force, F. Due to this compression, shear stress develops, within the compressed region, in different directions and rapidly increases in magnitude. As soon as the value of the shear stress exceeds the shear strength of the work material in the deformation region, yielding or slip takes place resulting in shear deformation (called the primary deformation), and metal is removed in the form of chip. The chip passes over the rake face of the cutting tool and receives additional deformation (called the secondary deformation) due to the shearing and sliding of the chip over the tool. These two plastic deformation processes [2] have mutual dependence. The chip that rubs the rake face of the tool has been heated and plastically deformed during its passage through the primary shear process; therefore, the secondary process is influenced by the phenomena on the shear plane. At the same time, the shear direction is directly influenced by the rake face deformation and friction processes. The shear direction influences the heating and straining of the chip in the primary process. However, the forces causing the shear stresses in the region of the chip quickly diminish and finally disappear while that region moves along the tool rake surface toward and then goes beyond the point of chip–tool engagement. As a result, the slip or shear stops propagating long before total separation takes place. In the meantime, the succeeding portion of the chip starts undergoing compression followed by yielding and shear. This phenomenon repeats rapidly resulting in formation and removal of chips in thin layer by layer. If A1 is the thickness of the material removed, the chip thickness is slightly lower, A2, due to compression. The lower surface becomes smooth due to further plastic deformation by the intensive rubbing with the tool at high pressure and temperature. The pattern and extent of total deformation

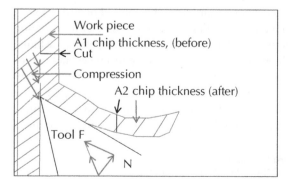

FIGURE 16.1 Chip formation during cutting operation.

of the chips due to the primary and secondary shear deformations of the chips ahead and along the tool face depend upon several factors such as:

1. Work material properties
2. Tool material and its geometry
3. Machining speed
4. Cutting fluid characteristics and application

The overall deformation process causing chip formation, however, is very complex. The function of MWF or cutting oil is to extract heat to reduce tool wear, lubricate to improve surface finish, reduce BUE, and modify shear flow properties of the material in the primary and secondary deformation zones [3, 4]. During drilling operations, these functions become more complex. The lubrication is occurring at the cutting area as well as between the margins of the drill and the side walls of the hole [5].

Various other metalworking operations are briefly discussed in the following lines.

Turning operations produce parts that are round, cylindrical, or conical in shape from rough cylindrical blank. The blank rotates about its longitudinal axis and against the cutting tool. The speed of the workpiece, the depth of the cut, and the feed rate of the tool are important parameters to obtain desired surface finish and dimensional accuracy.

Honing is a low-speed, surface-finishing process used to produce uniform components with high accuracy and fine surface finish. Metal particles are removed by the shearing action of a moving honing stone or stick.

Superfinishing is a polishing process that utilizes abrasive tapes or specialized superfinishing stones to provide a consistent finish.

Grinding is a process in which small particles of metal are removed from the workpiece surface using an abrasive wheel. It is used to produce components of precise geometry and surface finish. The type of grinding process and the choice of wheel material and grit size depend upon the shape of the component, the metal removal rate, and the required surface finish. Grinding fluid plays an important role in the process by reducing friction between the abrasive material and the workpiece in the contact zone. Fluid also takes away the heat generated during the process. Fluid viscosity and additives play a significant role.

Drilling is a machining process where a solid fluted tool is used to produce a hole in a solid component. The cutting tips are situated at the end of the drill, and the flutes allow coolant to enter the cutting zone and allow metal chips to evacuate through the hole. With deep hole and ejector drilling, the cutting tip is supplied with coolant at high pressure, through a hole at the bottom of the center of the drill, and the chips are ejected through a groove or a second hole down the length of the drill body. Deep hole drilling (gun drilling) is a complicated process and requires specific fluids for satisfactory operations.

Broaching is a machining process in which a cutting tool is pushed or pulled through a hole or across a surface to remove metal. The cutting teeth are set so that each tooth removes a small amount of metal and several teeth engage at the same time. Broaching tool is a complicated part, manufactured from a single piece, and the main thrust is to improve tool life and reduce tool wear. Neat oils containing active sulfur and sometimes chlorine additives are preferred in this operation to reduce buildup at the cutting edge. Synthetic emulsions with high extreme pressure (EP) additives are also used in broaching.

Gear hobbing is a multipoint machining process in which gear teeth are progressively generated by a series of cuts with a helical cutting tool (hob). Both the hob and the workpiece revolve constantly as the hob is fed across the face width of the gear blank.

Gear shaping uses a gear-shaped cutter that is reciprocated and rotated, in relation to a blank, to produce the gear teeth.

Gear shaving is applied to unhardened gears to improve their accuracy and uses a gashed rotary cutter in the shape of a helical gear that removes small slivers of material from the tooth profile.

All the three gear manufacturing processes utilize a single fluid based on neat oil containing EP additives similar to broaching oils.

CLASSIFICATION OF METALWORKING FLUIDS

There are several classifications for MWFs. The common systems used currently are described as follows:

DIN-51385 has classified MWFs (Code S) into the following seven categories:

1. Non-water-miscible MWFs—Code SN
2. Water-miscible metalworking concentrates—Code SE
   a. Emulsifiable metalworking concentrates—Code SEM
   b. Water-soluble metalworking concentrates—Code SES
3. Diluted MWFs—Code SEW
   a. Metalworking emulsions, oil-in-water—Code SEMW
   b. Metalworking solutions—Code SESW

Japanese standard JIS-2241 has classified MWFs into 29 grades with 23 neat cutting oils and 6 water-soluble oils. The neat oils are characterized by fatty oil content, chlorine content, viscosity, copper corrosion, and viscosity. The soluble oils are classified according to the appearance of the diluted emulsion and mineral oil to emulsifier ratio.

ASTM D-2881 has also classified MWFs into four categories: oil-based fluids, aqueous emulsions and dispersions, chemical solutions, and solid lubricants.

ISO has classified eight neat oils and nine water-miscible fluids, depending on the additives present in them (Table 16.1).

To sum up, MWFs would fall into the following categories:

1. Neat oils with and without additives.
2. Soluble oils and oil-in-water emulsions with or without additives, ranging from milky white emulsions to transparent microemulsions.
3. Semisynthetic oils, usually transparent emulsions, with small amount of mineral oil.
4. Synthetic oils without mineral oil. These are transparent emulsions of synthetic oils with synthetic emulsifiers and EP additives.
5. Chemical solutions in water.

TABLE 16.1   ISO 6743/7 metalworking lubricant classification L—family M

		Description
1.	**Neat oil category**	
	MHA	Mineral oil or synthetic oil, may have anticorrosion properties
	MHB	MHA oils with friction-reducing properties
	MHC	MHA oils with EP properties, chemically nonactive
	MHD	MHA oils with EP properties, chemically active
	MHE	MHC oils with friction-reducing properties
	MHF	MHD oils with friction-reducing properties
	MHG	Greases, pastes, and waxes, applied as such or diluted with MHA fluids
	MHH	Soaps, powders, solid lubricants, or their blends
2.	**Aqueous fluid category**	
	MAA	Concentrates giving milky emulsion with water and having anticorrosion properties
	MAB	MAA with friction-reducing properties
	MAC	MAA with EP properties
	MAD	MAB with EP properties
	MAE	Microemulsions with anticorrosion properties
	MAF	MAE with friction-reducing and/or EP properties
	MAG	Concentrates yielding transparent solutions with water and having anticorrosion properties
	MAH	MAG with friction-reducing and/or EP properties
	MAI	Greases and pastes blended with water

TABLE 16.2   Main application of ISO MWF categories

S.N.	Applications	Fluid category
1	Cutting operations	MHA–MHF and MAA–MAF, MAH
2	Abrasion	MHC,MHE,MHF, MAG,MAH
3	Sheet metal forming	MHB MHG and MAA,MAB, MAD, MAI
4	Wire drawing	MHB, MHG, MHH, and MAB,MAI
5	Forming and stamping	MHB and MAG,MAH
6	Rolling	MHA, MHB, and MAG

The major applications of ISO-category fluids (both neat and aqueous fluids) are represented in Table 16.2. These are only general guidelines, and actual products are used depending on the type of operation and the equipment design parameters.

MWFs require a large number of additives, and some of the typical additives belong to the following categories:

Antimicrobial/antibacterial or biocides

Antifoams

Alkalinity control and hard water tolerance

Coupling agents for stabilizing emulsion

Complexing agents

Corrosion and rust inhibitors

Dyes

Emulsifiers or wetting agents (surface-active compounds)

EP additives (corrosive and noncorrosive)

Lubricity additives

Metal deactivators

Synthetic esters

Under each of these classes of additives, there are several choices available and briefly indicated in Table 16.3.

TABLE 16.3 Some typical additives used in MWFs

S.N.	Additive type	Chemical compounds
1	EP	Elemental sulfur, sulfurized base oil, corrosive and noncorrosive sulfurized fats, sulfurized olefins, phospho-sulfurized terpenes, alkyl/aryl phosphate esters, amine phosphates, 400 TBN over-based Ca-sulfonate
2	Lubricity improver	Fats, fatty acids/ester, triethanol amine oleate, complex polyolesters, linear alcohols, phosphate esters
3	Corrosion and rust inhibitors	Imidazolines, sarcosines, triethanolamine, sodium nitrite, alkyl phosphonates, amine borate, sulfonates, naphthenates, polyetheramines
4	Coupling and complexing compounds	Linear alcohols and glycols, diethylene glycol, PEG-200/400, ethylene diamine tetra acetic acid (EDTA)
5	Metal deactivator	Sodium mercaptobenzothiazole, alkyl-benzotriazole
6	Emulsifiers	Various anionic, nonionic, and cationic surface-active compounds such as petroleum sulfonates, naphthenates, ethoxylated phenols, alcohols and fatty acids, EO/PO copolymers
7	Antifoam	Dimethyl siloxanes and some nonionic surfactants
8	Alkalinity control	Primary and tertiary amines
9	Hard water tolerance	Primary amines
10	Biocides	p-Chloro-m-cresol/xylenols, O-phenyl phenol, formaldehyde, phenoxyethanol/phenoxypropanol, sodium dimethyl dithio carbamate, isothiazolin derivates, triazine, admontate compounds, etc.

The use of water-based products in MWF is encouraged due to three reasons. Firstly, water is an excellent coolant due to its high specific heat and high thermal conductivity (water Sp. heat is 4.18 kJ/kg K; thermal conductivity 0.58 W/m·K, as against oil Sp. heat of about 1.8 kJ/kg K and oil thermal conductivity of 0.15 W/m·K), which is one of the prime requirements of MWF; secondly, water-based fluids are much less toxic as compared to neat oils; and thirdly, it is abundantly available at a very low cost. Water also has high latent heat of vaporization (2270 kJ/kg), which means that it takes away a large amount of heat on vaporization. These thermal properties make water and water-based products as one of the cheapest and best coolant for metalworking operations.

Earlier poorly refined low-viscosity-index oils were used in MWF formulation due to their easy emulsion-forming properties, mainly due to their polar characteristics. These oils, however, contain polycyclic aromatic hydrocarbons, and they have been slowly phased out due to the health hazard issues. The disposal of mineral oils used in MWF is also a concern. These problems led to the development of high water-based fluids, microemulsions, and semisynthetic and synthetic products for metalworking operations. Wherever mineral oil-based products are unavoidable, highly refined base oils are now preferred over the earlier less refined base oils. This shift has posed serious challenges to the formulation of soluble oils using high-quality base oils, due to their nonpolar nature.

EMULSIONS AND LUBRICANTS

In most of the lubricant applications, emulsions are undesirable, but in MWFs, oil-in-water emulsions are extensively used due to their superior cooling properties. Emulsions are a dispersion of small globules of a liquid into another liquid. Generally, one of the liquids in emulsions is water. These are classified as follows:

Oil-in-Water Emulsions

In these systems, oil is the dispersed phase and water is the continuous phase.

Water-in-Oil Emulsions

In these systems, water is the dispersed phase and oil is the continuous phase. Such emulsions are called invert emulsions and are used in fire-resistant hydraulic fluids in mining industry.

The two types of emulsion can be easily identified by simple tests. Oil-in-water emulsions can be easily diluted with water but not by oil. Similarly, water-in-oil emulsions can be diluted with oil and not with water. Oil-soluble dyes (Sudan red and dimethyl yellow) will easily spread on water-in-oil emulsions, and water-soluble dyes (methylene blue, brilliant green, and acid fuchsin) will spread on oil-in-water emulsions. Also, when water is in continuous phase, the emulsion is conducting. Electrical conductivity measurements can be used to distinguish the emulsion types provided, if electrolytes are not present in the systems.

TABLE 16.4   Oil particle size in different types of fluids

Emulsion type	Dispersed phase particle size, μm (10^{-4} cm)
Unstable emulsion for thin steel sheet rolling	>50
Meta-stable emulsion for sheet rolling	10–40
Milky emulsion	0.1–10
Microemulsion	0.01–0.1

Polar head   long chain hydrophobic (lipophilic) group

Hydrophilic group   oil solubilizer

FIGURE 16.2   Emulsifier or surface-active compound structure.

For preparing emulsions, an emulsifier or surfactant is required to stabilize it. Without the emulsifiers, the emulsion can be prepared by homogenizing two phases, but the system will be unstable and will separate on standing. The emulsifiers have affinity to both the phases and concentrate at the surfaces of the droplets to form a film at the interface. Different types of emulsions have varying particle sizes of the dispersed phase, and the size is also related to the stability of the emulsion. Table 16.4 provides approximate size distribution of dispersed phase in various emulsions.

Emulsifiers have been categorized into the following four types:

1. Anionic surface-active compounds dissociate in water and form a positively charged cation and a negatively charged anion. Since the anion possesses surfactant property, these are called anionic surfactant. Anionic emulsifiers, in which the oleophilic/lipophilic part is negatively charged, are soaps, sulfonates, phenates, salicylates, naphthenates, and carboxylates. These are generally soaps of sodium, potassium, calcium, magnesium, and barium.

2. Cationic surface-active compounds dissociate into a positively charged ion, which possesses surface-active properties. Cationic emulsifiers, in which the oleophilic part is positively charged, are quaternary ammonium salts.

3. Nonionic surfactants do not dissociate in water and are neutral. In these emulsifiers, the lipophilic part does not contain a charge and are prepared from fatty alcohols, fatty amines, and fatty acid esters or are ethylene/propylene oxide condensate of alcohols and phenols.

4. Amphoteric or zwitterion emulsifiers behave like either anionic or cationic surfactants depending upon the pH of the medium. At isoelectric point, there is no overall charge in these molecules.

All emulsifiers are characterized by the following typical structure (Fig. 16.2), having a hydrocarbon chain and a polar head. The hydrocarbon chain provides the oil

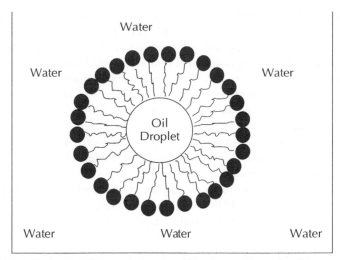

FIGURE 16.3   Oil droplet (dispersed phase) stabilized by surfactant in oil-in-water emulsion.

solubility and the polar head has affinity toward water. The ratio of the two is referred to as hydrophilic–lipophilic balance (HLB) value.

Each emulsifier is assigned an HLB value, which determines whether the emulsifier will behave as an oil-soluble or water-soluble type of emulsifier. The HLB value ranges from 1 to about 30. The lower HLB number indicates oil solubility, and the higher number is indicative of water solubility. These emulsifiers stabilize the emulsion by adsorbing at the oil–water interface and by reducing interfacial tension. In several cases, an electrical double layer is also established at the interface, which stabilizes the dispersed droplets. Since the emulsifiers form a film between oil and water, the interfacial tension between water–emulsifier and oil–emulsifier also assumes importance. If the tension between water–emulsifier is smaller than the oil–emulsifier tension, the former surface will tend to be larger, and oil-in-water emulsion will be formed (Fig. 16.3). Similarly, if the interfacial tension between oil and emulsifier is smaller than the water–emulsifier, the oil surface will be larger, and water-in-oil emulsion will result. The stability of the emulsion also depends upon the relative interfacial tensions or as a corollary upon the dispersibility of the emulsifier in the two phases. Often, more than one emulsifier is used to prepare stable emulsion: one having more affinity toward water and the other toward oil. There are large numbers of chemical compounds [6–8] in each category that act like surface-active agents. The list is so large that it is not possible to describe all the surface-active agents in this chapter and is beyond the scope of this book. Large numbers of products are industry specific. Some of the commonly used products are as follows:

Anionic products are based on sulfonate, carboxylate, and sulfate ions such as sodium dodecyl sulfate, ammonium lauryl sulfate, alkyl benzene sulfonate, sodium and calcium petroleum sulfonate, sulfated castor oil (turkey red oil), soaps, amine soaps, amine salts of phosphoric acids, naphthenic acids, and fatty acid salts. The sodium soaps of fatty acid like oleic and stearic acids are good emulsifiers for oil-in-water emulsion, but these are more effective at higher pH of about 10. Triethanolamine soaps of fatty acids are less dependent on pH. However, these soaps react with calcium or magnesium present in water as hardness to form insoluble soaps, which interfere with the emulsion and cause problems during machining operations. Both petroleum and natural sulfonates have been extensively used in oil-in-water emulsion in combination with other additives.

Cationic emulsifiers are based on quaternary ammonium cations such as cetyltrimethylammonium bromide (CTAB), cetylpyridinium chloride/bromide (CPC), polyethoxylated tallow amine, fatty amine salts, imidazolines, and sarcosines. These emulsifiers are used in special applications. Bitumen/asphalt emulsions are prepared by using cationic emulsifiers. Cationic products are not compatible with anionic emulsifiers. Quaternary ammonium compounds are stable in hard water and are also resistant to several microorganisms.

Nonionic emulsifiers are based on ethylene oxide condensates of alkyl alcohols, fatty acids and alkyl phenols, EO/PO condensates, ethoxylated fatty amines, various polyols, cetyl alcohol, oleyl alcohol, and alkanolamine salts of fatty acids, sorbitan mono-oleate, and glycerol mono-oleate. Nonionic emulsifiers form a large group of compounds and are used in a variety of applications. Both oil-soluble and water-soluble products can be produced by changing the ratio of ethylene oxide in the molecule. Such products can also be produced as mixed alkyl/aryl polyglycol ethers with different HLB values and solubility. These nonionic compounds exhibit high surface activity even in the presence of electrolyte and have been used in a variety of industries for different applications.

SURFACE-ACTIVE COMPOUNDS IN METALWORKING FLUIDS

In cutting, rolling, grinding, and drawing operations, oil-in-water emulsion is usually used. The primary function of emulsion is to cool the tool and remove heat from the point of contact and also lubricate surfaces to prevent seizure. For effective cooling, water must be the continuous phase. These oils in lubricant terminology are called soluble oils and are manufactured from a light mineral oil, emulsifying agent, and corrosion inhibitor. When this oil is mixed in water, emulsion is formed, where fine droplets of oils are dispersed through the continuous water phase. Conventional soluble oils are generally milky in appearance, but for certain applications, translucent emulsions are preferred. These translucent emulsions have higher level of emulsifiers in it, which reduces the oil droplet size, thus increasing the clarity.

For different applications, the oil content in the emulsion may vary from 1 to 10% volume. There are large numbers of metalworking operations to produce a variety of metal products, and different types of fluids are used. The choice of the emulsifier system in cutting oils is mainly governed by the cost. Conventionally, 12–18% of the emulsifiers in light base oil are used to produce the soluble cutting oil, although several advance emulsifier systems are available to produce the emulsion by using much smaller quantity of the emulsifier. The conventional emulsifier system consists of a mixture of petroleum sulfonate, rosin soap, and several auxiliary additives to impart stability and anticorrosion properties. These include glycols, tall oil esters, triethanolamine, oleic acid, naphthenates, and alkyl imidazolines. Nonionic surface-active compounds based on ethylene oxide condensates of alkyl alcohol or phenols have also been used in combination with the conventional additives to improve the emulsion stability. Although the science of emulsion

TABLE 16.5   Typical composition of soluble and synthetic emulsifying MWFs

S.N.	Self-emulsifying synthetic ester concentrate [6]		Conventional soluble emulsion concentrate	
	Components	Percent	Components	Percent
1	Water	55	Water	20
2	Synthetic ester	25	Mineral base oil	50
3	Alkanolamine	8	Alkanolamine	10
4	Oleyl alcohol-2EO	1	Emulsifier anionic	7
5	Boric acid	5	Emulsifier nonionic	5
6	Iso-octyl oleate	6	Corrosion inhibitor	5
7	—		Biocide	3

[6–8] has advanced to a great extent, in petroleum industry, somewhat empirical methods are used in the selection of emulsifier systems for manufacturing commercial products. The main reason is the use of several other additives to impart anticorrosion, EP, antiwear, antifoam, and antimicrobial properties to the emulsion to meet the operational requirements. This makes the formulation complex and additive interaction modeling becomes difficult. In an emulsion, particle size and stability play an important role in the applications of these fluids. The type of emulsifier and its dosage in the oil are the controlling factors for these parameters. Table 16.4 provides the particle size requirement of different emulsions for different applications.

Details of two typical formulations of oil-in-water emulsion system used in MWFs are provided in Table 16.5. One is based on conventional additives and the other uses a self-emulsifying ester [9]. The product based on self-emulsifying ester is stable for more than 60 days in 880 ppm hardness water. The life of MWFs is prolonged by the use of a suitable biocide. This information is provided only as general guideline, since in each class of additives, large numbers of commercial products are available, and the choice of a particular additive is based on technoeconomic considerations.

EP and antiwear additives are generally used in EP neat cutting oils and not in soluble-type oils. The anionic emulsifier in the conventional system may be a mixture of sodium sulfonate, sodium naphthenate, naphthenic acid, rosin soap, and oleic acid.

The corrosion inhibitors may be alkanolamine salts of carboxylic acid, boric acid amide, sodium borate, and sodium nitrite. The combination of sodium nitrite and triethanolamine, which forms a good anticorrosion combination, is now discarded due to carcinogenicity.

Microemulsion-based products have also been used in the metalworking industry, where it is desirable to have clarity and watch the tool surfaces during the use. A new additive chemistry-based product has been reported [10], and details are provided in Table 16.6. Machining fluids or MWFs are in use throughout the manufacturing industry for their coolant, lubricant, and corrosion-resistant properties during metal cutting, grinding, boring, drilling, and turning operations. Various patents disclose additives used in metal forming and machining fluids. Boric acid, alkali borates, and borate esters are conventionally known for their beneficial effects when included in lubricating oil formulations. Polyhydric alcohol and polyalkylene glycol have also

TABLE 16.6 Microemulsion-based MWF

S.N.	Components	Percent
1	Light neutral oil	21.2
2	Deionized water	5.50
3	Super amide alkoxylate[a]	4.45
4	Ether carboxylate	5.05
5	Alkyl phosphonate	1.00
6	Tallow fatty acids	17.20
7	Amine borate solution (55%)	44.60
8	Biocide	1.00

Above blend can be diluted to a fine microemulsion; Super amide alkoxylate provides hard water stability.

[a]Super amide alkoxylate has the following general structure:

$$\text{\\\\\\\\}\overset{\displaystyle O}{\overset{\|}{C}}\text{- NHCH}_2\text{CH}_2\text{- (O- CH}_2\text{ -}\overset{\displaystyle CH_3}{CH})_n - (OCH_2\text{- CH}_2)_m \text{ OH}$$

been used or added in oils to enhance their properties. Alkyl carboxylates, such as the ester glycerol mono-oleate, have also found uses in MWF compositions. Boron-based additives, when mixed with carrier fluids, such as water, cellulose, or cellulose derivatives, polyhydric alcohol, polyalkylene glycol, polyvinyl alcohol, starch, and dextrin, in solvated forms, provide improved MWFs [11]. These fluids effectively reduce friction, prevent galling, and reduce wear of cutting and forming tools.

Boron-containing compounds are known to be noncorrosive and to possess anti-oxidant, fire-retarding, and potential antifatigue characteristics. Such compounds also exhibit antiwear properties such as borated dihydrocarbonyl phosphonates. Borated phosphite additives may be synthesized by reacting dihydrocarbonyl phosphites with boron-containing compounds such as boric oxide, metaborates, alkylborates, or boric acid in the presence of a hydrocarbonyl vicinal diol. Organometallic boron-containing compounds are another class of additives useful in lubricants.

METALWORKING FLUID MONITORING

MWFs have to be constantly monitored for their in-service quality, since the active ingredient in these fluids is continuously being removed by the metal surfaces. Water also gets evaporated during the process. The following tests [12] are generally required to keep the fluid in good condition:

1. Concentration—A refractometer is a simple tool for checking the level of coolant and water mixture. Some soluble coolants may require acid splits or alkalinity measurements to maintain concentration. This ensures the correct level of lubrication and product constituent.

2. pH—Maintaining pH will reduce biological activity in the system, extend system life, and maintain corrosion protection. Cationic titration can be carried out to find out the anionic emulsifier's concentration.

3. Tramp oil—Excessive tramp oil can lead to biological concerns and weaken the emulsion stability of the soluble coolant.

4. Hardness—Hardness is the measure of inorganic salts (calcium, magnesium) dissolved in water. Increased hardness can lead to emulsion stability issues and cause *splitting* of the soluble oil emulsion. Oil companies, generally have full range of soluble coolants to meet any water conditions.

5. Bacteria/fungus—Dip slide is a simple method of checking the bacterial and fungal conditions of the soluble coolant. Bacteria/fungus may cause corrosion, poor filterability, reduction in pH value, and odor issues.

In addition to these tests, if required, four-ball EP and wear tests may also be used to monitor these properties. General characteristics of MWFs, their composition, and relative costs are provided in Table 16.7.

TABLE 16.7   Characteristics of different MWFs and their composition

Products	Composition	Characteristics	Cost
Neat oils	Mineral oil, EP/antiwear/ fatty oils, and biocide	Excellent lubricity and wetting, low cooling, microbial attack, and cleaning an issue	Highest
Water-soluble oil-milky emulsions	Mineral oil, EP/antiwear/ emulsifier/coupling agent/ anticorrosion, and biocide	Excellent cooling, good lubricity and wetting, fluid stability, and microbial attack is problem	Moderate
Semisynthetic solutions	Low concentration of mineral oil/synthetic ester, EP/antiwear/ emulsifier/coupling agent/ anticorrosion, and biocide	These have properties in between the synthetic solutions and milky emulsions due to low concentration of oil. Better cooling and microbial attack resistance	Low
Synthetic solutions	Mineral oil free, EP/antiwear/ emulsifier/coupling agent/ anticorrosion, lubricity, film-forming compounds, and biocide	Better cooling properties, resistant to microbial attack, good surface conditions	Lower
Chemical solutions	A water solution of chemical additives such as EP, corrosion inhibitor, and lubricity additives	Best cooling action, excellent rust preventive characteristics, suitable as grinding fluid	Lowest
Microemulsions	Mineral oil with high dosage of specific emulsifier to reduce oil particle size containing EP and corrosion inhibitor	Excellent cooling, good lubricity, and wetting. Transparency of the fluid is advantageous, stability associated with milky emulsions is overcome	Moderate

ROLLING OILS FOR STEEL

A conventional steel plant converts manufactured steel ingot into various products through the process of hot and cold rolling. The primary mills convert the ingot into blooms and slabs by hot rolling operation. These mills are of low speed and heavily loaded and generate peak loads for short duration. However, in a modern plant, continuous casting process is used to convert molten steel into semifinished billet, bloom, or slab. Depending on the product end use, various shapes can be cast. In recent years, the melting, casting, and rolling processes have been integrated to get the final finished product. The near-shape cast section has been applied to beams and flat-rolled products, and this results in a highly efficient operation. The complete process chain from liquid metal to finished rolling can be achieved within a few hours. Continuous casting process thus provides improved quality, productivity, cost efficiency, and yield as compared to the old process of forming ingot and then hot rolling the ingots into various products. These rolling operations are energy intensive and consume about 50% of the total electrical energy [13]. About 37% of this energy is lost as mechanical friction. This friction can be greatly reduced by the application of hot and cold rolling oils. The temperature in the hot rolling mill is quite high; therefore, many mills use only water during rolling. However, a lean dispersion of fatty oil in water is sprayed by other mills on the surface to reduce friction and energy consumption.

Thin steel sheets are produced by cold rolling of thicker metal sheets by spraying cold rolling oil, which is usually a lean emulsion/dispersion of synthetic or fatty oil combination. The primary function of cold rolling oil is to facilitate rolling process by controlling friction and temperature rise at the interface of work roll and steel strip. Therefore, cold rolling oil is a coolant to dissipate metal deformation heat and frictional heat and also to lubricate roll bite to optimum level. Most of the cold rolling oils used for reducing steel sheets to various thicknesses are oil-in-water emulsions or dispersions of varying droplet size of dispersed phase. The oil emulsion is sprayed onto the rolls and sheet during rolling operation. The emulsion used in cold rolling is not a stable emulsion, but is of metastable nature, and the oil separates out at the roll bite providing lubrication and consequent friction reduction between the rolls and sheet. With these emulsions, a phenomenon of oil pooling [14, 15] is observed at the predeformation zone, which causes an increase in oil film thickness at the work zone of roll bite. It has also been reported [16] that oil gets trapped in the micropits of the sheet and with the increase in pressure, oil viscosity rises, increasing oil film thickness. Oil pooling is an increase in oil volume fraction, and this can become so high that phase reversal of emulsion can take place at the surface. Phase reversal means that the oil-in-water emulsion becomes water-in-oil emulsion. With such a phase reversal, the volume fraction of the more viscous component oil will increase with the positive pressure gradient at the inlet zone. The diffusion process depends on the initial concentration of oil in the emulsion. This process of phase reversal is also induced by the evaporation of water at higher strip temperature of 120–200°C and the attraction of polar fatty oil or acids toward the metal surface [17].

Oil viscosity and oil particle size in the emulsion are also important factors. Smaller oil particle sizes form film at higher speed, while coarser particles will form film at lower mill speed [18]. These oil films reduce friction between the roll bite and metal sheet, but the friction must be optimum. The friction must be high enough to permit traction in the roll bite. However, if the friction is very low, skidding can take place. On the other hand, it is desirable to have low friction to minimize power consumption and maintain surface quality of the rolled sheet. When the roll bite friction is low, neutral plane shifts toward the exit of the bite, and there would not be enough traction to pull the strip. The roll will spin on the strip, and this phenomenon is called *skidding*. Conversely, when the neutral plane is ahead of the roll bite entry (with high friction), the strip is extruded through the roll bite, and this phenomenon is called *slipping*. Therefore, during cold rolling operation, the coefficient of friction should remain consistent. With sudden friction reduction, the forward slip gets reduced, and if the rolling speed is not reduced, it will lead to tension and the strip will break. Inconsistent friction is also responsible for shudder, chatter, poor surface finish, and shape. In cold rolling operation, selection of appropriate rolling oil with proper friction characteristics, emulsion properties, and oil droplet size are important factors. Each mill has its own characteristics, and thus, the oil also has to match the mill parameters.

TYPES OF COLD ROLLING MILLS

Rolling mills of different designs are used for cold rolling of steel. Generally, the following three types of mills are used, although there could be different configurations within these designs:

1. Single-stand reversing mills (4 hi/6 hi—4 hi means that two rolls are backed up, while in 6 hi, there are four backup rolls)
2. Multistand tandem mills (four stands or five stands)
3. Sendzimir mills (20-hi/Z-hi mills)

ROLLING OILS

As discussed earlier, each mill is specific and requires specially formulated rolling oil to suit its design, operating parameters, sheet thickness reduction per pass, oil circulation, purification, and emulsion formation system. Due to such variations, there are no guiding specifications for cold rolling oils. Rolling oils generally consist of mineral oil, synthetic esters, fats, fatty acids, EP additives, biocide, emulsifier, and coupling agent to stabilize the emulsion. Oils are distinguished on the basis of emulsion formation characteristics and friction modification properties. For low-speed mills, oils of strong emulsion stability with moderate friction-reducing characteristics are used. Metastable emulsions are capable of rolling sheets up to 0.4-mm thickness. For thinner sheets, oil dispersions (unstable emulsions) with strong lubricity characteristics are employed. Oils with strong emulsion-forming tendencies can be used for longer duration, but their antifriction properties are inferior. Rolling oils with weak emulsion stability are capable of reducing roll wear in the multistand mills due to superior antifriction properties but have the tendency to separate out in the oil-holding tanks. A strong agitator in such systems would be useful. ASTM D-3342 test method has been developed to assess the emulsion stability of rolling oil dispersion in water. Another method of assessing emulsion stability index (ESI value) has also been developed and is useful in determining tightness or looseness of the emulsion [19].

Emulsion stability is also affected by the water hardness, temperature, and its pH value.

PERFORMANCE EVALUATION OF STEEL ROLLING OILS

Many laboratory test methods such as Falex, SRV, LFW-1, and optical EHD have been attempted to simulate rolling mill conditions. Small experimental rolling mills have also been used to simulate oil performance, but none of these correlate with the actual mill operations. At best, these methods can be used for screening various rolling oil formulations during the development stages.

There are several rolling mill parameters that affect the performance of the oil. Roll condition, rolling load, steel strip cleanliness, spray pattern of the oil through nozzles, oil cleanliness, concentration of the oil in emulsion, all contribute to the performance of oil in the mill. Any deviations in these parameters can lead to high roll bite temperatures, which indirectly mean low oil viscosity and consequently higher friction. These parameters are therefore required to be controlled precisely to avoid roll spalling, high roll wear, strip breakage, and rolled sheet quality.

Formulation of MWFs is a challenge to the oil developers, since there are no well-defined standard laboratory tests for evaluation. The available antiwear, EP, and friction tests, at the best, offer screening opportunity to the candidate blends. Each operation is unique and requires careful selection of base fluid and chemical additives to obtain satisfactory performance in the intended equipment.

REFERENCES

[1] Jean C. Childers, metal working fluids—an industry analysis. Lubr Eng 1985;41 (4):214–220.
[2] ASM International Handbook Committee. *ASM Handbook*. Volume 16, Materials Park: ASM International; 1989. Machining; p 7–12.
[3] Shaw MC. *Metal Cutting Principals*. Oxford: Clarendon Press; 1984. p 461–481.
[4] Cassin G, Boothroyd G. Lubricating action of cutting fluids. J Mech Eng Sci 1965;7 (1):67–81.
[5] Haan DM, Olson WW, Sutherland JW. The mechanism of metal cutting fluids in machining aluminium alloy. AAMA Metal Working Fluid Symposium; March 1996; Dearborn. p 301–306.
[6] Becher P. *Encyclopedia of Emulsion Technology*. Volume 2, New York: Marcel Dekker; 1985. Applications of emulsions.
[7] Surfactant science series, Volumes 1–132. London: CRC/Taylor & Francis.
[8] Tadros TH, Vincent B. In: Becher P, editor. *Encyclopedia of Emulsion Technology*. Volume 1, New York: Marcel Dekker; 1983.
[9] Hoogendoorn R. New emulsifiable esters in metal working fluids. Current trend in industrial tribology. Proceedings of the XI NCIT; January 22–25, 1995; New Delhi: Tata McGraw Hill. p 531–536.
[10] Le Helloco JG, et al., New generation of additives for high performance metal working fluids. 13th LAWPSP Symposium, Mumbai, Volume 13, No 13, 2002. p AD. 04.01-06.
[11] Erdemir A, Sykora F, Dorbeck M. Metalworking and machining fluids. US patent 7,811,975. October 12, 2010.

[12] Byers J. Laboratory evaluation of MWFs. In: Byers J, editor. *Metalworking Fluids*. Boca Raton: CRC Press; 2006. p 148–173.

[13] Cichelli AE, Poplawski JV. Tribology and energy considerations in the rolling of steel. ASLE Conference; February 3–7, 1980; Anaheim.

[14] Pathak P, Ranjan M, Marik AK, Sengupta PP, Murthy GMD, Jha S. In: Srivastava SP, editor. Improvement in effectiveness of cold rolling lubrication in steel mill. Proceedings of the International Symposium on Fuels and Lubricants (ISFL 97); 1997; New Delhi. New Delhi: Tata McGraw Hill. p 74–79.

[15] Wang AZ, Rajgopal KR. Lubrication with emulsion in cold rolling. J Tribol 1993;115 (3):523–531.

[16] Fudanoki F, Inoue S, Araki J, Yanai K. Development of model for formation of surface properties in cold rolling of stainless steel and application to the actual mill. Nippon Steel Technical Report No. 99; September, 2010.

[17] Vergne P, Kamel M, Querry M. Behaviour of cold rolling oil in water emulsion, a rheological approach. J Tribol 1997;119 (2):50–58.

[18] Makino Nakahara T, Koyogoku K. Observation of liquid droplet behaviour and oil film formation. J Tribol 1988;110 (2):348–353.

[19] Takakusaki T, Iwasaki M, Fukuyama S, Misonoh K. Development of rolling oil for high speed cold strip rolling. Trans ISI J 1982;22:233–244.

BLENDING, RE-REFINING, MONITORING AND TEST METHODS

PART V

BLENDING,
RE-REFINING,
MONITORING AND
TEST METHODS

CHAPTER 17

LUBRICANTS: BLENDING, QUALITY CONTROL, AND HANDLING

A finished lubricant is a mixture of one or more base oils with several oil-soluble performance chemical additives. Production of lubricant thus involves blending these components together with the desired end properties. Blending of solids and liquids is an art and has been practiced in different fields such as food and pharmaceutical industry since ages. The blending formula for lubricants is arrived at by carrying out a series of laboratory experiments, rigs/engine testing, and field testing to prove the performance in actual engine/equipment. A typical blending plant usually manufactures 300–500 products in different quantities depending upon the market need. The quantity may range from a few kiloliters of special oil batch to several hundred kiloliters of fast-moving engine oil batch. For blending complete range of these lubricants, blending plant would need about 10–15 grades of base oils and about 100–150 chemical additives in their inventory. A manufacturing document (formulation sheet) is issued by the laboratory to the plant along with quality control parameters. A typical such document would be on the following format. Usually, the base oil and additive details are coded for security purpose.

Formulation

Brand Name: Lubricant ABCD

Components	Date Issued
	Percent by wt. or vol.
Base oil (B_1)	W1/V1
Base oil (B_2)	W2/V2
Additive (A_1)	W3/V3
Additive (A_2)	W4/V4
Additive (A_3)	W5/V5
Additive (A_4)	W6/V6
Total	**100**

Developments in Lubricant Technology, First Edition. S. P. Srivastava.
© 2014 John Wiley & Sons, Inc. Published 2014 by John Wiley & Sons, Inc.

Alternate base oils and additives permitted in the blend are also indicated, if any, in the formulation sheet. Quality control tests such as viscosity at 40 and 100°C, pour point, viscosity index, total acid and base number, and elements are specified with a maximum or minimum range. Wherever necessary, performance tests such as rust, corrosion, oxidation, water separation, and foaming tests are also specified in the formulation sheet.

The quality control tests indicated are only typical for the purpose of understanding. These will vary depending upon the quality of oil and its control requirements. There could be several new control tests for specialty oil. The base oil control is usually difficult. There are batch-to-batch viscosity variations in base oils. Blending plants also do not have control on the source of these base oils. Usually, the most cost-effective base oils are processed, and they could be from multiple sources with different viscosity, viscosity index, and pour point. It is therefore advisable to prepare a laboratory blend with the available base oils and additives and optimize the base oil ratio and the quantity of additives (pour point and viscosity modifiers) to meet the finished product requirements. Other performance additive package, however, has to remain unchanged.

It is also necessary to check the quality of all the incoming base oils and additives before attempting a blend. These have to be checked for their main/specification and nonconforming products to be segregated for appropriate replacement or disposal. Usually, during bulk storage and transfer, water content in the base oil goes up, and this must be removed from the base oil by heating or dry air blowing before blending.

CHOICE OF BLENDING

Majority of small plants use batch blending process ranging from 1 to 10 kl or more batch size. The blending kettle is mounted on a load cell, and each component is entered sequentially and weighed into kettle. Smaller additives like antifoam or pour point depressants can be added manually. There are, however, several choices available for a larger blender to achieve the most cost-effective process. Several oil companies have established fully automatic blending plants with computerized and robotized inventory and warehouse management. Following different types of blending arrangements and controls are available for a modern blending plant operation.

In-Line Blending

This blending method offers high production rate of 1000 l/min and is suited for large-volume blends. The product is blended in the line and is sent directly to the filling line or bulk loading. In this system, no blending kettle is required and the products are produced on specification. All components, base oils and additives, are introduced into the line through real-time measuring of liquid components using a series of flowmeters. The flow rate is automatically controlled at the desired ratio and is moved to a common header, which employs an in-line mixer. With this system, the output of the blend header is homogeneous and on specification. In-line blender requires a minimum batch size of approximately 6000 l. The in-line blender has been extensively used by large oil companies for the production of large-volume finished products such as engine oil or hydraulic oil.

Automatic Batch Blending

In automatic batch blending (ABB) system, all the components are transferred to the blending kettle in the desired ratio by volume or mass measurements. Measurement is typically done by load cell or meter. The base stocks and additives are transferred by dedicated piping to the top of the blending kettle to avoid contamination. The main advantage for this blending method is that a basic system can easily be automated. The whole mass is then agitated or recirculated at a temperature of about 60°C. The product is then tested for essential parameters and transferred to storage tank or filling lines. In this process, small- to medium-sized batch is accurately produced. However, this is a three-stage process of filling, mixing, and discharging, which makes the production slow. Small and medium local blenders generally use this method of blending. But larger blenders can use this method to produce specialty products.

Simultaneous Metered Blending

Simultaneous metered blending (SMB) is a term used to describe measuring liquids with a flowmeter and transferring these known quantities into blending tanks so that all the liquid components are in the correct ratio. Small-volume additives are preblended together and then transferred to the kettle. The mixing of the component is typically carried out through recirculation in the blending tank. This operation can quite easily be automated. The SMB is generally used in the small basic blending unit.

Integrated Blending

The integrated blending system is designed and customized by combining in-line blending, ABBs, and SMBs to meet the low-volume and high-volume product requirements. The integrated system takes advantage of the common components and operates under a common control system integrating all operator stations. These systems are employed in the modern new blending plants. The complexity of the control system depends on the particular application. Completely automated system can be designed to integrate liquid asset management, product transfer, tank gauging, product bulk loading, pigging, valve manifolding, and others. Several benefits can be obtained from automation:

1. Blend accuracy is improved resulting in fewer reblends.
2. Both throughput and efficiency are improved.
3. Just-in-time delivery would require less tankages.
4. A larger number of blends and products can be handled.
5. Improved safety and security.

OTHER FACILITIES

Drum Decanting

Most new batch blending operations incorporate an automatic drum decanting system. The system is designed to minimize product loss and contamination when additives are pumped directly into the blending kettles. This system along with the blending control system selects proper drum and setup sequence of transfer process. All other operations can be done automatically, including withdrawal of the exact amount of additive, rinsing of lines between additives to prevent contamination, rinsing emptied drums, and pigging of the lines connected to the blending kettles.

Pigging Systems

Pigging is the system to clean the wall of a pipe after the transfer of additives, base oils, and finished products. This has several benefits:

1. There is minimal product loss, and the lines are cleaned and emptied after the pigging operation.
2. The product contamination is avoided, eliminating the need for flushing.
3. Common lines can be used for multiple products. These systems can be completely automated, and pigging can be carried out after the completion of each batch.

Manifolding

There are many situations in lube plants where multiple input lines must be connected to blending systems or to direct product to various locations. A common solution to this problem in the past has been the hose manifold where the operator physically makes these connections using a flexible hose or short pipefitting. These installations are often referred to as snake pits and are hazardous working areas. The operators stand on a floor that is often slippery due to spilled products and are required to lift heavy hoses to make each connection. In addition, there is also a possibility of product loss and product contamination. There are a variety of manifolds that can be used to overcome these problems. These manifolds range from manual manifolds to fully automatic manifolds that are easily integrated into the blending control system. When incorporated into the blending control system, component connection to the blender or outlet connection to the product tanks is set up automatically when the formula is selected in the setup process.

CONTAMINATION CONTROL IN BLENDING PLANT

Contamination in blending plant usually takes place from the residue of the previous blend either in the tank or piping. These can be minimized by the proper design of the tanks and pipes. Pipeline pigging is a convenient method to remove the residual product. Another method adopted by most of the plants is to follow a sequential blending pattern to avoid product contamination. This, however, requires proper planning of blending. For example, to begin with, engine oils of reducing TBN values are blended, followed by straight mineral oil (bearing oils). This can then be followed by hydraulic oils, compressor oils, turbine oils, etc. This sequence can also be reversed by first starting blending turbine oils, compressor oils, and hydraulic oils followed by engine oils of increasing base number. The idea is to select the next oil in such a way that the contamination of the previous oils can be tolerated to a limited extent. Each plant can develop its own blending sequences to minimize cross contamination. A typical blending arrangement of a modern plant is depicted in Figure 17.1. The plant has base oil tanks B1–B5 and final blended product holding tanks P1–P4. The number of tanks will depend on the requirement of the individual plant. There is a separate handling area for additives. Some additives are solids and need to be dissolved in base oil or in other solvent for easy handling. Some additives

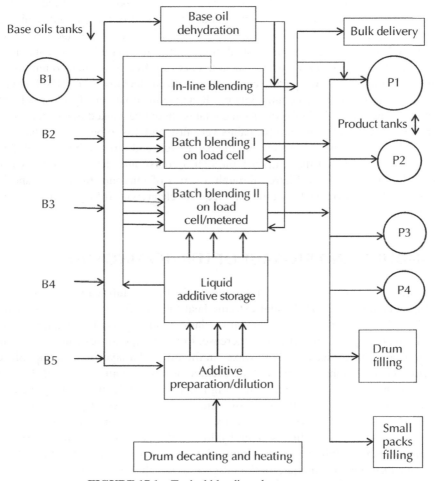

FIGURE 17.1 Typical blending plant arrangements.

like antifoam additives are used in such a small dosage of ppm level that it is difficult to meter them properly. It is, therefore, necessary to dilute these with a solvent like kerosene or light base oil. Most base oils during transit from refinery to the blending plant pick up moisture, and it is necessary to remove it before blending operation. Plant may have an in-line dehydrator or a batch dehydrator for both base oils and additives. Bulk products such as engine oils can be blended in the in-line blender, and smaller batches are prepared in batch blenders. The batch blenders could have both system of weight blending and volume (metered) blending system. The entire system can be integrated with central control and automation system. The extent of automation will depend on the individual plant requirements.

CONTAMINATION FROM CONTAINERS (PACKAGES)

Containers in which lubricants are filled can sometimes become a source of contamination if not properly selected. New steel drums and their bungs sometimes contain a protective oil coating (rust preventive), and direct filling of a high-quality oil like turbine or clean hydraulic oil will deteriorate when it comes in contact with such coatings. Dust, dirt, and sometimes small insects may also get into these containers. It is, therefore, necessary to ensure that containers are free from such contaminants before the oil is filled. In specific cases, phosphatized steel drums or HDPE drums may be used to avoid such contamination. Reactive materials or glycol or solvent-containing products like brake fluid and coolants should be packed in an appropriate container. These materials dissolve paints very fast, and it is to be ensured that such materials do not come in contact with the paint. Dust and dirt can be controlled by filling finished oil into the container through a course and the fine filters. Clean hydraulic oils need to be filtered through a series of micronic filters in a special facility. Filling of aviation hydraulic fluid requires clean room-filling facility to avoid particulate contamination.

HANDLING AND STORAGE OF LUBRICATING OILS

The storage environment can affect the shelf life of lubricants and greases. Extreme cold (below $-15°C$) and extreme heat (above $45°C$) can affect lubricant stability. Heat increases the rate of oil oxidation and degradation, leading to the formation of deposits and viscosity increase. Lower temperatures can result in the separation of otherwise dissolved waxes in the oil, and a layering can take place. During day and night, the low and high temperatures may alternate, and this will result in the breathing in of air containing moisture. Slowly, the oil may then get contaminated with water. Presence of water can affect several properties of oil promoting microbial growth (will affect water separation properties) and promote hydrolysis of several additives, thereby losing the performance characteristics. After opening the lube or grease drum, if it is left open in a dusty atmosphere, abrasive dust may accumulate in the oil/grease, which will cause excessive wear of the equipment.

Greases are prone to oil separation over a period of time. If a certain amount of grease is taken out from the drum, the surface should be smoothened to prevent oil separation into the surface pits.

Proper storage and handling methods are, therefore, important for obtaining maximum service life from the lubricant. It guarantees proper product quality and eliminates product contamination. Proper storage also eliminates the chances of use of incorrect grades, mixing up of grades and also reduces leakage and other losses. The best lubricant will also fail to serve its purpose efficiently, if stored and handled in a wrong way. The ideal system would be to store all lubricants in a warehouse that would be easily accessible, well ventilated, well lighted, spacious, away from heat sources, and well protected from rain.

OUTDOOR STORAGE METHOD

- Drums should be placed on wooden planks in a cool drum indoor area.
- Layers should be limited to two/three numbers.
- Drums should be kept in a horizontal position with the bungs at 3 o'clock and 9 o'clock positions.
- Wooden ramps should be used for removing barrels from the layers.
- Drums should not be stored near steam/gas lines.
- Precautions should be taken against fire hazards.
- Direct sun rays should not fall on drums.

GUIDELINES FOR INDOOR STORAGE

- Warehouses should be located away from possible sources of industrial contamination as coke, dust, ore dust, cement dust, textile mill fly, and similar forms of grit or soot.
- Sealed drums should be placed separately marked by fences.
- Sufficient space should be there between rows of drums for easy movement.
- Grease drums should be kept in upright position to avoid spillage.

HANDLING OF LUBRICATING OILS AND GREASES

- Preferable to use fork lift or chain hoist for stacking drums as well as moving out.
- Should be issued on first-in first-out basis.
- Oil should be withdrawn from the drums by semirotary hand pump.
- Drums should have two bung holes, one large and one small. Small is used for lighter grades, and big is for heavier grades.

- Metal trays should be provided below the containers during withdrawal to avoid wastage by spillage.
- Separate containers/oil cans and grease guns should be used for different grades of oil or grease.
- Galvanized containers should never be used for transporting oil.
- Grease should be withdrawn by metal paddles from the drums.
- Dispensing equipment should bear a label that matches the container from which it was filled.
- Grease should not be stored for a long time as it may result in oil separation.

SHELF LIFE OF OILS AND GREASES

Usually, oils have long shelf life, if stored properly, but certain blended products are thermodynamically unstable, and it is important to understand the life of these products. The following table serves as a guideline to indicate the average shelf life. However, it is suggested that if a product in the field is found to have lost its shelf life, it should not be summarily discarded. It is advisable to test the product, and if found satisfactory, it can be used with caution or can be diverted to a secondary noncritical application.

Products	Shelf life in years
Base oil	Above 5
Industrial/engine oils	5
Greases	2–5
Soluble cutting oils	1–5
Steel mill rolling oils	2–5
Rust preventive	2
Open gear oils	2–5
Compounded oil	2–3

Greases are soap-thickened oil and are prone to oil separation or surface cracks. When this happens, grease becomes thicker and nonhomogeneous. Care must be taken to use such a product in critical equipment. Similarly, soluble cutting oils have very high percentage of soap-based polar additives and tend to settle down over a period of time. Steel mill rolling oils, rust-preventive oils, and compounded oils contain high percentage of fatty oils, and their quality may also deteriorate over a period of time. Such lubricants must be tested before use, if these have been stored for more than a year.

CHAPTER 18

REREFINING AND RECYCLING OF USED LUBRICATING OIL

USED OIL AND ITS COMPOSITION

All lubricating oils in a system would become unfit for further use after a certain period, and oil needs to be replaced. This duration depends upon the quality of oil, operating parameters, and system maintenance practices. Used lubricating oil is defined as that charge of oil filled into the system that cannot perform the original function in the same equipment due to changes in oil properties. There are four main reasons for this oil degradation:

1. Chemical changes in the oil due to thermal and oxidative degradations
2. Contaminants from fuel combustion such as water, sludge, soot, carbon, and unburnt fuel in fuel-fired engines
3. Wear particles and airborne impurities like dust and silica
4. Depleted and degraded lubricant additive components and metals therefrom

These changes in oil properties could vary considerably. In some cases, only minor changes such as contaminants like water, dust/dirt, or other grade of oil may get into the oil, while in other cases major changes like oil oxidation, high viscosity increase or decrease, excessive soot level, excessive contamination and increase in acidity, or decrease in total base number can take place. Due to these minor or major changes, the oil has to be removed from the system, and new oil is charged. The loading of each of these impurities determines when the oil has to be changed. Equipment manufacturers generally set out rejection limits for the lubricant in a large system based on oil analysis. However, in a smaller system such as a passenger car, the rejection limit is determined by the number of miles/kilometers run by the engine and coupled with oil duration in the engine.

Terminologies such as reconditioning, recycling, and rerefining are used to describe the processes for the recovery of useful oil from the discarded used oil. The waste oil can be utilized in three ways, that is, reuse by rerefining, by thermal cracking to produce fuels, and by direct incineration in the furnaces (use as a fuel). The first one is the best option. The second one, although produces acceptable cracked products, only converts high-value lubricant into low-value fuels. The third option produces a

Developments in Lubricant Technology, First Edition. S. P. Srivastava.
© 2014 John Wiley & Sons, Inc. Published 2014 by John Wiley & Sons, Inc.

high amount of ash containing heavy metals and pollutes the environment. Recycling of used lubricants is now attracting more attention due to dwindling world oil reserves and environmental concerns.

Wherever industrial oils can be segregated and collected efficiently, reclamation/reconditioning of these oils is easier. Most industrial oils contain a small percentage of additives (usually 1–4%), and contamination from carbon, soot, and sludge is avoided since these do not come in contact with fuel combustion process. A large user of lubricating oil such as steel plant can take out used hydraulic, compressor, turbine, and gear oils with minor changes and recondition the oil by means of simple techniques such as settling/centrifugation, clay contacting, and filtration. With these simple treatments, water and minor impurities are removed from the used oil. After testing, the oil may be added with necessary additives such as antioxidant, antirust, or antiwear additive and can be used in the same system or in another noncritical equipment. This is possible with several grades of turbine, hydraulic, and bearing oils, provided there is no intermixing of grades during the collection of oils. This is called reconditioning of oil.

Used oil would generally contain the following impurities, and rerefining processes aim at removing these in a stepwise manner:

1. Water
2. Solid particles, dust, and wear debris
3. Fuels and fuel fractions
4. Carbon, soot, sludge, etc.
5. Oil-oxidized products
6. Lube oil additives and their degraded by-products

All engine oils during the use get contaminated heavily with soot, fuel combustion products, and other contaminants, and it is not possible to prolong their life by means of the simple process described earlier. These oils also generate several hazardous ingredients, which call for careful disposal or rerefining. Unfortunately, all used oils cannot be collected. According to a European CONCAWE report, 49% of the lubricants are collectable, and only 28% of these are actually collected in Europe. About 40% of the new oil charge is lost to the environment. In Germany, 48% of the total oils are recovered. Out of these, 41% motor oils are rerefined, 35% are burnt in cement plant kiln, and 24% are burnt in other applications. In the United States, only 17% used oils are rerefined and 83% are burnt. Rerefining of used oil saves 8.1% energy content as compared to burning [1]. All process oils and total loss system oils are also lost into the environment. Used oil is a valuable resource since it has both lubrication value and heat value. It becomes unusable due to various contaminations. If these contaminations can be removed by certain processes and treatments, the used oil can yield a product similar to virgin lubricating base oil. To achieve maximum energy conservation and environmental benefit, it is preferable to rerefine used oils into base oils, which can be blended with additives to produce finished lubricants of high value as compared to combustion for only heat recovery. However, the environmental benefit of combusting used oil depends on the fuel that is displaced. For example, displacement of high environmental impact fuels like coal would make combustion of used oil a better proposition from an environmental point of view. The cost difference between burning and rerefining of used oils would always decide the best way to handle the used oil.

The main problem in used oil rerefining processes is the segregation and collection of used oil. The cost of collection, transportation, and rerefining process makes the rerefined oil sometimes costlier than the virgin oils. Due to this, rerefining does not find favor with large oil companies. In such a situation, the only application of used oil is to burn it in furnaces.

Used oil rerefining involves some of the following steps:

Collection, segregation, and storage of used oil

Dehydration to remove water

Fuel stripping

Vacuum distillation/thin-film evaporation

Lube oil recovery

Extraction/hydrotreatment

Filtration and finishing

Additive treatment to produce finished lube oil

Use of the residue obtained as road binder and utilization of fuel fractions recovered as internal fuel in rerefining units

The following important processes have been developed for lube rerefining over a period of time. There are several other processes that are variations or combinations of these processes:

1. Acid refining (H_2SO_4, Meinken process)—no longer used due to environment hazards
2. Propane extraction process
3. CEP—Mohawk
4. Kinetics Technology International (KTI) process
5. PROP technology (Phillips Petroleum)
6. Safety Kleen process
7. DEA technology

Table 18.1 compares various steps involved in these processes.

Thus, rerefining of used oil involves removal of water and fuels by distillation followed by vacuum distillation or thin-film evaporation to recover base oil streams and then subjecting the recovered lube oil streams to hydrotreatment. If the processing is being carried out in a refinery setup, then after vacuum distillation, entire lube stocks can be fed to the refinery system where solvent extraction and hydrofinishing can be carried out. Some processes prefer propane deasphalting to recover asphalt binder for road construction and heavy base oil such as bright stock.

TABLE 18.1 Comparison of steps involved in rerefining processes

Propane extraction ↓	KTI ↓	PROP-Phillips ↓	Safety Kleen ↓	DEA technology ↓
Atmospheric distillation	Atmospheric distillation	Chemical demetallization with diammonium phosphate	Atmospheric distillation	Atmospheric distillation
First, propane extraction	Fuel stripping		Remove fuels by vacuum stripping	Thin-film evaporation
Vacuum distillation	Thin-film evaporation	Distillation	Thin-film evaporation	Lube feed to lube refinery
Second, propane extraction of the vacuum residue	Hydrofinishing	Clay treatment and filtration	Hydrotreatment at high temperature, pressure,	Solvent extraction
Hydrofinishing of raffinates	Vacuum distillation	Hydrotreatment	and catalyst	Hydrofinishing
Light, medium base oils and bright stock	Light, medium, and heavy base oils	Base oils	Base oils	Base oils

Meinken Process

Sulfuric acid refining is the earliest rerefining process developed by Meinken. The acidic sludge and spent clay disposal problems have now made this process obsolete. This process is based on chemical pretreatment. First the solid impurities are removed by filtration and the used oil is stored into the waste oil storage tanks to separate water by gravity. The dewatered oil is treated with sulfuric acid (96%) followed by vacuum distillation and clay treatment to separate lube base oils, low-boiling spindle, and gas oil.

Propane Extraction Process

This process developed by IFP, France, uses a combination of distillation, two-stage propane extraction, and hydrofinishing to separate out the oil into light, medium, and heavy oils (bright stock). The process is relatively costly due to several steps. In this process, water-free waste oil from atmospheric distillation is transferred to an extraction column with liquid propane at 75–95°C. Dirt and insoluble sludge settle out. After extraction, the oil-containing propane is removed from the extractor. Snamprogetti (Italy) has further improved IFP process by including a propane extraction step before and after vacuum distillation and adding hydrofinishing step. In the first stage of the Snamprogetti technology, light hydrocarbons and water are removed by atmospheric distillation. In the second stage, all the impurities in the engine oil, including the additives and partly degraded polymers, are removed by extraction with propane. In the next stage, the extracted oil is fractionated by vacuum distillation. The vacuum residue is then submitted to a second extraction stage in which metal content and resinous asphaltic components are further reduced. After hydrotreatment, bright stock is obtained.

CEP: Mohawk

The Mohawk process (subsequently CEP—Mohawk) using high-pressure hydrogenating was introduced in the United States at the end of the 1980s. The first stage of the process removes water from the feedstock. In the second stage, light hydrocarbons are removed by thin-film vacuum distillation. In the third stage, evaporation vaporizes the base oil, separating it from additives, metals, sediments, etc. This residue is used in asphalt industry. The distillates are then subjected to hydrogenation over a standard catalyst. Catalyst life of 8–12 months has been obtained, which makes the process economical. The Mohawk process is a continuous operation with low maintenance and longer catalyst life.

KTI Process

The KTI process combines thin-film evaporator and hydrofinishing to remove most of the contamination and additives. The key to the process is the thin-film vacuum distillation to minimize thermal stress through mild temperatures of 250°C. The hydrofinisher then removes sulfur, nitrogen, and oxygen. The yield of finished base oils is high (82% on a dry waste oil basis).

PROP Process

This has been developed by Phillips Petroleum and uses chemical demetalizing by mixing used oil with a solution of diammonium phosphate. This is followed by distillation, clay treatment, and hydrogenation over Ni/Mo catalyst steps to get base oil.

Safety Kleen Process

A modern rerefining technology has been developed by Safety Kleen based on vacuum distillation and hydrotreatment. In this process, distillation is carried out in three stages. The first stage removes water and light hydrocarbons such as gasoline and solvents. A separate fractionation unit will then recover fuels from water–hydrocarbon mixture. Recovered fuel can be used as process fuel or can be sold off. Water obtained contains several impurities like sulfur compounds, ammonia, alcohols, and ethylene glycol from antifreeze coolant. Some of the valuable products can be recovered from this water and sold off. The dehydrated used oil is now subjected to vacuum fuel stripping, and the recovered fuel can be utilized as process fuel. In the third stage distillation, oil passes through vacuum flash tower and thin-film evaporators. Two grades of lube oil fractions are collected at this stage. The residue is utilized as asphalt binder for paving. The two lube oil fractions are now subjected to hydrotreatment at high temperature and high pressure and in the presence of catalyst. This process removes sulfur, chlorine, oxygen, unsaturated organic compounds, and other impurities. This step also improves oil stability, color, and odor. Oil thus obtained can be used as normal base oil in different automotive and industrial oil formulations.

DEA Technology

This process developed by Mineralol, reffinierve, Germany, is reported to provide very-good-quality base oils. This is achieved by a combination of thin-film distillation followed by selective solvent extraction. In this process, the distillate from vacuum thin-film distillation towers is finally treated in a lube refinery solvent extraction plant followed by hydrofinishing. After this extraction process, the polyaromatic content in the recovered base oil is lower than that of virgin solvent neutrals.

Other Technologies

Vaxon (Enprotec fabrication facilities in Denmark) uses three or four vacuum cyclone evaporators and finishing treatment with chemicals for rerefining of lubricating oils. The key step in the ENTRA technology is the special vacuum evaporation in a vacuum linear tubular reactor (single tube). After continuous evaporation by means of rapidly increasing temperature, vapor condensation is performed by fractional condensation. Complete dechlorination can be achieved with metallic sodium. Clay finishing is used to obtain final base oils.

TDA Technology

The thermal deasphalting (TDA) process has been developed by Agip Petroli/ Viscolube using the technology of PIQSA Ulibarri in Spain. The process is based on chemical treatment to facilitate subsequent deasphalting. The Viscolube technology, also known as TDA, is an improvement of a deasphalting process, which has been operated for several years by Viscolube Italiana SpA. TDA is followed by vacuum distillation and clay finishing.

PetroTex Hydrocarbon also claims to have developed a rerefining process based on distillation and hydrotreatment, which yield API groups I and II quality base oils.

An environmentally friendly laboratory process has been reported [2] based on the coagulation of sludge and other impurities with the use of triethanolamine and ethylene glycol monobutyl ether in the dehydrated used oil. The purified solvent–oil mixture is subjected to vacuum distillation to separate solvent, light hydrocarbons, and lubricating oil fractions. A small amount of high-temperature antioxidant is added in the solvent–oil mixture to avoid oxidation of oil during distillation. This lube base stock is treated with activated clay to get polychlorinated biphenyls (PCBs) and metal-free lubricating base oil. The residue can be used for making soft asphalt and in other applications.

The lubricant industry has upgraded product quality over a period of 20 years substantially. This has happened quite extensively in automotive oils where the quality levels of both engines and lube oils have been upgraded (e.g., transition from API CF4 (1991) to CG4 (1995), CH4 (1998), CI4 and CI4 plus (2002), and CJ 4 (2010)). Gasoline engine oils have been similarly upgraded (API SG in 1988 to SH in 1993, SJ in 1997, SL in 2001, SM in 2004, and SN in 2010). Consequently, quality of base oils utilized has also gone up. This resulted in the availability of higher-quality used oils, containing higher level of additives, sometimes synthetic PAOs, or other synthetic materials and polymeric viscosity modifiers. This situation will pose new problems in rerefining. This trend indicates that the quality of used oil will constantly change with the quality of finished oil, and rerefiners will have to adjust their processes to accommodate this change. The rerefining of synthetic oils or the mixture of synthetics and mineral oil will pose another challenge. They will, however, get improved rerefined base oil that will fetch them higher value. Different countries have evolved separate strategies for waste oil management.

France: Seventy-eight percent of used oils are collected. Forty-two percent of used oil is rerefined by government-directed associations.

Germany: Ninety-four percent of used oil is recovered. Of the recovered oil, 41% of the recovered used oil is refined, 35% used oil is burnt in cement plant kilns, and 24% burnt in other applications. Used oil is treated as hazardous waste, and oil marketers have to provide facilities for used oil collection near retail outlets.

Japan: Rerefining is very limited. There is no national-level recycling program. Whatever recovered is burnt for heating value.

Italy: Use of rerefined oil is mandatory in engine oils. Only 10% can be directed for burning. The use of rerefined oils in motor oil is mandatory.

India: The Hazardous Wastes (Management and Handling) Rules [3] has been amended with effective from May 23, 2003, and it is mandatory that rerefining/recycling of the used oil/waste oil is done only through application of environmentally sound technologies. Existing industries based on acid/clay process are required to switch over to the new technologies. Approved processes have to be based on vacuum distillation with clay treatment, thin-film evaporation process, and vacuum distillation followed by hydrotreatment. Any industry generating more than 10 kl/annum of waste oil has to sell it to registered rerefiners. Table 18.2 provides permissible limits in used oil for recycling and rerefining processes [4].

Australia: There are high subsidies for rerefining and 81% of used oil is collected.

United States: Different states have implemented a range of recycling programs. There is no central body coordinating the efforts. Some states classify used oil as hazardous waste and restrict dumping. Most used oil is burnt.

United Kingdom: Government policies make used oil burning an attractive proposition. The United Kingdom imports large quantities of waste oil from Europe and burns them.

Thus, different countries around the world follow different strategies for waste oil management. Several publications have appeared on this subject discussing different aspects of lube rerefining [5–13].

ENVIRONMENTAL IMPACT

It is estimated that globally about 20 million tons of new lubricating oils are lost in the environment, and a similar amount is collectable, which could be rerefined for further use. This uncontrolled loss and improper disposal of used lubricants is a

TABLE 18.2 Waste oil properties for recycling and rerefining processes in India

Parameter	Waste oil suitable for recycling	Used oil suitable for rerefining
Properties	Maximum limit	Maximum limit
Sediments	5% max.	—
Color	—	8 hazen units
Water	—	15% max.
Density	—	0.85–0.95
Kinematic viscosity cSt at 100°C	—	1–32
Diluents	—	15% vol.
Neutralization No.	—	3.5 mg KOH/g
Saponification value	—	18 mg KOH/g
Total halogens	4000 ppm max.	4000 ppm
PCBs	Below detection limit	Below detection limit
Lead	—	100 ppm max.
Arsenic	—	5 ppm max.
Heavy metals	(Cadmium + chromium + nickel + lead + arsenic) 605 ppm max.	Cadmium + chromium + nickel 500 ppm max.
Polyaromatic hydrocarbons (PAHs)	6% max.	6% max.

potential threat to the natural resources by contaminating soil, water, and air [14–16]. Even the uncontrolled burning of used oil for fuel value produces hazardous emissions, which affect humans, wild life, and vegetation. It has been reported that environmental biodegradation of used engine oils is 24–84% in soil after 1 year of application [17] or 20% in lakes after 100 days of spillage [18]. Oil, in any form, is harmful to the environment. Post oil spill studies indicate that it takes up to 20 years for an aquatic environment to return to normal healthy condition. In aquatic community, oil residue tends to settle at the bottom, coating the substrate and organisms. When poured on the ground, oil quickly migrates through the soil. In both instances, bacteria, plants, invertebrates, and vertebrates experience physiological stress. Oil film on water can reduce the penetration of light into the water and, consequently, reduces the rate of photosynthesis, which in turn reduces oxygen production. The oil film may also inhibit the movement of oxygen from the air through the surface of the water. The reduction of dissolved oxygen in the water stresses animals living in it. Thus, it is important that oil is not disposed of into water or soil and used oil is rerefined and used again and again. Used engine oil may also pose health hazard to humans due to several contaminants entering into oil:

1. PAHs as fuel and oil combustion products
2. Contamination with halogens and PCBs (these have now been restricted)
3. Contamination with heavy metals such as arsenic, cadmium, chromium, and lead (lead has now phased out from gasoline)
4. Presence of nitrosamine—produced between the reaction of sodium nitrite and ethanolamine in coolants (the use has now been restricted)

With some of the restriction imposed, PAHs and heavy metals are the main contaminants in used oils, which pose health hazards. One hundred and forty PAHs have been found in used engine oils, which are the outcome of the combustion process in the engine [17]. Most standards of used refined oil have restricted PAHs, PCBs, halogens, and heavy metals [14]. Even the used oil that can be recycled has restriction on these parameters (Table 18.2). It is therefore important that the used oils are properly segregated, collected, and refined through an appropriate technology. Rerefining of used oils (industrial hazardous waste) is a good approach to reduce burden on environment [19, 20].

REFERENCES

[1] Used oil re-refining study to address energy policy Act of 2005 Section 1838, U.S. Department of Energy, July 1–7, 2006.

[2] Nagarkoti BS. Eco-friendly process for re-refining used crankcase engine oils, Indian patent 209949. September 2007.

[3] Hazardous Wastes (Management and Handling) Rules, 2004, as amended. INDIA and Directives of the Supreme Court of India in Writ Petition 657 of 1995 dated 14.10.2003.

[4] Boralkar DB. Implementation of environmentally sound technologies for re-refining/recycling used oil and waste oil. Report. Maharashtra: Maharashtra Pollution Control Board; 2005.

[5] Jhanani S, Joseph K. Used oil generation and management in the automotive industries. Int J Environ Sci 2011;2 (2):648.

[6] Hamad A, Al-Zubaidy E, Fayed ME. Used lubricating oil recycling using hydrocarbon solvents. J Environ Manage 2005;74:153–155.

[7] Boughton B, Horvath A. Environmental assessment of used oil management methods. Environ Sci Technol 2004;38 (2):353–358.

[8] Bridjanian H, Sattarin M. Modern recovery methods in used oil re-refining. Pet Coal 2006;48:40–43.

[9] Nakaniwa C, Graedel TE. Life cycle and matrix analyses for re-refined oil in Japan. J Life Cycle Assess 2002;7:95–102.

[10] Jha MK. Re-refining of used lube oils: an intelligent and eco-friendly option. Indian Chem Eng Sect B 2005;47 (3):209–211.

[11] Ostrikov VV, Matytsin GD, Nagornov SA. Regeneration of used lubricating oils. Chem Pet Eng 2002;38:3–4.

[12] Pawlak Z, Rauckyte T, Oloyede A. Oil, grease and used petroleum oil management and environmental economic issues. J Achiv Mater Manuf Eng 2008;26 (1):11–17.

[13] Shiung Lam S, Russell AD, Chase HA. Microwave pyrolysis, a novel process for recycling waste automotive engine oil. Energy 2010;35:2985–2991.

[14] Schieppati R, Winter DD. Re-refining in Europe-quality and performance. In: Srivastava SP, editor. Proceedings of the 1st International Symposium on Fuel and Lubricants (ISFL-97); 1997; New Delhi. New Delhi: Tata McGraw Hill. p 435–438.

[15] Vazquez-Duhalt R. Environmental impact of used motor oils. Sci Total Environ 1989;79:1.

[16] Donkelaar PV. Environmental impact of crankcase and mixed lubrication. Sci Total Environ 1990;92:165.

[17] Raymond RL, Hudson JO, Jamison VW. Oil degradation in soil. Appl Environ Microbiol 1976;31:522.

[18] Higgins IJ, Gilbert PD, Wyatt J. Microbial degradation of oil in environment. Stud Environ Sci 1981;9:85.

[19] Singh MP, Rawat BS, Srivastava SP, Bhatnagar AK. *Recycling of Lubricants, Environmental Health Concerns and Future Strategies*. New Delhi: Hydrocarbon Technology, Centre of High Technology; August 1993.

[20] Betton CI. Lubricants and their environmental impact. In: Mortier RM, Orszulik ST, editors. *Chemistry and Technology of Lubricants*. New York: Springer; 1992. p 283.

CHAPTER 19

IN-SERVICE MONITORING OF LUBRICANTS AND FAILURE ANALYSIS

Monitoring of lubricant condition in running equipment provides useful information about the health of the equipment as well as oil itself. Changes in both oil and equipment take place continuously with time. Oil gets oxidized and additives are gradually depleted. On the other hand, there is normal wear and tear of metal surfaces. There would also be ingress of impurities in the oil system from atmosphere and other leaking fuels, oils, and greases. These wear debris, oxidized products, degraded additives, and other contaminants are then present in the oil. The oil analysis would thus provide information about the changes taking place in the system. Any abnormal change in these conditions would result in faster oil degradation and higher equipment wear. Monitoring of these changes along with equipment performance analysis, vibrational analysis, visual inspection, and oil analysis is known as condition monitoring and provides useful tribological parameters [1–3].

LUBRICANT CONDITION MONITORING

Condition monitoring by oil can be divided into two main categories:

1. Debris monitoring
2. Lubricant condition monitoring

Debris monitoring measures the quantities of wear particles carried away from the wearing surfaces by the lubricant. Lubricant condition monitoring determines if the lubricant itself is fit for service based on physical and chemical tests. These techniques together, when combined with statistical analysis, provide a complete program of condition monitoring by oil analysis.

The useful life of oil in a particular system is limited by the process of degradation and contamination. Oil degradation is defined as those irreversible changes that happen to the oil. The reasons for oil degradation are many, but some of the important parameters are as follows:

Developments in Lubricant Technology, First Edition. S. P. Srivastava.
© 2014 John Wiley & Sons, Inc. Published 2014 by John Wiley & Sons, Inc.

- Thermal and oxidative degradation leading to oil thickening, sludge and deposit formation, acidity increase, and viscosity increase
- Loss of additives such as EP, antiwear, antifoam, and rust inhibitors
- Viscosity loss due to shear of polymeric viscosity modifiers and fuel dilution
- Fuels and fuel combustion products in the case of engine oils
- Solid contaminants such as wear metal particles, dirt, and dust
- Liquid contaminants such as water, fuel, coolant, and other grade of fluids
- Oxidation and nitration level in engine oils specially diesel and gas engine oils

The used oil analysis program provides an opportunity for monitoring the health of the machine and also solves operational problems. The oil change interval can also be determined precisely irrespective of original equipment manufacturers' (OEMs) recommendations. There are three basic options for lubricant monitoring:

1. Continuous online monitoring
2. Routine monitoring using portable instruments without sampling
3. Taking regular samples for analysis in a laboratory

The main advantage of online monitoring is that these instruments operate continuously and can be linked to a computerized condition monitoring system that processes other condition-related signals such as vibrational signals. The system may be set in such a way that it provides initial warning when limits are crossed and further investigations are needed. The disadvantage of online monitoring is that this is dedicated to a system and therefore requires high capital outlay. Furthermore, online monitors are not available for all the properties.

The portable equipment are simpler to use and suitable for routine checkup and do not require bottle sampling. The limitation of portable equipment is that these are also not available to test all the properties. Several commercial portable kits are available for used oil analysis.

Taking regular oil sample in sample bottles and sending to the laboratories for condition monitoring are more labor intensive in comparison to online monitoring but has various advantages. It does not require any modification to system/machine and provide a great deal of information about the condition of oil and health of the machine. Extensive laboratory tests can be carried out to assess oil condition and equipment performance.

DEBRIS MONITORING

Debris monitoring pertains to the detection and analysis of metallic wear particles. The most common techniques and devices used for condition monitoring include atomic absorption spectroscopy (AAS), atomic emission spectroscopy, ferrography, magnetic plugs, magnetic chip detector, and microscopic examination of filter debris.

Debris monitoring is the backbone of oil analysis condition monitoring program. It helps us to determine if the system is approaching or has already reached failure stage. The further damage can be avoided through corrective actions. In many cases, component failure may be identified by using effective lubricant management. The microscopic examination of wear particles can provide information regarding the type of wear and causes of wear such as overloaded components, high-temperature operation, inadequate lubrication, inappropriate metallurgy, and assembly fault. Spectroscopic analysis is the most widely used for debris monitoring, and the following techniques may be utilized for this analysis [4, 5]:

Atomic Absorption Spectroscopy

Atomic absorption spectrometer has been in use for wear metal analysis since long. In this technique, energy absorbed by elements at a characteristic frequency is determined and converted to concentration with reference to established calibration curve. The technique is sequential and ideal for limited number of elements.

Rotating Disk Emission Spectroscopy

In rotating disk emission (RDE) spectrometer, the elements in the oil sample are excited by spark/DC arc source, and emitted light by all the elements is measured at characteristic wavelength simultaneously. These measured light intensities are converted into concentration with respect to established calibration curves. In this way, 20–30 elements may be analyzed simultaneously. The techniques is suitable for used oil analysis for wear metals such as Fe, Al, Cr, Cu, Sn, Pb, Ag, Ti, and Ni; contaminants like Si and Na; and additive elements Ca, Zn, P, Mg, and Ba. However, larger particle analysis is missed in this test due to the technique limitation.

ICAP Emission Spectrometer

The ICAP emission spectrometer is the most modern technique for metal analysis. In this technique, argon is allowed to pass through a quartz tube under high-frequency magnetic field causing partial ionization of argon gas to form plasma. Plasma is a high-temperature (8000 K) source and is most suitable for the analysis of trace and additive elements. The instrumentation part is the same as RDE spectrometer, except source. This instrument is much more versatile and useful over RDE spectrometer. In ICAP, wear particles of all sizes get analyzed.

Spectrographic analysis measures the levels of wear metals and the concentration of additive elements. The results, usually reported in parts per million (ppm), provide an indication of the rates of wear of engine components and the depletion of additives. Emission spectrograph analyzes wear particles that are about 5–10 µm in size. Bigger particles are not excited in this method. Inductively coupled plasma is a better method for analyzing all the wear particles in used oil. Therefore, the interpretation of spectrographic test results must be carried out with caution. Moreover, similar equipment running under similar conditions will not generate identical wear pattern. Each equipment is an individual piece and has to be monitored independently. Wear rate trends can be established after the interpretation of at least three to four oil samples taken at the same sample interval such as after 500-h operation or 5000-mile service in an engine.

Ferrography

Ferrography separates the debris under the influence of an inclined high-intensity magnetic field. In this technique, a ferrogram is prepared by passing a diluted oil sample across a specially prepared glass microscopic slide, which is subjected to a strong magnetic field gradient. The developed ferrogram is monitored under microscope, which indicates the size and shape of the particles. Heat treatment of the ferrogram can further distinguish between steel, cast iron, and high alloy steel between nonferrous metal and organic/inorganic materials. For elaborated investigation, the ferrogram is examined under SEM equipped with XRF.

LUBRICANT CONDITION MONITORING

Lubricant condition is determined by carrying out following basic tests. Further tests can be planned, if these tests indicate the need.

Appearance and Odor

This is the first test to be conducted on used oil. Oil sample is placed in a narrow glass test tube and examined for clarity; haziness; cloudy, milky, or opaque appearance; and odor.

The hazy or cloudy appearance generally indicates water contamination. A dark to brown appearance and characteristic burnt smell indicate that oil has undergone oxidation and thermal degradation. Any other color of oil indicates some sort of chemical reaction with the oil additives and needs further analysis.

Viscosity

Viscosity is an important property of the oil, and any change in viscosity has to be considered seriously. Increase in viscosity is caused by contaminations such as soot, dirt, and glycol from coolant, oxidation, and water. Decrease in viscosity results from fuel dilution, wrong top-up from a lighter oil, and shear of viscosity index (VI) improver. It is also possible that there is no change or very little change in viscosity due to the opposing effects; for example, the combined effect of fuel dilution, oxidation, and polymer shearing may show no viscosity change or very little change. These conditions need to be examined further by carrying out other tests. Fuel dilution can be measured by distillation, soot content can be determined by differential scanning calorimetric method, and oxidation level can be measured by Fourier transform infrared spectroscopy (FTIR) techniques.

Viscosity Index

This is an arbitrary value indicating viscosity changes with temperature and is calculated from the viscosities obtained at 40 and 100°C.

Presence of fuels in used oil increases the VI value of the original oil. Low shear stability of the VI improvers used in the engine or hydraulic oil causes lowering of VI of used oil. Again, wrong top-up with a lower VI product could also be responsible for lower VI.

Total Base Number

This is a measure of the total amount of alkalinity in the oil. Most oils, especially engine oils, contain alkaline additives such as overbased detergents and dispersants to neutralize acidic combustion products. The total base number (TBN) of used oil indicates the amount of neutralizing additive left. ASTM D-2896 procedure determines TBN by titration with perchloric acid and expressed as mg KOH per gram of oil.

Total Acid Number

This is a measure of the amount of acid in the oil. It is measured by following ASTM D-664 method and expressed as mg KOH per gram of oil. On oxidation, oil forms acidic products, which can be corrosive to metal components. The greater the oil degradation, the higher will be the value of the total acid number (TAN). Generally, new oil also has a positive value of TAN due to the presence of certain acidic components and additives derived from organic acids. Such additives are rust inhibitors and EP/antiwear additives. Therefore, the difference between the TAN of used oil and new oil will indicate the extent of oxidation.

Flash Point

The flash point of the used oil is determined to detect the fuel dilution. It is measured by ASTM D-92 procedure in an open cup. Closed-cup flash point is determined by ASTM-93 (PMCC) procedure. The low flash point of the used oil may be due to fuel dilution or cracking of the oil. The fuel dilution reduces viscosity and lubrication properties of oil.

Water Content

Water reduces the lubricity of the oil and leads to the corrosion of metal parts. Higher level of water may emulsify oils containing detergents and dispersants such as engine oils. In circulating system, water may enter from oil cooler leakages. In steam turbines, water can enter as a result of steam condensation, and in automotive oils, water may come due to condensation of flue gases containing water from combustion of fuels. In most of the cases, water content is determined by following ASTM D-1744 using Karl Fischer reagent. High amount of water can be determined by Dean and Stark method of distillation. A simple crackle test on a hot plate also qualitatively indicates the presence of water.

Insolubles

Hexane-insoluble material consists of oxidation products such as resins, matter like carbon, soot, wear particles, dust, and dirt. Toluene insoluble dissolves the oil oxidation products and consists of only extraneous matter. Therefore, difference in hexane and toluene insoluble indicates the amount of the oxidized products. Low hexane insoluble indicates that the oil is in good condition. If hexane insoluble is high, the subsequent toluene insoluble indicates whether contaminant is due to oil oxidation or extraneous matter. Insolubles are determined by following ASTM D-83 method.

Rust Test

Lubricating oils in turbine, hydraulic, or other system having oil cooling often gets contaminated with water, resulting in rust. Particles of rust in the oil can act as a catalyst and tend to increases the rate of oxidation of the oil. Rust particles are abrasive in nature and can cause wear and clogging of valves. This property is determined by ASTM D-665 A or B method for 24 h at 60°C. The used oil should pass the test. In the case of a failure, if other oil condition is favorable, an antirust additive in appropriate dosage may be added in consultation with the oil supplier.

Nitration and Oxidation

Nitration and oxidation in engine oils depend upon conditions such as air-to-fuel ratios and oil operating temperatures. Oxidation is caused by the reaction of oil with oxygen in combination with catalysts such as copper wear particles. Oxidation occurs in all lubricated systems and results in an increase in viscosity and acidity of the oil. The oxidation process increases with the increase in system temperature. Nitration, on the other hand, occurs in diesel and natural gas engines and, if left uncontrolled, can deteriorate oil quickly. Nitration is a chemical reaction with the oil and nitrogen dioxide (NO_2) formed during combustion, causing premature thickening of the oil. Nitration of oil depends on the temperature and air-to-fuel ratio. Both oxidation and nitration can be easily be monitored by IR or FTIR spectroscopic method.

FTIR Analysis

FTIR spectroscopy is a reliable method to find out used oil condition. In FTIR analysis, the IR spectrum of used oil is subtracted from the IR spectrum of fresh oil [6]. The changes observed at various frequencies in the differential spectrum provide accurate information of used oil condition. The oxidation ($1700 \, cm^{-1}$), nitration ($1630 \, cm^{-1}$), sulfation ($1150 \, cm^{-1}$), soot content ($2000 \, cm^{-1}$), water content ($3400 \, cm^{1}$), fuel dilution ($750–800 \, cm^{-1}$), and depletion of antiwear additive ($960 \, cm^{-1}$) can be monitored from this spectrum [7]. Fourier transform Raman spectroscopic techniques [8] are also now being developed for the used oil analysis.

Glycol and Coolant Contamination

Testing for glycol leaks in accordance with ASTM D-2982 must be part of any oil analysis program. Any amount of glycol in the analysis can indicate a coolant leak into the engine and will cause catastrophic failure by promoting corrosive acids, sludge, and varnish. If present, the ingress of glycol into the lubricant must be checked. Some industrial systems use water-based coolant containing a mixture of salts and chromium compounds. In these systems, elemental analysis will indicate coolant leakage.

Particulate Count

Accurate count of particulate matter in lubricants is important in many applications where a clearance between the moving surfaces is very small and critical. Any ingress of larger particle in the system is not tolerated and can lead to wear and equipment failure. Such systems are provided with micronic filters to continuously filter oil and keep it clean in the system. This is true in aviation hydraulics, electrohydraulic servo valves, and numerically controlled machine tools. There are several methods to determine particulate matter by filtration and gravimetric methods. Presently, accurate particulate counting can be carried out by automatic particle counter equipment based on photoelectric sensors. For such counting sampling of lubricant from the equipment, storing it in clean sample bottle attain importance for accurate analysis. Atmospheric contamination of the sample has to be avoided.

LOCAL AREA NETWORKING AND DATA MANAGEMENT IN OIL ANALYSIS LABORATORY

Oil condition monitoring can become very fast by applying local area networking and data management system [9]. Oil condition monitoring report can be generated within a few hours.

An automated modern lube oil condition monitoring laboratory equipped with LAN and data management system can be set up to meet the large testing requirement of a field trial for monitoring a new product in service or for monitoring the health of the equipment. Generally, five most essential sophisticated computer-controlled equipment, namely, auto viscometer, emission spectrometer, FTIR, autotitrator for TAN and TBN, and insoluble measurement, are set up. The viscometer and autotitrator for TAN and TBN are capable of analyzing large number of samples unattended for viscosities at 40 and 100°C and TAN, TBN simultaneously. The condition of the used oil is automatically evaluated for its oxidation, nitration, sulfation, soot content, contaminants, water, and fuel dilution in percent with the help of FTIR. The buildup of metal contents in the oil, as a result of wear and/or external contamination, is monitored through emission spectrometer. These equipment can be integrated with computer networking to speed up the analysis by autotransferring of the test results into a database and to generate analytical reports along with remarks and recommendations automatically. The analytical reports may be then transmitted to the customer through computer networking to the remote locations. The advantage of this system is that any number of instruments can be integrated at any given time, depending on the analytical need. Several laboratories operate such a system for speedy analysis.

SAMPLE HANDLING FOR CONDITION MONITORING

Analysis by an automated laboratory requires a relatively small sample of oil, and great care is to be exercised in taking the sample so that it is representative of the system. The following steps and care must be taken:

1. Have dedicated clean, dry sample bottles with a good stopper cap.
2. The sample point should be a point where the oil is flowing freely.
3. Sample should be taken from the same point each time.
4. Machine must be running at the time of sampling.
5. Initial quantity of the oil should be discarded before filling in the sample bottle. This would allow the removal of entrapped water or wear debris.

Some examples of good sampling points are as follows:

Storage tanks	Sample from top and near bottom and then mix them in equal ratio
Large stationery engine	Discharge side of the circulating oil pump
Gears	Return line before the filter
Hydraulic pumps	Return line before the filter
	Bearing return line
Compressors	Midpoint of crankcase or bearing return line
Turbine	Bearing return line

Sample Label

The sample label must have the following details:

Name of the oil

Name of equipment

Location

Oil service hours

Consumption, liters/hour

Oil top-up history

Date of sample

TESTS TO BE CONDUCTED ON DIFFERENT OILS

Table 19.1 indicates the initial tests to be carried out on the oil sample. Detailed testing may be required depending on the problem and test results.

Laboratory equipped with FTIR also provides additional information about the oil condition with respect to oxidation, nitration, sulfation state, and soot content of the oil.

INTERPRETATION OF LABORATORY RESULTS

Appearance

Visual appearance of the sample is the first test carried out on the sample, and this simple step provides (Table 19.2) useful information.

TABLE 19.1   Typical tests to be conducted on used oil sample

Oil type	Tests
Engine oils	Appearance and odor, kinematic viscosity 40 and 100°C, VI, TBN, water content, pentane insoluble, flash point, elements[a] (wear metal and additives)
Gear oils	Appearance and odor, kinematic viscosity 40 and 100°C, VI, TAN, water content, and elemental analysis. Water separation characteristics
Industrial oils	Appearance and odor, kinematic viscosity 40°C, VI, TAN, water content, pentane insoluble, elemental analysis[a], particle count (for clean hydraulic oils)

[a]Elements include Al, Sb, Ba, B, Ca, Cr, Fe, Pb, Mg, Mn, MO, Ni, P, Si, Na, K, Sn, V, and Zn.

TABLE 19.2   Visual appearance and oil condition—indicator

Visual appearance	Oil condition
Clear and bright	Oil in good condition
Hazy	Presence of water
Brown colored	Oil oxidized
Dark black	Oxidized, contains soot
Brownish milky	Emulsified water
Burnt smell	Oil oxidized

Water Content

A simple hot plate crackle test indicates the presence of water in used oil. Water in oil in engine could be from two sources. It could come from the leakage of the coolant, or it could be from the combustion of the fuel. In industrial oils, presence of water indicates leakage into the oil system. Coolant leakage can be identified from the elemental analysis of the oil. In either case, the leakage should be traced and corrected. If centrifuge is in the system, the equipment should be checked and corrected for proper functioning. Water in oil is undesirable and will affect lubrication. This will also lead to rusting and corrosion of metal parts. Steel rolling mill bearing oils and steam turbine oils usually get contaminated by process oil and condensed steam. Such oils are supposed to possess high demulsibility characteristics, and the oils are regularly cleaned through centrifuge to control water content.

Abrasive Matter

This can be identified easily by diluting one part of used oil with four to five parts of cleaning solvent or naphtha or gasoline and allowing it to stand for 24 h in a clean glass jar or test tube. If solids are present in the used oil, it will settle down at the bottom and can be further characterized by elemental analysis or microscopic examination. Abrasive or insoluble matter can be removed by centrifugation or filtration. Some elements such as silica are quite abrasive and can cause high wear. Large systems generally have a bypass filtration/centrifugation system to continuously clean oil.

Viscosity

Low   Viscosity of oil in use can be lowered by fuel dilution or by the ingress of lighter oil. In the case of multigrade oils, viscosity can also be lowered by polymer (VII) shear. If fuel dilution is the main cause, then flash point will also come down.

High   Viscosity increase is the outcome of oil oxidation or soot dispersion in the case of engine oils. Viscosity may also go up due to wrong top-up of higher-viscosity oil. If the soot levels in oil are low but the viscosity is still high, this could be due to oil oxidation, which will show up in insoluble matter and IR spectrum.

Sometimes, the combined effect of oil thickening and fuel dilution/polymer shear may give misleading results. In such cases, detailed investigation can reveal the actual cause.

Total Base Number

In used oil, TBN of used engine oil will gradually come down due to its depletion by reaction with acid formed during the combustion. It should not be allowed to come down below the rejection limit set for the oil or as per the OEM requirement. If low, the oil should be topped up with higher TBN oil. The rate of TBN depletion depends on the type of fuel used, engine severity operation, and top-up rate. TBN reduction is the indication of additive depletion and the oil needs to be changed, if the value has reached rejection limit.

Acid Number

Any increase from the original value indicates oil oxidation and decrease means depletion of additives. Effective oil purification and oil makeup can check acidity increase in turbine/hydraulic oils and can prolong oil life. If acidity goes beyond a specified value, oil needs replacement.

Pentane Insoluble

In an engine oil, partial fuel combustion and oil oxidation may generate products that are insoluble in pentane. These can lead to the formation of sludge, varnish, and lacquer. If pentane insolubles are beyond the set limit, oil needs to be changed.

Elemental Analysis

This provides a very good insight into the oil quality. Depletion of elements present in the oil additive directly shows the oil degradation. Presence of wear metals shows the wear rate of the related equipment (refer to the table for origin of elements). Presence of silica indicates ingress of dust/dirt and nonfunctioning of air filter.

Presence of silica is also responsible for overall high wear rate. Contamination by other oils can be checked by the presence of elements.

Particulate Matter

This is specially required for clean hydraulic oil system with servo controls. High particulate matter will lead to chocking of fine micronic filters. Clean hydraulic oils should be monitored by checking particle count through an automatic counting device. In such case, the oil should be cleaned through a bypass filter, and fine filter should be changed regularly.

Origin of Elements

The presence of elements in oil comes from multiple sources such as equipment metallurgy, environment, oil additives, fuels, and greases. Table 19.3 provides approximate information about the possible source of the elements in used oil, which can be taken into account while interpreting wear metal analysis.

Presence of Higher Level of Si and Fe

Silica comes from air and shows the malfunctioning of air filter. In such case, the air filter needs to be changed. The engines might be operating under dusty environment. Presence of iron indicates wear and requires further investigation, and the source must be located. Iron could also be from wear due to the presence of abrasive silica.

TABLE 19.3   Possible sources of element in used lubricating oil

Equipment	Cooling system	Environment	Lubricating oil	Grease	Fuel oil
Aluminum				Aluminum	Aluminum
Antimony				Antimony	
	Barium		Barium	Barium	
	Boron		Boron	Boron	
Calcium	Calcium	Calcium	Calcium	Calcium	
Chromium	Chromium				Chromium
Copper					
Iron	Iron				
Lead		Lead		Lithium	
Magnesium	Magnesium		Molybdenum	Molybdenum	
Manganese			Manganese	Manganese	
Nickel					
	Phosphorus		Phosphorus	Phosphorus	Phosphorus
	Potassium	Potassium		Potassium	
		Silicon			Silicon
	Sodium	Sodium		Sodium	Sodium
Tin					
					Vanadium
Zinc			Zinc	Zinc	

TABLE 19.4 Typical rejection limits for lubricants

Engine oils	Turbine oils	Hydraulic oils	EP gear oils	Compressor oils	Heat transfer oils
—	Appearance satisfactory	Appearance satisfactory	Appearance satisfactory	Appearance satisfactory	—
Viscosity at 100°C, ±one grade max.	Viscosity at 40°C, +10 to −5% change	Viscosity at 40°C, ±10% change	Viscosity at 40°C, ±one VG grade	Viscosity at 40°C, ±20% of the original	Flash point lowering not more than 25°C
Water content 0.5% vol max.	Water content 0.1% vol max. after purification	Water content 0.1% vol max. after purification	Water content. 0.1% vol max. after purification	Water content 0.2% vol max.	—
Excessive value of Si and Fe to be reported. Hexane insoluble 2.5% max.	Sediments % wt. 0.1 if abrasive, otherwise 0.5	Sediments % wt. 0.1 if abrasive, otherwise 0.5	Hexane insoluble 0.5% max.	Sediments % wt. 0.1 if abrasive, otherwise 0.5	—
TBN mg KOH/g 2 min	TAN mg KOH/g 1.0 max. Rust test Annual sample to pass the test	TAN mg KOH/g 1.5 max.	Phosphorous 50% of the original value	TAN mg KOH/g 1.5 max. Zinc 40% of the original	—

Field Problems

Large numbers of field problems are encountered with the use of lubricants [9, 10]. Customers immediately doubt the oil. In majority of these cases, it is found that while the oil is in good condition, problems arise due to equipment malfunction. In order to satisfy the customers, the problems need to be properly investigated with detailed oil analysis and equipment performance analysis should be carried out. The test results can be analyzed and interpreted in the light of information provided in this chapter. This can then be discussed with the customer, and a mutually agreed solution can be found out. In case, there is a genuine problem with the oil, either the oil recommendation is changed or the oil charge has to be changed. If the oil has completed its service life, it should be immediately changed. Some of the typical rejection limits of oils are provided in Table 19.4 for guidelines only. OEMs' recommendations, wherever available, must be followed.

REFERENCES

[1] Coats JP. Used oil analysis: an integral approach. In: Srivastava SP, editor. Proceedings of the International Symposium on Fuels and Lubricants (ISFL 97); December 8–10, 1997; Tata McGraw Hill, New Delhi. p 252.

[2] Nadkarni RA. *A Review of Modern Instrumental Methods of Analysis of Petroleum Related Materials.* Philadelphia: ASTM STP; 1991. p 19–51.

[3] Jacques J, Jean B, Hipeaux J-C. *Lubricant Properties Analysis and Testing.* Paris: I F P Publication; 2000.

[4] Patel MB, Christopher J, Jain SK, Srivastava SP, Bhatnagar AK. Determination of calcium, zinc and phosphorous in multigrade crankcase oils by XRF, DRES and ICP-AES. Tribotest J December 1998,5 (2).115–120.

[5] Bansal V, Sastry MIS, Sarpal AS, Jain SK, Srivastava SP, Bhatnagar AK. Characterization of nitrogen and phosphorous compounds by NMR and IR techniques. Lubr Eng April 1997;53:17–22.

[6] Murray W, Allen A. Monitoring oil degradation using FTIR analysis. Lubr Eng 1992;48:236.

[7] Socratese D. *Infrared Characteristics Frequencies, Tables and Charts.* New York: John Wiley & Sons, Inc; 1994.

[8] Hendra P, Jones C, Warnes G. *Fourier Transform Raman Spectroscopy, Instrumentation and Chemical Applications.* Chichester: Ellis Horwood; 1991.

[9] Jain MC, Negi MS, Taneja GC, Srivastava SP, Bhatnagar AK. Industrial tribology management system and its role in establishing the life of lubricants. 3rd International Petroleum Conference and Exhibition, Petrotech 99; 1999; New Delhi. p 275–280.

[10] Jain MC, Srivastava SP, Bhatnagar AK. Tribotesting and its role in establishing the service life of lubricants. Proceedings of the Conference on Industrial Tribology; December 1–4, 1999; Hyderabad. p 352–360.

CHAPTER 20

LUBRICANT TESTS AND THEIR SIGNIFICANCE

In order to test the performance of lubricating oils, a large number of physicochemical, mechanical, and engine tests are carried out. The significance of these tests with respect to the performance of the lubricant in the actual equipment should be understood. Different countries and their standardization institutes have defined the tests and written down methods for carrying out such tests. For example, in the United States, these are defined by the American Society for Testing Materials (ASTM), in the United Kingdom by the Institute of Petroleum (IP), in Germany by the Deutsche Industrie Norm (DIN), in Japan by the Japanese Industrial Standards (JIS), in France by the Association Francais Petroles de Normalisation (AFNOR), in Russia by the GOST, and in India by the Bureau of Indian Standards (BIS). There has been effort by the International Organization for Standardization (ISO) to consolidate and arrive at a mutually agreed test method.

During these tests, large numbers of common terms are used, which are provided in Table 20.1.

Some details of these tests are also provided in this table. It has not been possible to provide an exhaustive list. For details, we can refer to relevant standards for petroleum products.

Most of the lubricants can be characterized by the following initial tests. These tests provide enough information about the type of the products and their possible applications:

1. Viscosity at 40 and 100°C
2. Viscosity index (VI) calculated
3. Total acid number (TAN)
4. Total base number (TBN)
5. Pour point
6. Ash or sulfated ash percent
7. Color ASTM
8. Copper corrosion at 100°C for 3 h
9. Elemental analysis Calcium, zinc, phosphorus, barium, magnesium, molybdenum, etc.
10. Flash point °C

Developments in Lubricant Technology, First Edition. S. P. Srivastava.
© 2014 John Wiley & Sons, Inc. Published 2014 by John Wiley & Sons, Inc.

TABLE 20.1 Common tests, terms, and their significance

Test	Significance
Acid number	The milligram of KOH per gram of oil required to neutralize all or part of the acidity of a petroleum product is known as acid number. Neutralization value is a measure of the combined organic and inorganic acidities. Total acidity, inorganic acidity, and organic acidity terms are used to indicate corrosive property, mineral acid in the sample, and the acidity from organic compounds (obtained by deducting the inorganic acidity from the total acidity).
Additive	Any chemical compound added in small quantity to lubricating base oils or their blends to change its properties or performance is defined as additive.
Air release value	This is a measure of the oil tendency to release air when air is blown into the oil at a specific temperature. Air release value is the time at which the density is lower by 0.002 gm/ml of the density of neat oil.
Aniline point	The lowest temperature at which equal volumes of aniline and hydrocarbon fuel or lubricant base oils are completely miscible. It is a measure of the aromatic content of a hydrocarbon blend or base oil. It indicates the solvency of a base stock. High aniline point indicates that the oil is highly paraffinic. In case of aromatics, the aniline point is low.
Antifoam agent	A chemical additive to suppress the foaming tendency of petroleum products. Silicone oil and acrylic polymers are used to break up surface bubbles of air to reduce foaming.
Antistatic additive	An additive that increases the conductivity of a hydrocarbon fuel and lubricant to improve the dissipation of electrostatic charges during high-speed dispensing thereby reducing the fire/explosion hazard.
Antiwear agents	Chemical compounds that form thin tenacious films on highly loaded parts to prevent metal-to-metal contact. These compounds contain sulfur, phosphorous, or other reactive atoms in their molecules.
Apparent viscosity	A measure of the viscosity of a non-Newtonian fluid under specified temperature and shear rate conditions.
Ash	Ash is the inorganic material left out after burning of the organic compounds of the oil. This gives an idea of the ash forming impurities in the oil, mostly metallic or inorganic contaminants.
Ash (sulfated)	The ash content of oil, determined by charring the oil, treating the residue with sulfuric acid, and evaporating to dryness, provides sulfated ash value of the oil. Sulfated ash reduction/depletion in used oil indicates additive depletion. If it runs higher than that of new oil, contamination with dirt or wear metals is suspected, and further analysis is required to identify the source of foreign material.
Automatic transmission fluid (ATF)	Fluid for automatic, hydraulic transmissions in motor vehicles.
Bactericide	Additive to inhibit bacterial growth in the aqueous component of fluids, preventing foul odors. These are used in metal working fluids, coolants, and aviation turbine fuels.

TABLE 20.1   (Continued)

Test	Significance
Base number	The amount of acid (perchloric or hydrochloric) required to neutralize all or part of a lubricant's basicity, expressed as mg of KOH equivalents.
Base stock	The base fluid, usually a refined petroleum fraction or a selected synthetic material, into which additives are blended to produce finished lubricants.
Bitumen	Also called asphalt or tar, bitumen is the brown or black viscous residue from the vacuum distillation of crude petroleum. It also occurs in nature as asphalt *lakes* and *tar sands*. It consists of high-molecular-weight hydrocarbons and minor amounts of sulfur and nitrogen compounds.
Black oils	Lubricants containing asphaltic materials, which impart extra adhesiveness. These are used for open gears and steel cable lubrication.
Blow-by	Passage of unburned fuel and combustion gases through the piston rings of internal combustion engines, resulting in fuel dilution and contamination of the crankcase oil.
Boundary lubrication	Lubrication between two rubbing surfaces without the development of a full fluid lubricating film. It occurs under high loads and requires the use of antiwear or extreme pressure (EP) additives to prevent metal-to-metal contact.
Bright stock	A heavy residual lubricant base stock with low pour point, used in finished blends to provide good bearing film strength, prevent scuffing, and reduce oil consumption. Usually identified by its K. viscosity at 100°C.
Brookfield viscosity	Measure of apparent viscosity of a non-Newtonian fluid as determined by the Brookfield viscometer at a controlled temperature and shear rate.
Cams	Eccentric shafts used in most internal combustion engines to open and close valves.
Carbon residue	Carbon residue can be defined as the amount of carbon residue left after burning an oil and is intended to provide some indication of relative coke-forming tendencies of the lubricant.
	This can be determined either by Conradson method or by Ramsbottom method.
Catalytic converter	An integral part of vehicle emission control systems since 1975. Oxidizing converters remove hydrocarbons and carbon monoxide (CO) from exhaust gases while reducing converters control nitrogen oxide (NOx) emissions. Both use noble metal (platinum, palladium, or rhodium) catalysts.
Cetane index	A value calculated from the physical properties of a diesel fuel to predict its cetane number (CN).
CN	A measure of the ignition quality of a diesel fuel, as determined in a standard single-cylinder test engine. The higher the CN, the easier a high-speed, direct injection engine will start and the less *white smoking* and *diesel knock* after start-up.
	CN of diesel fuels is determined in a single-cylinder CFR engine by comparing the ignition delay characteristics of the fuel with that of reference fuels (normal cetane (100 CN) and heptamethylnonane (HMN)). HMN has now been replaced by alpha-methylnapathalene, which has a CN of 0.
CN Improver	An additive (usually an organic nitrate) that boosts the CN of a fuel.

(*continued*)

TABLE 20.1 (Continued)

Test	Significance
Cloud point	The temperature at which a cloud of wax crystals appears when a lubricant or distillate fuel is cooled under standard conditions. Cloud point indicates the tendency of the material to plug filters or small orifices under cold weather conditions. It provides a rough idea of the temperature above which the oil can be safely handled without any problems of congealing or filter clogging.
Cold cranking simulator (CCS)	An intermediate shear rate viscometer that predicts the ability of oil to permit a satisfactory cranking speed to be developed in a cold engine.
Color	This is determined either by Saybolt Chromometer or by Lovibond Tintometer. The color has no significance on the performance of the product. It is an indication of the degree of refining of the product. Used oils are of darker color and indicate oil oxidation and contamination with soot.
Compression ratio	In an internal combustion engine, the ratio of the volume of combustion space at bottom dead center to that at top dead center.
Copper strip corrosion	A qualitative measure of the tendency of a petroleum product to corrode pure copper. A cleaned and smoothly polished copper strip is immersed in the sample at 100°C for three hours. This strip is washed with sulfur-free petroleum spirit and examined for evidence of etching, pitting, or discoloration and then compared with ASTM copper strip corrosion standard color code to measure the degree of corrosion. This test indicates the corrosive effect of oil on copper, brass, or bronze parts.
Corrosion inhibitor	Additive that protects lubricated metal surfaces from chemical attack by water or other contaminants.
Demulsibility	A measure of a fluid's ability to separate from water. ASTM D-1401 and ASTM D-2711 have been the standard test methods for measuring demulsibility of industrial oils and gear oils. In many applications, oil is exposed to contamination by water condensed from the atmosphere. Turbine oils are exposed to condensed steam, which can then form emulsion. IP-19 test method measures the steam emulsion number of steam turbine oils. Water promotes the rusting of ferrous parts and accelerates oxidation of the oil. For effective removal of water, the oil must have good demulsibility characteristics.
Density	Mass per unit volume. It is used for calculating the mass when volume of the bulk is known or vice versa (density = mass/volume).
Detergent	A substance added to a fuel or lubricant to keep engine parts clean. In engine oil formulations, the commonly used detergents are overbased Ca/Mg sulfonate with a reserve of basicity to neutralize acids formed during combustion.
Detergent/dispersant	An additive package that combines a detergent with a dispersant. Dispersants are usually PIB succinimide/amides.
Dilution of engine oil	Contamination of crankcase oil by unburned fuel, leading to reduced viscosity and flash point. May indicate component wear or fuel system malfunctioning.

TABLE 20.1   (Continued)

Test	Significance
Dispersant	An additive that helps keep solid contaminants in crankcase oil in colloidal suspension, preventing sludge and varnish deposits on engine parts. Usually, dispersants are ashless and used in combination with detergents.
Distillation	This is the basic test used to characterize the volatility of a gasoline or distillate fuel.
	Petroleum products do not boil at a particular temperature but boils over a range of temperature. This range is of importance in fuels and solvents and is measured by distillation tests.
	In case of crude oil, the ASTM distillation data provide some idea of the fractions that could be collected below 300°C. In a true boiling point (TBP) distillation, the TBP curve provides detailed characteristics useful for the refinery design. The 10% vol. of distillation for gasoline is an indication of the ease with which the engine can be started.
Elastohydrodynamic (EHD) lubrication	A lubricant regime characterized by high unit loads and high speeds in rolling elements where the mating parts deform elastically due to the incompressibility of the lubricant film under very high pressure.
Emissions (mobile sources)	The combustion of fuel leads to the emission of exhaust gases that may be regarded as pollutants. Water and CO_2 are not included in this category, but CO, NOx, and hydrocarbons are subject to legislative control. All three are emitted by gasoline engines; diesel engines also emit particulates that are controlled.
Emissions (stationary sources)	Fuel composition can influence emissions of sulfur oxides and particulates from power stations. Local authorities control the sulfur content of heavy fuel oils used in such applications.
Emulsifier	Additive that promotes the formation of emulsion of oil and water is called emulsifier.
	These are surface-active polar compounds that concentrate at the interface and reduce surface tension/interfacial tension.
End point	Highest vapor temperature recorded during a distillation test of a petroleum stock.
Engine deposits	In engine, sludge, varnish, and carbonaceous residues due to blow-by of unburned and partially burned fuel or the partial breakdown of the lubricant get deposited at different places and are jointly called engine deposits. Water from the condensation of combustion products, carbon, residues from fuel or lubricating oil additives, dust, and metal particles also contribute to engine deposits.
EP additive (EP agent)	Lubricant additive that prevents sliding metal surfaces from seizing under EP conditions.
Exhaust gas recirculation (EGR)	System to reduce automotive emission of nitrogen oxides (NOx). It routes exhaust gases into the intake manifold where they dilute the air/fuel mixture and reduce peak combustion temperatures, thereby reducing the tendency for NOx to form.

(continued)

TABLE 20.1   (Continued)

Test	Significance
Flash point	Minimum temperature at which a fluid will support instantaneous combustion (a flash) but before it will burn continuously (fire point). Flash point is the lowest temperature at which the test flame causes the vapor above the sample to ignite. Flash points are determined by Abel Apparatus and also by Pensky–Martens. Flash point is an indicator of the fire and explosion hazards associated with a petroleum product.
Fire point	Fire point is the lowest temperature at which the oil ignites and continues to burn for 5 s.
Foaming	Foaming in an industrial oil system is a serious service condition. This may interfere with satisfactory system performance and even lead to mechanical damage of the equipment. While good-quality straight mineral oils do not foam, the presence of additives changes the surface properties of the oils and increases their tendency to foaming in the presence of air. Additives such as silicones and acrylate polymers impart foam resistance to the oils and enhance their ability to release trapped air quickly. Foaming consists of air bubbles that rise to the surface of the oil. Foaming is different from air entrainment, consisting of slow-rising bubbles dispersed throughout the oil. Contamination of the oil with surface-active materials such as rust preventives and detergents can also cause excessive foaming. Antifoaming additives are generally used to reduce foaming tendencies.
Fluid friction	Occurs between the molecules of a gas or liquid in motion and is expressed as shear stress. Unlike solid friction, fluid friction varies with speed and area.
Friction	Resistance to motion of one object over another. Friction depends on the smoothness of the contacting surfaces, as well as the force with which they are pressed together.
Gaseous fuels	Liquefied or compressed hydrocarbon gases (propane, butane, or natural gas), which are finding increasing use in motor vehicles as replacements for gasoline and diesel fuel.
Gasoline	A volatile mixture of liquid hydrocarbons, containing small amounts of additives and suitable for use as a fuel in spark-ignition, internal combustion engines.
Gasoline/ethanol blend	A spark-ignition automotive engine fuel containing denatured fuel ethanol in a base gasoline.
Hydrofinishing	A process for treating raw extracted base stocks with hydrogen to saturate them for improved stability.
Hydrolytic stability	Ability of additives and certain synthetic lubricants to resist chemical decomposition (hydrolysis) in the presence of water.
Induction period	In an oxidation test, the time period during which oxidation proceeds at a constant and relatively low rate. It ends at the point where oxidation rate increases sharply.
Inhibitor	Additive that improves the performance of a petroleum product by controlling undesirable chemical reactions, that is, oxidation inhibitor, rust inhibitor, etc.

TABLE 20.1   (Continued)

Test	Significance
Insolubles	Contaminants found in used oils due to dust, dirt, wear particles, or oxidation products are measured as pentane or benzene insolubles. ASTM D-893 test method used for lubricating oils plays an important part in the analysis of contamination. The tests for pentane and benzene insolubles provide three values: 1. *Pentane insoluble*. The total amount of materials insoluble in pentane is oxidative resins plus extraneous matter. 2. *Benzene insoluble*. It is extraneous matter. 3. Pentane insoluble minus benzene insoluble is resins/oxidation products. Pentane- and benzene-insoluble data are useful in determining the used crankcase oil service condition.
Kinematic viscosity	Measure of a fluid's resistance to flow under gravity at a specific temperature (usually 40 or 100°C).
Lubrication	Control of friction and wear by the introduction of a friction-reducing film between moving surfaces in contact. This film-forming compound may be a fluid, solid, or plastic substance.
Multigrade oil	Engine or gear oil that meets the requirements of more than one SAE viscosity grade classification and that can be used over a wider temperature range than single-grade oil. For example, 10W-30 or 0W-30. W stands for winter grade.
Naphthenic oil	A type of petroleum fluid derived from naphthenic crude oil, containing a high proportion of closed-ring methylene groups.
Neutralization number	Neutralizing number is the measure of the acidity or alkalinity of oil and is expressed as milligrams of the amount of acid (HCl) or base (KOH) required to neutralize 1 g of oil. Depending on its source, additive content, refining procedure, or deterioration in service, oils may exhibit certain acid or alkaline (base) characteristics. Since acidity and alkalinity have opposing characteristics, an acid solution can be neutralized by addition of a base and vice versa. The acid number is expressed in milligram of potassium hydroxide (KOH) required to *neutralize* a gram of sample. TBN is total base number, which indicates the basicity of the oil, and TAN is total acid number, which indicates the acidity of the oil. TBN is significant for engine oils, whereas TAN is important for industrial oils like turbine oils and hydraulic oils.
Neutral oil	The basis of most commonly used automotive and diesel lubricants, they are light overhead cuts from vacuum distillation.
Newtonian flow	Occurs in a liquid system where the rate of shear is directly proportional to the shearing force, as with straight grade oils, which do not contain a polymeric viscosity modifier. When the rate of shear is not directly proportional to the shearing force, flow is non-Newtonian, as it is with oils containing viscosity modifiers.
Nitration	The process whereby nitrogen oxides attack petroleum fluids at high temperatures, often resulting in viscosity increase and deposit formation.

(*continued*)

TABLE 20.1   (Continued)

Test	Significance
Octane number	The octane number of a gasoline is a measure of its antiknock quality, that is, its ability to burn without causing the audible *knock* or *ping* in spark-ignition engines.
	Octane number is measured in a standard single-cylinder, variable-compression-ratio engine by comparison with primary reference fuels. Under mild conditions, the engine measures research octane number (RON), and under severe conditions, motor octane number (MON). Antiknock index (AKI) is the arithmetic average of RON and MON, $(R+M)/2$. It approximates the road octane number, which is a measure of how an *average* car responds to the fuel.
	Octane number requirements of gasoline engines depend on their compression ratio, and if the fuel meets the minimum requirements in respect to octane number, it ensures trouble-free operation.
Octane requirement (OR)	The lowest octane number reference fuel that will allow an engine to run knock-free under standard conditions of service. OR is a characteristic of each individual vehicle.
Octane requirement increase (ORI)	As deposits accumulate in the combustion chamber, the ORI of an engine increases, usually reaching an equilibrium value after 10,000–30,000 km. ORI is a measure of the increase, which may be in the range of 3–10 numbers.
Oxidation	Occurs when oxygen attacks petroleum fluids. The process is accelerated by heat, light, metal catalysts, and the presence of water, acids, or solid contaminants. It leads to increased viscosity and deposit formation.
Oxidation inhibitor	Substance added in small quantities to a petroleum product to increase its oxidation resistance, thereby increasing its service or storage life; also called antioxidant.
Oxidation stability	Oxidation stability of petroleum product is its ability to resist changes when coming in contact with air/oxygen at higher temperatures. It is also a measure of its potential service or storage life.
	Oxidation is a chemical reaction that occurs between oil and oxygen. The oxidation of lubricating oils is accelerated by high temperatures, catalysts, and the presence of water, acids, or solid contaminants. The rate of oxidation increases with time. Oxidation tends to increase the viscosity and acidity of oil. Hence, oxidation is determined by an increase in the oil's acidity. Acids formed by oxidation may be corrosive to metals with which the oil comes in contact. Sludge may deposit on sliding surfaces, causing them to stick or wear, or they may plug oil passages.
	Oxidation stability is an important factor in the prediction of oil's performance. Without adequate oxidation stability, the service life of oil may be limited unless the oil is constantly replaced. There are a large number of oxidation tests for each oil with different air flow rates, temperatures, and time durations.

TABLE 20.1   (Continued)

Test	Significance
Oxygenate/ oxygenated fuels	An oxygen-containing ashless organic compound such as alcohol or ether that can be used as a fuel or fuel supplement in internal combustion engine. Term also applies to blends of gasoline with oxygenates, for example, gasohol, which contains 10% by volume anhydrous ethanol in gasoline.
Paraffinic oils	A type of petroleum fluid derived from paraffinic crude oil and containing a high proportion of straight-chain saturated hydrocarbons. These oils have high pour points.
Percentage temporary viscosity loss (PTVL)	Difference between the viscosities of an oil measured at low and high shear stresses, divided by viscosity measured at low shear stress, multiplied by 100.
Permanent viscosity loss (PVL)	Difference between the viscosity of fresh oil and that of the same oil after engine operation or special shear stability test conditions of polymer degradation.
Poise (P)	Measurement unit of a fluid's resistance to flow, that is, viscosity, defined by the shear stress (in dynes per square centimeter) required to move one layer of fluid along another over a total layer thickness of one centimeter at a velocity of 1 cm/s. This viscosity is independent of fluid density and directly related to flow resistance $$\text{Viscosity} = \frac{\text{shear stress}}{\text{shear rate}}$$ $$\frac{\text{dynes}/\text{cm}^2}{\text{cm}/\text{s}/\text{cm}}$$ $$\frac{\text{dynes}/\text{cm}^2}{\text{s}} = 1 \text{ poise.}$$
Polishing (bore)	Excessive smoothing of the surface finish of the cylinder bore or cylinder liner in an engine to a mirror-like appearance, resulting in the reduction of ring sealing and oil consumption performance.
Pour point	An indicator of the ability of an oil or distillate fuel to flow at cold operating temperatures. It is the lowest temperature expressed in multiples of 3°C at which the oil is observed to flow when cooled and examined under prescribed conditions. Pour point indicates the waxy nature of oil and serves as a guide to its pumpability.
Pour point depressant	Additive used to lower the pour point or low-temperature fluidity of a petroleum product.
Preignition	Ignition of the fuel/air mixture in a gasoline engine before the spark plug fires. Often caused by incandescent fuel or lubricant deposits in the combustion chamber, it wastes power and may damage the engine.
Pumpability	The low-temperature, low shear stress–shear rate viscosity characteristics of oil that permit satisfactory flow to and from the engine oil pump and subsequent lubrication of moving components.
Refining	A series of processes to convert crude oil and its fractions into finished petroleum products, including thermal cracking, catalytic cracking, alkylation, hydrocracking, hydrogenation, hydrogen treating, solvent extraction, dewaxing, deoiling, and deasphalting.

(*continued*)

TABLE 20.1 (Continued)

Test	Significance
Rerefining	A process of reclaiming used lubricant oils and restoring them to a condition similar to that of virgin stocks by filtration, clay adsorption, or more elaborate methods such as vacuum distillation and hydrogen treatment.
Ring sticking	Sticking of a piston ring in its groove in the engine or reciprocating compressor due to heavy deposits in the piston ring zone.
Rings	Circular metallic elements that ride in the grooves of a piston and provide compression sealing during combustion. Also used to spread oil for lubrication.
Rust inhibition	For turbine/hydraulic oils, this is determined by immersing a steel rod in 300-ml oil with 300-ml distilled or synthetic seawater at 60°C for 24 h. For a pass, no rusting should be observed; usually, ASTM D-665 A/B methods are used. Additives used to control rusting are called rust inhibitors.
Rust preventive	Compound for coating metal surfaces with a film that protects against rust. Commonly used to preserve equipment in storage and transit.
Scuffing	Abnormal engine wear due to localized welding and fracture. It can be prevented through the use of antiwear, EP, and friction modifier additives.
Shear stability index (SSI)	The measure of a viscosity modifier's contribution to oil's percentage kinematic viscosity loss, when the oil is subjected to engine operation or special shear test conditions.
Sludge	A thick, dark residue, normally of mayonnaise consistency, that accumulates on nonmoving engine interior surfaces. Generally removable by wiping unless baked to a carbonaceous consistency. Its formation is associated with insolubles' overloading of the lubricant.
Solvent extraction	Refining process used to separate aromatic components from lubricant distillates to improve oxidation stability and VI of the base oils. Selective solvents such as furfural/phenol/NMP are used.
Stoke (St)	Kinematic measurement of a fluid's resistance to flow defined by the ratio of the fluid's dynamic viscosity to its density.
Sulfur	This is determined by lamp method or Wickbold procedure for volatile petroleum products and by bomb method for heavier products. Sulfur in the sample is oxidized by combustion and is estimated volumetrically after absorption in H_2O_2 or by gravimetric methods after converting to barium sulfate. Sulfur in any form creates corrosion and environmental problems and is undesirable.
Synlube	Lubricating fluids synthesized from chemical feedstocks rather than refined from oil.
Synthetic lubricant	Lubricating fluid made by chemically reacting materials of a specific chemical composition to produce a compound with known structure and predictable properties.
Temporary shear stability index (TSSI)	The measure of the viscosity modifier's contribution to oil's percentage viscosity loss under high shear conditions. Temporary shear loss results from the reversible lowering of viscosity in high shear areas of the engine, an effect that can influence fuel economy and cold cranking speed.
Temporary viscosity loss (TVL)	Measure of decrease in dynamic viscosity under high shear rates compared to dynamic viscosity under low shear.

TABLE 20.1   (Continued)

Test	Significance
Tribology	Science of the interactions between surfaces moving relative to each other, including the study of lubrication, friction, and wear.
Varnish	A thin, insoluble, nonwipeable film occurring on interior engine parts; can cause sticking and malfunction of close-clearance moving parts; called lacquer in diesel engines.
Viscosity	Viscosity of a liquid is a measure of its resistance to flow. It is expressed in centistokes (kinematic viscosity).
	Viscosity is an important characteristic of a lubricant, and different equipment need different viscosity grades of oils.
VI	Liquids have a tendency to thin out when heated and to thicken when cooled. The property of resisting changes in viscosity with respect to temperature is called VI. VI is an empirical number. The higher the VI of an oil, the less its viscosity changes with changes in temperature. Lubricating oils are subjected to a wide range of temperatures in service. At high temperatures, the viscosity of oil may become low, and the lubricant film may be broken, resulting in metal-to-metal contact and wear. However, at lower temperatures, oil may become very viscous for proper circulation. Several applications require oil with high VI. VI of oil can be improved by adding VI improver additives or viscosity modifiers. These are usually oil soluble/dispersible high-molecular-weight polymers.
	High VI fluids tend to display less change in viscosity with temperature than low VI fluids.
Viscosity modifier	Lubricant additive, usually a high-molecular-weight polymer that reduces the tendency of the oil viscosity to change with temperature.
White oil	Highly refined lubricant stock used for specialty applications such as cosmetics and medicines.
Zinc (ZDP) or ZDDP	Commonly used name for zinc dialkyl/aryldithiophosphate, an antiwear/oxidation inhibitor chemical.

Further tests would be necessary only to identify the exact performance level of the product.

These tests will reveal whether the oil is automotive type or industrial. If the oil has TBN exceeding 5–6, then in combination with viscosity, one can decide what tests need to be conducted to identify the type of automotive oil. Similarly, for industrial oil, further tests can be organized after finding out the ISO viscosity range and the possible applications. Oxidation and rust tests would be useful for turbine and compressor oils. Additional wear tests may be required for antiwear hydraulic or gear oils. For superclean hydraulic oil, a particle size analysis is useful.

For the analysis of used oil, additional metals like iron, silica, chromium, nickel, Cu, Ti, and Si are also determined to establish the wear pattern. If wear is high, a ferrographic analysis can reveal the nature of wear.

Table 20.2 provides equivalent chart of ISO, ASTM, IP, and DIN test methods. These are technically equivalent methods. However, there may be slight variation in

TABLE 20.2   Standard test methods for lubricating oils: equivalent chart

Test	ISO	ASTM	IP	DIN
Acidity				
Inorganic (bromophenol blue)	—	—	182	—
Total, color indicator (*p*-naphthol benzene)	—	D:974	139	51,558[a] part 1
Total, potentiometric	DIS:6619	D:664	177	—
Air release value	—	D:3427	313	51,381
Aniline point	2977	D:611	2	51,775[a]
Apparent viscosity by Brookfield viscometer	—	D:2983[a]	267	—
Ash content	6245	D:482	4	EN 7
Ash, sulfated	3987	D:874	163	51,575
Asphaltene content	—	—	143	51,595
Barium in lube oils	—	—	110	—
Barium by emission	—	—	187	—
Base number total	3771	D:2896	276	—
Borderline pumping temperature	—	D:3829	—	—
Bromine number	—	—	129	—
Calcium in lube oils	—	—	111	—
Calcium emissions	—	—	187	—
Carbon residue				
Conradson	6615	D:189	13	51,551
Ramsbottom	4262	D:524	14	—
Channel point	—	—	—	—
Chlorine	—	D:808	—	51,577
Cloud point	3015	D:2500	219	51,597
CCS	—	D:2602	350	51,377
Color ASTM	2049	D:1500	196	51,578
Copper corrosion	2160	D:130	154	51,579
Demulsibility	—	D:2711	—	—
Density and relative density	3675	D:1298	160	51,757
Dielectric strength	—	—	295	—
Distillation	—	D:1160	—	51,356[a]
Elastomer compatibility	—	D:3604	—	—
Emulsion characteristics	6614	D:1401	—	51,599
Evaporation loss (Noack)	—	—	—	51,581
Falex film strength	—	D:2670	—	—
Fire point	2592	D:92	36	51,376
Flash point (COC)	2592	D:92	36	51,376
Flash point (PMCC)	2719	D:93	34	51,758[a]
Floc point	—	—	—	51,351
Foaming characteristics	DP:6247	D:892	146	51,566
Fuel oil dilution (gasoline)				
By GC	—	D:3525	—	—
By distillation	—	D:322	23	51,565
Fuel oil dilution (diesel)	—	D:3524	—	—
FZG gear rating	—	—	334	51,354
Hexane insolubles	—	D:893	—	—
HTHS viscosity	—	D:4624	—	—
		D:4683		
Hydrolytic stability	—	D:2619	—	—
ISO viscosity classification	3448	D:2422	—	—

TABLE 20.2 (Continued)

Test	ISO	ASTM	IP	DIN
Kinematic viscosity	3104 3105	D:445	71	51,550
Nitrogen	—	D:3228	—	—
Oxidation characteristics of EP lube oils	—	D:2893	—	51,586
Oxidation characteristics of inhibited steam turbine oils	—	D:943	—	51,587
Oxidation characteristics of inhibited steam mineral oils	DIS:4263	—	280	—
Oxidation stability of steam turbine oils by rotating bomb	—	D:2272	229	—
Oxidation test for lubricating oils	—	—	48	51,352
Oxidation stability of mineral turbine oils during use assessment	—	—	328	—
Particulate count				
Automatic	—	—	327	—
Microscopic	—	F—312	275	—
Pentane insolubles	—	D:893	—	—
pH value	—	D:664	177	51,369[a]
Phosphorus	—	D:4047	149	—
Phosphorus by emission	—	—	187	—
Precipitation number	—	D:91	—	51,586[a]
Pour point	3016	D:97	15	51,597
Refractive index	—	D:1747 D:1218[a]	—	51,423 part 2
Rust and corrosion test (humidity cabinet test)	—	D:1748	366	51,359[a]
Rust test	DIS:7120	D:665	135	51,585
Saponification number	6293	D:94	136	51,559
Sediments by centrifuge	3734	D:1796	75	51,793
Sediments by extraction	3745	D:4763	53	51,789
Shear stability	—	—	294	51,382
Stable pour point	—	—	—	—
Standard viscosity Temperature chart	—	D:341	—	51,563
Steam emulsion number	—		19	51,589[a]
Sulfur				
By bomb	—	D:129	61	—
By XRF	—	D:2622	336	51,400 part B
Timken OK load	—	D:2782	240	—
Toluene insolubles	—	D:893	—	—
Unsulfonated residue	—	D:483	—	51,362
VI	2909	D:2270	226	51,564
Water content by Dean and Stark	3733	D:95	74	51,582
Water content by Karl/Fischer	DIS:6296	D:1744	—	51,777 part 1
Wear test Vickers vane pump	—	D:2882	281	51,389
Wear test, four-ball	—	D4172	—	51,350[a] part 1
Weld load, four-ball	—	—	239	—
Yield stress and app. vis at low temp	—	D:4684	—	—
Zinc in lube oils	—	—	117	—

[a] Indicates that the particular test method differs in some details from the other corresponding methods.

each of these equivalent methods. This table is very useful in conducting a test when some customer has asked, for example, a DIN test, and if exact facility for this test is not available, then a corresponding ASTM or other equivalent test can be found out, and the test can be organized.

RIG TESTS

In lubricant terminology, rigs are small machines on which performance evaluation of lubricant additive and finished products is carried out. It is useful to understand these tests since they are referred in many lubricant specifications:

Timken EP test—ASTM D-2782

Four-ball EP test—ASTM D-2783

Four-ball wear test—ASTM D-4172

FZG load-carrying capacity test—IP-334/DIN 51354

FZG low-speed test—ASTM D-4998-89

FZG micro pitting test

Denison T6C hydraulic vane pump test—Denison specification TP-30283

Vickers vane pump test for hydraulic fluids

IP high-torque test—IP-232

There are various other friction and wear tests used in lubricants such as Amsler, Falex, SRV, SAE, IAE gear, and other machine tests.

ABBREVIATIONS OF ORGANIZATIONS

In lubricant specifications, various institutional organizations are involved, and these are usually referred in abbreviated form. Appendix 20.A provides the full form of these abbreviations.

APPENDIX 20.A

	Abbreviation of Organizations
AAMA	American Automobile Manufacturers Association (Formerly MVMA)
ACEA	Association des Constructeurs Europeens de I' Automobile (Association of European Automotive Manufacturers)
ACS	American Chemical Society
AFNOR	Association Francais Petroles de Normalisation
AGMA	American Gear Manufacturers' Association
ANSI	American National Standards Institute
APE	Association of Petroleum Engineers (USA)
API	American Petroleum Institute
ASME	American Society of Mechanical Engineers

APPENDIX 20.A (Continued)

Abbreviation of Organizations	
ASTM	American Society for Testing and Materials
ATC	Technical Committee of Petroleum Additive Manufacturers (Europe)
ATIEL	ATIEL Association Technique de I' Industries Europeenne des
BIS	Bureau of Indian Standards
BLF	British Lubricants Federation
BNP	Bureau de Normalisation des Petroles
CARB	California Air Resources Board
CCMC	Comite des Constructeurs d' Automobiles du Marche Commun (replaced by ACEA)
CEC	Conseil Europeen de Coordination pour les Developments des Essais de Performance des Lubrifiants et des Combustibles pour Moteurs (Coordinating European Council)
CEN	Conseil Europeen de Normalisation
CIMAC	International Council on Combustion Engines
CLR	Cooperative Lubrication Research
CMA	Chemical Manufacturers Association
CONCAWE	Conservation of Clean Air and Water (Europe)
CRC	Coordinating Research Council (USA)
DIN	Deutsche Industrie Norm
ECE	Economic Commission for Europe
EFTC	Engine Fuels Technical Committee (of CEC)
ELTC	Engine Lubricants Technical Committee (of CEC)
EMA	Engine Manufacturers Association
EPA	Environmental Protection Agency
FZG	Forschungsstelle für Zahnrader und Getriebebau
IS	Indian Standards
IFP	Institute Francais du Petrole
ILSAC	International Lubricant Standardization and Approval Committee
IP	Institute of Petroleum (UK)
ISO	International Organization for Standardization
JAMA	Japan Automobile Manufacturers Association Inc.
JARI	Japan Automobile Research Institute
JASO	Japan Automobile Standards Organization
JIS	Japanese Industrial Standards
JSAE	Society of Automotive Engineers (Japan)
LRI	Lubricants Review Institute (USA)
MITI	Ministry of International Trade and Industry
NLGI	National Lubricating Grease Institute (USA)
NMMA	National Marine Manufacturers Association
NPRA	National Petroleum Refiners Association
PAJ	Petroleum Association of Japan
SAE	Society of Automotive Engineers
STLE	Society of Tribologists and Lubrication Engineers

INDEX

Developments in Lubricant Technology, First Edition. S. P. Srivastava.
© 2014 John Wiley & Sons, Inc. Published 2014 by John Wiley & Sons, Inc.

Printed and bound by CPI Group (UK) Ltd, Croydon, CR0 4YY

27/10/2024

14580336-0003